Introdu
Natural R
Plann

Introduction to Natural Resource Planning

Charles Yoe

CRC Press
Taylor & Francis Group
Boca Raton London New York

CRC Press is an imprint of the
Taylor & Francis Group, an **informa** business

CRC Press
Taylor & Francis Group
6000 Broken Sound Parkway NW, Suite 300
Boca Raton, FL 33487-2742

First issued in hardback 2017

ISBN 13: 978-1-138-40150-1 (hbk)
ISBN 13: 978-1-4398-9261-9 (pbk)

Library of Congress Cataloging-in-Publication Data

Yoe, Charles E.
Introduction to natural resources planning / Charles Yoe.
p. cm.
Includes bibliographical references and index.
ISBN 978-1-4398-9261-9 (pbk. : alk. paper)
1. Natural resources--Management. 2. Environmental management--Planning. I. Title.

HC85.Y64 2012
333.7--dc23 2012028460

Visit the Taylor & Francis Web site at
http://www.taylorandfrancis.com

and the CRC Press Web site at
http://www.crcpress.com

Contents

Preface

Here are three things you should know about this book: First, it was written to help keep the art and practice of planning alive. Second, it is not terribly academic. If it is terribly anything, I hope it is terribly practical. Third, I did not want to write it alone and as it turns out I did not.

Here's how I became a planner. Ed and Dave were setting up a keg of beer in the office as I walked to my job interview in the evaluation and economics branch of the Planning Division of the Baltimore District of the U.S. Army Corps of Engineers. It was the morning of the office Christmas party way back when people had kegs at office Christmas parties. I thought I might like it there even before the interview began. I did. That job was the most fun I ever had in an office. I signed on as an economist but within a year I was a planner; not a good one, but an enthusiastic one. I like to think I got better as I went along.

If I did, it was because of Bob, Mr. T, Nels, Harry, Stu, Ken, Milt, Dan, Raleigh, Ed, Dave, Ron, Steve (God rest you, friend), Phil, John, Mike, George, Cliff, T, Tom, Johnny, Pearl, Al, and Ed. It was the seventies, and there were few women planners. Women were still secretaries, and Paula, Patty, and Sandy were as good as the best planners in the office. They are all, in many ways, the coauthors of this book. I learned from them and I learned with them and it was exciting, important, and fun.

I wanted my good friend, Ken Orth, to write this book with me. He is a world-class planner. He has thought about the "stuff of planning" throughout his career, and he is smarter about how to actually do planning than anyone I know. We wrote a planning manual together many years ago. It was good work. I thought we might have a little more good work left in us, so I lured him to my house in Maryland from his house in Georgia one day and plied him with some very good black lagers. He turned me down. The Beatles were his first love, and he has been documenting Beatle album (plus) parody art for years. That project was more important than the tedium of a textbook—go figure.

The truth is this book would not exist without him. His voice is heard throughout it. There is no part of this book he has not touched. Perhaps the only good thing about his declining to join me formally is if he had we'd likely still be arguing the value of the all possible combinations formulation method or whether compare and contrast are two separate activities or one and the same. He's the Beatles, I'm the Stones. He read the draft and told me what sucked and what was good. I almost always took his advice, but I did keep some of the parts he said sucked, and I hope he was wrong about them. I cannot thank him enough for his lists of things, the arguments, the agreements, the collaboration and support, but most of all the friendship. I am most grateful for the opportunity to have worked with him and hope to do so again in the future.

This is not a very academic book. You will be appalled at the paucity of references you'll find, so let me try to explain that. When I decided to write this book, I thought I'd start by defining planning. About 20 pages in, with dozens of articles cited, I still didn't have a satisfactory definition. It was then I decided I didn't want to write that carefully researched book and I figured you didn't want to read it either. My experience says planning is (or ought to be) standing in the present and envisioning the possible futures for the purpose of finding a route to one that appeals to us. That same experience also says there is no one right way to do planning, but there are a whole bunch of wrong ways and if you don't know for sure exactly how you're doing it you're likely to end up doing it one of those wrong ways. And so, I wrote from experience, but not necessarily my experience. It is more the experience of all the planners I benefited from working with.

I have tried to keep this text practical, almost, but not quite, a "how to" manual. The examples draw heavily on my water resources planning background. I am sorry if I missed your interest area, natural resources planning is just too broad, and I hope that does not get in the way of your understanding. I should leave planning history to the scholars, but it seems to me planning was once going to solve a lot of society's problems. There was a great period of planning that seemed, in the final quarter of the last century, to give way to brute force politics and the economic and other exigencies of globalization. During that time, planning often felt devalued to me and I fear, but cannot prove, that we began to lose some ability to practice the art of planning. This book was written to help prevent that from happening.

In 2007, Congress asked the U.S. Secretary of the Army to revise the nation's water resources planning guidance. Several draft versions were created with significant changes. Missing from the drafts was the how to do it piece. Leaving a variety of agencies to figure out how to implement the policy was a bit unnerving coming off a few decades during which planning practice seems to have regressed. That got Ken and me to thinking about how our own thoughts on the art of planning had evolved and changed and it looked, to me at least, to be a good time to put it all in writing.

This feels to me like an exciting time to be doing planning again, especially natural resources planning. There are amazing technologies available to planners, and there is a wide and growing array of fascinating resources problems and opportunities that need our attention. Resource planning issues are being taken up by private and public interests all over the world and at all levels of involvement. The only thing that seems missing is some of that communal experience we had begun to lose. Keeping that flame alive is the real purpose of this book.

I want to repeat, there is no one way to do planning but there are some better ways and they should not get lost. Planning policy and procedures sometimes seem to threaten the art of planning by strangling the fun out of it. That ought not to happen. So it seems to be a good time for an art and practice of planning textbook. Here, you'll find a process I have named "Five By Two" (an oblique Rolling Stones reference) that I hope planners from outside the natural resources arena will find useful as well. Good planning is just good planning.

Planning is important work. It is social decision making. It changes lives. It forms the future. It is a career where you ought to have fun. Every planner, but especially young planners, ought to be able to feel excited about the important work they can do. This book is by the past keepers of that flame, and it is for the future keepers of that flame.

Charles Yoe
Professor of Economics
Notre Dame of Maryland University
Baltimore, Maryland

Author

Charles Yoe, PhD, is a professor of economics at Notre Dame of Maryland University and an independent risk analysis consultant, but at heart he is a planner and educator. He began his planning career with the U.S. Army Corps of Engineers and has since worked with a number of U.S. and other government agencies as a planner, consultant, risk analyst, and trainer in a wide range of areas touching natural resources, including food and feed, natural disasters, public works infrastructure, homeland security, ecosystem restoration, resource development, biotechnology in crops, all manner of water resources, ecosystem services, and the like.

CHAPTER 1

Envisioning the Future

1.1 INTRODUCTION

We live in an uncertain world. Even in a world of perfect certainty—where we knew the truth of climate change, the future price of energy, or the outcome of the next election—planning for that future would be a difficult challenge. We would still have to deal with differing social values. But the world is not certain and planning is all the more difficult, important, and necessary because of this simple fact. Planning is one of our most underrated means of social intercourse and collective action. Planning gives us the ability to envision the future through all the uncertainty. Done well, it empowers us to choose a better future.

This is a book about planning, and it's written for planners and people who don't yet know they are or are going to become planners. Some readers know they are planners because they work for capital "P" Planning organizations. Other readers think they are engineers, environmentalists, economists, lawyers, and interested citizens when they are really planners with a little "p." Still others are sitting in a classroom wondering, "When I am ever going to need to know this?" Just wait—many of you are going to be planners, too.

The genesis for this book is steeped in the water resources planning policy of the U.S. government. For much of my career that policy was articulated in something called "the P&G" (Water Resources Council 1983). The P&G articulated principles and standards for water resources planning. They also offered a rather serviceable six-step planning process that I have spent years unpacking with friends and colleagues for ourselves and others (Yoe and Orth 1996; Orth and Yoe 1997). In Section 2031 of the Water Resources Development Act of 2007, Congress directed the Secretary of the Army to revise the P&G. This book is written, in part, to try to get planning right in case the bureaucrats in Washington don't.

This book is about small "p" planning. This includes the principles, standards, process, and practice of planning; it's what to do and how to do it. It is the art and practice of planning. Capital "P" Planning is the policy guidance of organizations, the budget process, the necessary administrivia that is sometimes important but not nearly as much fun as planning. This book is not about Planning.

Why natural resources planning? The simple answer is that there are a lot of people all around the world doing it with varying degrees of success. We care about it, and it is important.

Another reason for a natural resources planning book is that it is good to be green these days. I came of age as a planner during an exciting time when the National Environmental Policy Act of 1969 was just beginning to have some impact. Planning was exciting and vital, and it was fun to do. Then planning entered its Muppet Age, when Kermit spoke for many natural resource planners when he sang, "It's not easy being green." The economic exigencies of the 1980s devalued planning at the national level.

The beautiful thing about planning though is that we really do need it. The streamlined devalued version of the last decades of the twentieth century did not solve all the problems and lead us to a greener future. Now we see people who don't even know they are planners springing up all over the nation. They are planning at the national, state, and local levels. They are planning in nongovernment organizations (NGOs) and in their communities.

Natural resources planning is where the past, the present, and the future all converge, with no less than the future of the planet in the balance. That makes it vital and important. This is an area with more sexy issues than the TMZ website. Just think about it. We've got climate change, sustainability, biodiversity, oil spills, tsunamis, nuclear disasters, brownfields, carbon footprints, a growing population, people wanting to eat locally, debates about drilling, and concerns about natural resource adequacy and rates of resource use, and that is just for starters. Managing natural resources is a concern of every national, state, and local government in the world. Land use and conservation, water use, pollution, energy issues, natural hazards, and development of natural areas fill headlines somewhere every day. Fascinating new issues continue to emerge as we seek to identify and preserve natural DNA, struggle with invasive and nonindigenous species, and worry about the well-being of native and managed pollinators. Our awareness of the value of unique assets continues to grow and convert new followers, whether it is generic topics like agriculture, beaches, fish, groundwater, habitat, livestock, rivers, sanitation, wastewater, and wetlands or specific treasures like the Everglades, the Gulf Coast, Puget Sound, salmon in the Columbia River Basin, Asian carp headed toward the Great Lakes, or the woods a few blocks from your home. These are among the most important issues that current and coming generations face.

Natural resource issues intersect with many technology issues. How will mini-chromosome technology affect our environment? Genetically modified food and feed are already staples in many countries. Nanotechnology will enable us to create legions of new solutions to once-intractable problems. Particles and filters can bind with or remove inactive pollutants. Renewable energy is changing the landscape. More-efficient uses of resources are being discovered every day. Natural resources are one of the hottest topics on the planet.

How will society respond to these challenges? By what means shall we organize to answer them? Natural resource issues will be nudged off the front pages by geo-political events and the latest celebrity meltdown or athlete screwup, but they are not going away. Neither will these problems stay the same. Something is going to happen, and if we want a say in what that something is, we need to begin planning. That is why this book exists.

Planning is a ubiquitous human endeavor. We have urban planning, strategic planning, military planning, financial planning, and don't forget wedding planning, just to name a few. So if you read with one of these interests, welcome to planning. Although the examples here come from the world of natural resources planning, you'll find the methods presented as applicable to your interests as they are to natural resources, with the possible exception of wedding planning, where we yield to the experts. Planning is standing in the present and peering off into a vast array of uncertain futures that stretch ahead of us, for the purpose of choosing the direction that best suits the needs of those for whom we plan. It's time for us to begin to envision the future.

1.2 UNCERTAINTY

One of the emerging constants in the modern world is uncertainty. Growing social complexity and an increasingly rapid pace of change are normal parts of the decision-making landscape, and they contribute a great deal to the uncertain environment in which we all operate. Planning for an uncertain future presents a viable alternative to drifting through an uncertain environment.

The world grows more complex. Think of complexity as a social phenomenon. It refers to the size of a society, the number of its parts, the distinctiveness of those parts, the diverse specialized social roles it incorporates, the number of distinct social personalities present, and the variety of mechanisms for organizing these into a coherent, functioning whole.

As a species, we have spent virtually all our history living as low-density foragers or farmers in egalitarian communities of no more than a few dozen persons and even fewer distinct social roles. In the twenty-first century, we live in societies with millions of different roles and personalities. Our social systems grow so complex they defy understanding. Solving the ill-formed problems of natural resource use and preservation in such a world grows correspondingly complex.

Added to this growing complexity is an increasingly rapid pace of change in almost every arena of human endeavor. Scientific breakthroughs make things once impossible to conceive commonplace. These "things" include disasters and challenges as well as delights. Much of this change is driven by rapid advances in technology. Technology changes social values and beliefs as well as the way we live, work, and die. The level of complexity in our social, economic, and technological systems is increasing to a point that is too turbulent and rapidly changing to be predicted by human beings. The impacts of this rapid change on natural systems are often poorly understood, if they are understood at all.

Rapid increases in social, economic, and technological connectivity are taking place around the world, shrinking, uniting, and dividing the globe at a dizzying pace. Social movements, for example, environmentalists, women's rights, World Trade Organization opposition, the Occupy movement, and the like are global in their pervasiveness. We are increasingly a global economy. Fashions are designed in New York, approved in London, patterns are cut in Hong Kong, clothes are made in Taiwan, and sold in Europe and North America. A computer virus spreads around the world in hours. A human virus spreads in weeks or months. An oil spill in the Gulf of Mexico, a tsunami in Japan, or a flood on the Mississippi River reverberate through our global society and its ever-more-integrated economy.

Social values change unpredictably and constantly. The resource fixes of the past are often the problems of the present. Efforts to drain the swamplands in Southern Florida, in part to increase agricultural land to enable us to support our European allies during a century of World Wars, have now become a threat to one of the world's most productive ecosystems. Taming the Wild West, straightening our meandering rivers to improve navigation, and clear cutting forests for farming and housing are values of the past that have contributed to the problems of the present. No doubt many of the values of the present are already seeding the resource problems of the future. Fields of wind turbines produce renewable energy but threaten the migratory patterns of birds and vistas undisturbed since creation. Who knows how they will be judged by society in 30 years?

The United States can no longer as confidently lay claim to the Judeo-Christian ethic. Families are more diverse than ever. The Census Bureau said, "Married couples represented just 48 percent of American households in 2010." Politics remains as divisive as ever. Wildlife and environmental resources now have numerous interest groups supporting both their harvest and their preservation. We are better educated than ever, with a greater divide between the most and least educated among us. Values are in conflict even when consensual views of problems may exist, and that is not nearly as frequently as we'd like. Discerning the uncertain values that decide issues is one of planning's greater challenges.

Relentless pressure on costs is a fixture in all public decision making. Faster, cheaper, better is the mantra of far too many organizations that have not thought carefully about what that means for the present, much less for the future. We have entered a world where irreversible consequences, unlimited in time and space, are now possible. This is or should be extremely important to resource managers in both the public and private sectors. Decades after the accident at Chernobyl, some of its victims have yet to be born. Many of the wicked resource problems* we face can have a long latency period.

Many of our country's landscape-scale ecosystem restoration problems, like those in the Florida Everglades, coastal Louisiana, and the Columbia River Basin, as well as global concerns like greenhouse gases and climate change, provide clear examples of problems that took decades to emerge and be recognized. The implications of the solutions being formulated may similarly take decades to be understood.

A new phenomenon of "known unawareness" has entered our lexicon. Donald Rumsfeld in November 2006 summarized this truth to scattered laughter when he said:

> There are known knowns. These are things we know that we know. There are known unknowns. That is to say, there are things that we now know we don't know. But there are also unknown unknowns. These are things we do not know we don't know.

No one is laughing anymore. As a society, we are beginning to realize that despite all we know, the unknown far outweighs what is known. Knowledge is as much to create more questions as it is to provide definitive answers.

Clearly scientists now know much more about bovine spongiform encephalopathy (BSE, or mad cow disease) than when it was first found in cattle in 1986. Even now, decades after the disease's discovery, its origins, its host range, its means of transmission, and the nature of the infectious agent and its relation to its human-counterpart variant Creutzfeldt-Jakob disease remain mostly unknown. We have begun to suspect that some problems are so complex there may be no narrative closure, no ending by which the truth is recovered and the boundaries of the problems established.

Although most of us live and work in nations, our interactions and our problems are increasingly global in nature. It becomes more difficult to affix responsibility for resource problems and their solutions. Who is destroying the ozone, causing global warming, and spreading BSE? What is happening to polar ice caps and aquifers? What is the future of our resource deposits as once sleeping giants like China, India, and Indonesia grow in affluence and expectations of a better standard of living? Meanwhile, familiar old problems like flooding continue to claim mounting numbers of casualties around the world.

Despite the world's rapid advances in all kinds of sciences, we are increasingly dominated by public perception. Public perception is a palpable force and in some circumstances it is an irresistible one. It can also be an ignorant force. When those

* Wicked problems are complex but ill-formed problems that lack right and wrong solutions. Instead, there are many candidate solutions where some are better and some are worse than others.

who benefit from a situation differ from those who pay the costs, the conditions are often considered unacceptable regardless of the facts of the matter. All resource issues have a social context. It is folly to regard social and cultural judgments as things that can only distort the perception of reality. Social and cultural judgments shape reality, and neither they nor the people who hold them can be ignored.

One result of this not uncommon disconnect between the scientific facts of an issue and its social context is that possibility is often accorded the same significance as existence in the public's view. This view has found its way into some public policy. This is in part because many things that were once considered certain and safe, and often vouched for by authorities, turned out to be deadly. The BSE experience in Europe, the SARS experience in Asia and elsewhere, the failed levees in New Orleans, the melamine contamination from Chinese products, the size of the Gulf oil spill of 2010, and the 2011 nuclear disaster in Japan provide vivid examples of this. Applying knowledge of these experiences to the present and the future devalues the certainties of today. This is what makes conceivable threats seem so possible and what fuels our fears of uncertainty. It is also what makes criticism of a decision that masquerades as certain embarrassingly easy.

Responsibility in this more connected world has become less clear. Who has to prove what? What constitutes proof under conditions of uncertainty? What norms of accountability are being used? Who is responsible morally? And who is should pay the costs? These questions plague decision makers nationally and transnationally.

We all live and operate in this uncertain reality. Yet many organizations and individuals cling stubbornly to a short-sighted, present-oriented, and deterministic approach to decision making that belies the experience of public and private sectors worldwide. Decision making needs a future-oriented "culture of uncertainty." Risk-informed planning* helps to build such a culture.

The future is fundamentally unknowable. There must be recognition of the central importance of demonstrating the collective will to act responsibly and accountably with regard to our efforts to grapple with this fundamental uncertainty and the inevitable wrong decisions and losses that will occur despite every best effort to account for this uncertainty. In an uncertain world we cannot know everything and we will make mistakes despite our best efforts to the contrary. This is the challenge that invites planners to reclaim an important social role as guardians of the future.

Planners operate in this complex and uncertain global environment, where values are fungible and tomorrow is unpredictable. Now add to this the more pragmatic concerns that challenge planners. There is never enough data. The theory, scenarios, and models they rely on are rarely complete. It is precisely because the world of planning is so uncertain that we need a systematic process that is intentional about identifying critical uncertainties and taking them into account in decision making. Planning, when it incorporates the best practice of risk analysis, offers society a rational approach to envisioning the uncertain future.

* This is planning that is intentional in honestly assessing and addressing the uncertainty that exists in the planning context.

1.3 NATURAL RESOURCES

Natural resources are physical assets provided by nature. Biotic natural resources include all living things. Plants, animals, and fossil fuels comprise this category. Abiotic natural resources are nonliving things. Land, water, air, and minerals in all their forms make up this category.

Most definitions of natural resources include some reference to their occurrence in a natural state as well as their economic value. Natural resources provide a flow of services essential to human life. Natural resources include the assets themselves, their amenities and attributes, and the services that flow from them. They can have functional value, seen in their usefulness in producing some other goods or services; aesthetic value in their mere existence; and moral values that require no further justification (Eriksson and Andersson 2010). These values benefit humans by improving human welfare and developing human capabilities (Edward-Jones et al. 2000). They can benefit all sentient beings that can experience pleasure or pain (Singer 1976). Teleological biocentrism argues that anything with an organic development path can be considered a benefit. Taylor (1986) has called this a "theory of respect for nature." An ecocentric view considers concern for entire species or ecosystems (Eriksson and Andersson 2010). This book embraces all these views of natural resources for planning purposes.

Natural resources offer a staggering array of commercial, recreational, aesthetic, and ecological services. Figure 1.1, adapted from the World Wildlife Federation's Living Planet Report 2010, shows the interconnections among people, natural

ECOSYSTEM SERVICES

Ecosystem services are defined as the benefits that people obtain from ecosystems (Millennium Ecosystem Assessment 2005). Four major categories of services are usually identified. *Provisioning services* are provided by the goods obtained directly from ecosystems. These include food, minerals, water, medicines, timber, fossil fuels, and the like. *Regulating services* are provided by ecosystems when waste is decomposed, carbon is sequestered, water is filtered, crops are pollinated, and climate is regulated. *Supporting services* necessary for the provision of all other ecosystem services are another major benefit category. These support services include such basic ecological functions and processes as nutrient cycling, photosynthesis, and soil formation, which make many of the other services possible.

The *cultural services* of ecosystems provide the psychological and emotional benefits humans realize when they interact with and relate to ecosystems. These benefits include recreational experiences as well as aesthetic and spiritual experiences.

Source: World Wildlife Federation. 2010. *Living Planet Report 2010 Biodiversity, Biocapacity and Development.* Gland, Switzerland: WWF International.

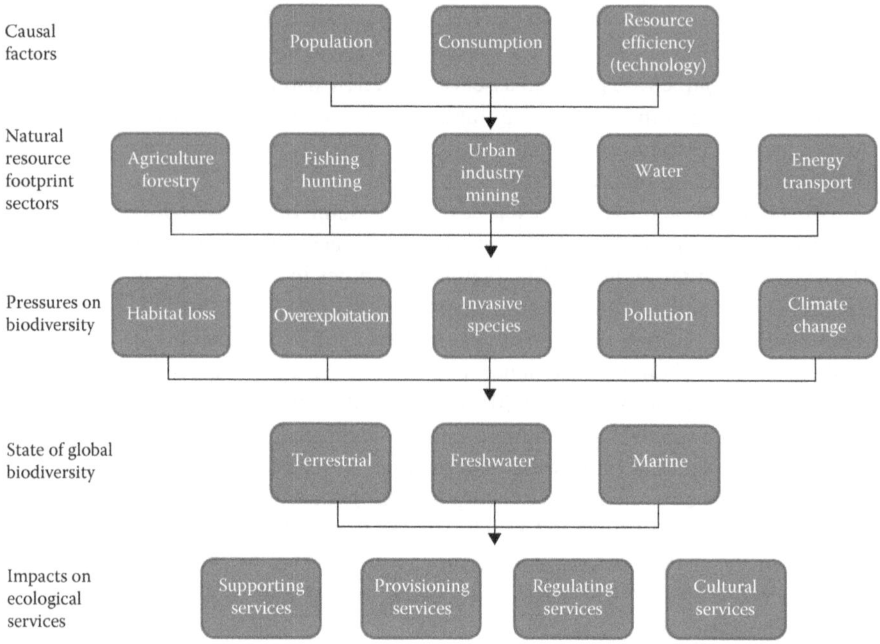

Figure 1.1 The role of natural resources in the production of ecosystem services. (Adapted from World Wildlife Federation. 2010. *Living Planet Report 2010 Biodiversity, Biocapacity and Development.* Gland, Switzerland: WWF International. With permission.)

resources, and ecosystem services. Maintaining the flows of those services is a primary challenge for planners. Battelle identifies two major natural resource tasks for governments. The first of these is determining how to clean up legacy problems, restore natural resources, and achieve human health protection. The second is designing strategies to allow for future growth while protecting the environment, maintaining biodiversity, safeguarding human health, and preserving cultural/social values (Boehm 2003). These tasks are integral components of the natural resources planning tradition.

The Global Environmental Outlook 2000 (UNEP 1999) identifies a wide variety of global legacy problems, including:

- Impacts of anthropogenic emissions of greenhouse gases on the atmosphere.
- Ground-level ozone, smog, and fine particulates.
- Overexploitation of surface water resources and aquifers.
- Deaths from water-related diseases.
- Biological diversity is under increasing threat.
- Land degradation continues to worsen.
- Many remaining forest ecosystems have been degraded and fragmented.
- Crop- and livestock-related buildup of reactive nitrogen in the global biosphere, contributing to the acidification and eutrophication of ecosystems.

- Urban air pollution and deteriorating water quality have major health, economic, and social impacts.
- Natural disasters are increasing in frequency and intensity.

Global issues for this young century are many and have been described as clustered in three groups. These are unforeseen events and scientific discoveries; sudden, unexpected transformations of old issues; and already well-known issues to which the present response is inadequate (UNEP 1999). The Scientific Committee on Problems of the Environment (SCOPE) of the International Council for Science asked 200 scientists from 50 countries what environmental issues are likely to require attention in the twenty-first century. Most of their responses reflect existing problems that receive inadequate attention. Figure 1.2 shows selected issues identified by SCOPE.

Local issues mirror global concerns and add countless variations of other concerns, including timber management, resources in urban areas, park management, forest management, recreation, biodiversity, wildlife, habitat, water quality, fisheries, flood risk management, grizzly bear recovery, invasive species management, water rights, reallocation studies (reservoirs and treaties), greenways, bike trails, nature

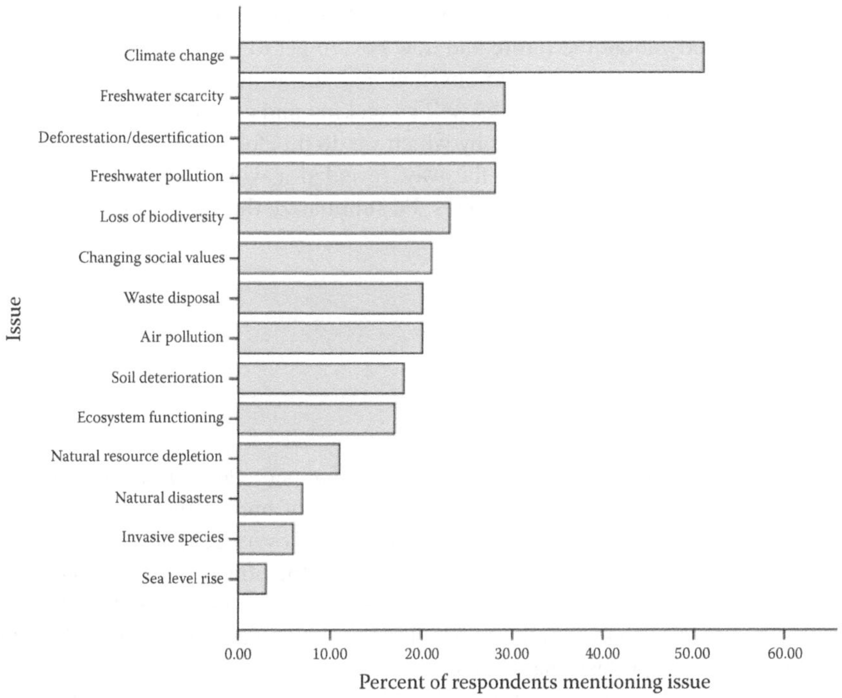

Figure 1.2 Selected environmental issues of the twenty-first century from the Scientific Committee on Problems of the Environment (SCOPE) survey. (Derived from United Nations Environmental Programme (UNEP). 1999. *Global Environment Outlook 2000*. Accessed July 23, 2011, http://www.grida.no/publications/other/geo2000/.)

walks, natural resource management, cultural resource management, energy security, poverty reduction, peregrine falcon management, adapting to climate change, food production, burro management, hazardous trees, fire management, freshwater, provision, and on and on the list goes.

1.4 PLANNING IS ...

Planning is not easy to define in a way that will please many, much less all people. Wildavsky (1973) wrote when central planning was very much a world force to be reckoned with and when national planning gave planning much more cache than it has had in recent years. In his article "If Planning Is Everything Maybe It Is Nothing," he described planning as future control, as cause, as power, as adaptation, as process, as intention, as rationality, and as faith. But he does not drive a stake into the shifting sand and define it. It seems easier to wax on about planning than it is to define it.

So let me define planning for natural resources and get that nasty task out of the way. In 1996, a friend and I defined planning as the deliberate social or organizational activity of developing an optimal strategy for solving problems and achieving a desired set of objectives (Yoe and Orth 1996). Today I prefer to say planning is thinking carefully about the future and how best to get to the most desirable future from the present. Its purpose is to envision and shape the future by anticipating, identifying, and solving problems as well as seeking and capitalizing on opportunities. This book is about the process by which we do this for natural resource issues.

A review of the literature, by the way, revealed a wide range of opinion and very little consensus on what planning is. I'd summarize the views as roughly these. Planning is

- Basic human activity
- Rational choice
- Control of future action
- Complex problem solving
- What planners do

Some see planning as a basic human activity that pervades our behavior at every level of society. In this view, planning is a process of human thought followed by action based upon that thought. This makes planning a very general human activity. You plan what to wear to work, the route you take to the office, and what you will do during your day. This also makes planning very ordinary. Those who do not plan in these ways might seem to drift through their days. Although everyone plans like this, few do it as well as professional planners.

Planning can be seen as a matter of deliberate choice. It is a process for determining appropriate future actions through a sequence of choices. It is a structured rational approach to achieving desired ends.

Closer to my own favored view is planning as an attempt to control future consequences through present actions. This view fuses planning and action together, for

PLANNING IS NOT ...

Planning is not budgeting, scheduling, and related reporting, but these are project management activities. They are important, but they are not planning. Neither is planning the legally mandated coordination and consultation with other agencies and groups. When these are proscribed exchanges of correspondence, reports, meetings, and the like within proscribed time frames, it is procedural administration. Planning is not conducting technical studies. The hydrology, design, geotechnical, biology, water quality, cultural resources, and all the other analyses are all vitally important support functions, but they are not planning either.

if we do not implement a plan, there can be no control exerted over the future. Some would measure the success of planning by the future consequences we are able to influence or control.

Another line of thought is that planning is problem solving that is aimed at very particular kinds of problems. Planning theorists have often defined the problems they deal with as ill-structured or wicked problems, that is, problems with no clear answers. Most natural resource problems are wicked problems.

Planners help decision makers identify their problems, conceive solutions to them, and compare the importance of the inevitable conflicting values inherent in any solution. This is a simple and intuitive definition with which many natural resource planners can identify. The job is unique. It differs so much from day-to-day that it defies a definition more precise than planning is what planners do.

Each definition offered here is like a point in a pointillist painting. No one of them alone is very enlightening, but taken together they provide a reasonable representation of what planning is.

1.5 ORGANIZATION OF THIS BOOK

This text is about how to plan. It is applied planning, that is, good planning practice. It is based largely on what others have taught me, mistakes I have made, and occasional good ideas I think I have had. It is not the academic prototype that reviewed every bit of literature on every topic. There are books and articles that do that far better than I ever could. What you do get here are these 16 chapters:

Chapter 1: Envisioning the Future
Chapter 2: A Team Sport
Chapter 3: The Planning Process
Chapter 4: Establish the Decision Context
Chapter 5: Gather Evidence
Chapter 6: Formulate Solutions
Chapter 7: Evaluate Solutions

Chapter 8: Compare Solutions
Chapter 9: Make a Decision
Chapter 10: Public Involvement
Chapter 11: Tell Your Story
Chapter 12: Uncertainty
Chapter 13: Scenario Planning
Chapter 14: Economics for Planners
Chapter 15: Fast Planning
Chapter 16: Practical Tips

Chapter 2 begins with the people who are indispensable to planning: the planners. Planning is indeed a team sport, and good teams do not occur by accident. Chapter 3 discusses the notion of a planning process and offers several examples, including the P&G mentioned earlier in this chapter. The P&G planning process is the parent of the planning process presented in this text, which I call "Five By Two" for the five steps and two ongoing processes that comprise it.

Chapters 4 through 10 constitute the heart of this book by devoting a chapter to each of these seven components. Step 1, the decision context, is described first in Chapter 4. This is followed in Chapter 5 by a description of the evidence gathering process, which begins in Step 1 and continues throughout the planning process. Chapter 6 is devoted to plan formulation, one of the most important and least discussed tasks in any good planning process. Chapters 7 through 9 are the solution-assessment chapters. Chapter 7 describes how plans are evaluated or found worthy of consideration as potential solutions. Chapter 8 discusses the process of comparing the evaluated plans to highlight the differences among them that may make a difference to decision makers. Chapter 9 describes the magic of decision making. Public involvement, which is so essential to each of the five planning steps, is the topic of Chapter 10.

Telling your story is the focus of Chapter 11, which argues against the death by pagination planning reports of the past in favor of stories that people can understand and care about. Chapter 12 discusses uncertainty, which as you may already suspect is an important element of the planning process. Good planning must be honest about what it does and does not know. Chapter 13 introduces scenario planning, a variation on the planning process that has proven especially useful in situations where uncertainty is too great to say with confidence what the future may look like without any action from your plan.

I am an economist; I came to planning through economics, so Chapter 14 is about economics. This is an important chapter, no matter my background, because people always care about costs and costs are a little trickier than you might think. Sometimes you may need to speed up the planning process. Chapter 15 describes why and how you might plan fast. The book ends with the catch-all Chapter 16, which is full of things I just had to say even if you didn't have to know them. The hope is you'll find the tips are practical and useful too.

Planning is exciting, challenging, important, and fun. At least it can be, so if you are planning and it is not, maybe this book will help. Having a way to bring all that theory and good knowledge you have acquired in school and life into practical focus to solve problems and realize opportunities is what this book is all about.

1.6 TAKE AWAYS

Lesson 1: The world is uncertain, and the future is the most uncertain part of it.

Lesson 2: Natural resource issues are among the sexiest issues on the planet.

Lesson 3: Planning is one of humankind's most effective forms of collective action.

Lesson 4: This book is about how to do planning.

Lesson 5: There is no one way to plan, but there is a better way and a lot of worse ways.

REFERENCES

Boehm, P. 2003. *Gearing up to Meet Global Environmental Challenges*. Columbus, OH: Battelle Corporation.

Edward-Jones, G., B. Davies, and S. Hussain. 2000. *Ecological Economics: An Introduction*. Oxford: Blackwell Science.

Eriksson, R. and J. O. Andersson. 2010. *Element of Ecological Economics*. London: Routledge.

Millennium Ecosystem Assessment. 2005. *Ecosystems and Human Well-Being: Biodiversity Synthesis*. Washington, DC: World Resources Institute.

Orth, K. and C. Yoe. 1997. *Planning Primer*. IWR Report 97-R-15. Alexandria, VA: Institute for Water Resources.

Rumsfeld, D. 2006. Press Conference at NATO Headquarters. U.S. Department of Defense. http://www.defense.gov/transcripts/transcript.aspx?transcriptid=3490.

Singer, P. 1976. Not for humans only: The place of nonhumans in environmental issue. In K. E. Goodpaster and K. M. Sayre (Eds.), *Ethics and the Problems of the 21st Century*. Notre Dame, IN: Notre Dame University Press.

Taylor, P. W. 1986. *Respect for Nature: A Theory of Environmental Ethics*. Princeton, NJ: University of Princeton Press.

United Nations Environmental Programme (UNEP). 1999. *Global Environment Outlook 2000*. Accessed July 23, 2011, http://www.grida.no/publications/other/geo2000/.

Water Resources Council. 1983. *Economic and Environmental Principles and Guidelines for Water and Related Land Resources Implementation Studies*. Washington, DC: Council on Environmental Quality.

Wildavsky, A. 1973. If planning is everything maybe it's nothing. *Policy Sciences*, 4(2): 127–153.

World Wildlife Federation. 2010. *Living Planet Report 2010 Biodiversity, Biocapacity and Development*. Gland, Switzerland: WWF International.

Yoe, C. and K. Orth. 1996. *Planning Manual*. IWR Report 96-R-21. Alexandria, VA: Institute for Water Resources.

CHAPTER 2

A Team Sport

2.1 INTRODUCTION

Planning is done by people and they plan best in teams. There are two primary reasons why planning must be done by a team. The first and simplest reason is that two heads are better than one and many heads are better than two. No one person, no one discipline, no one group has all the answers. Many disciplines are needed for planning. No one has the background, experience, and knowledge necessary to plan alone. Planning is a social phenomenon that requires people to work together to solve problems.

The second and more enduring reason is that the world is extremely complex, and only a team structure can respond to social needs in this uncertain environment. Individuals do not know enough to develop good solutions to wicked problems on their own. Teams are needed. Large bureaucratic organizations are too rigid, too slow, and too costly to respond effectively in today's environment. Teams are a sensible and efficient compromise between the limitations of individuals and the inefficiency of large organizations. They make sense (Yoe and Orth 1996).

Good teams don't just happen. They must be built, and they must work well together. Oddly enough, team building and teamwork are responsibilities that often remain beneath the planning radar. Teams are important enough that before we begin to talk about the principles, standards, processes, and practices of planning, we need to talk about teams. Who needs to plan? What does a team look like? What are the stages a team passes through as it matures? What processes does a team need, and what roles do team members fill? What does a team do? The answers to these questions are the subject matter of this chapter.

2.2 WHO NEEDS TO PLAN?

Individuals and organizations need to plan. The stewards of any public trust need to plan. Those charged with the care and keeping of public resources, whether natural resources or social resources like public health, public safety, homeland security, freedom of speech, and the like, would seem to have a social obligation to plan for their trusts. This would include government organizations at the federal, state, and local levels.

Nongovernmental organizations (NGOs) are often involved in planning. Sometimes they are the planning organization and other times they are major stakeholders in the planning efforts of government and private industries. Many private organizations with a social obligation or social conscience need to plan. Corporations that profit from natural resource use and exploitation have an obligation to stockholders and future generations to plan effectively for resource use, preservation, and reclamation. Trusts and foundations have assumed an active role in financing and planning for significant natural resources around the world.

Contractors, consultants, and architect and engineering firms all need to know how to plan. They are often paid to conduct the analytical aspects of the planning process and sometimes the deliberative ones as well.

Interested citizens need to know how to plan. Many local resource management efforts have flourished or floundered depending on the citizens' knowledge of and insistence on a planning process. Citizens who understand planning and insist on their role in the process are the best safeguard against bureaucratic processes replacing planning. Endangered and threatened species would plan if they could, so might abused waterways and other exploited resources. Planners include those who speak for those and that which cannot speak for itself.

2.3 PLANNING IS DONE IN TEAMS

Planning is a team sport. Social decision making should never be done in private or by individuals. Planning teams are more or less organized and more or less structured. They are collections of diverse people, essential roles, and necessary expertise; the best teams are all of these.

CHARACTERISTICS OF A GOOD TEAM

1. Team members are in close physical proximity and able to meet regularly.
2. The appropriate skills are represented on the team.
3. The appropriate levels of organizational authority are present within the team.

4. Team members are involved in the setting of objectives.
5. Objectives are understood by all members.
6. All individuals support the objectives.
7. Objectives are set and met within realistic time frames.
8. Roles are clearly defined and do not overlap.
9. Team members and their leaders know their assignments.
10. Roles are understood by all and are supported.
11. There is a strong, effective leadership with clearly defined responsibilities.
12. Members and leaders are accessible to each other.
13. Decisions are made by consensus.
14. Meetings are efficient and task-improvement oriented.
15. Emphasis is on problem solving versus blaming the individual responsible for the problem.
16. All members participate in discussions and meetings.
17. Minutes of meetings are promptly distributed.
18. Members listen well.
19. There is frequent feedback to individuals regarding performance.
20. All members are kept informed.
21. Deadlines and milestones are clearly established and agreed to by the team.
22. There is team identity or esprit de corps and pride.
23. There is tolerance for conflict with an emphasis upon resolution.
24. Conflict is openly discussed, often resulting in growth or learning.
25. Members enjoy each other.
26. Team members support each other.

Source: Yoe, C. and K. Orth. 1996. *Planning Manual. IWR Report 96-R-21*. Alexandria, VA: Institute for Water Resources.

Transdisciplinary teams are better than interdisciplinary teams are better than multidisciplinary teams are better than an uncoordinated group's efforts, which may be better than individual effort, which may be better than nothing. Your goal is to get your team as close to the beginning of that sentence as you can.

Doing nothing is easy. Ignoring things is one of the things we do best. Let's be clear about natural resource planning: it is not a one-person job. The natural systems are too complex, the social complexity is increasing, and no one person can do it well. Uncoordinated group efforts are, sadly, still relatively common. Many nominal planning efforts might well fall into this category.

Planning needs teams. A multidisciplinary team assembles a group of diverse experts who tackle complex problems together. The integration of their various disciplines is never a focus of the effort, and the work of such teams often has the flavor of a series of more or less well-connected analyses that we hope adds up to something meaningful to someone somewhere. Assembling the right disciplines is sometimes regarded as a hurdle to be leaped. Although this marks a substantial improvement over an uncoordinated group's effort, coordination of the efforts of the various disciplines falls well short of their integration.

Multidisciplinary teams are limited by the fact that disciplines have, over time, developed their own specific and occasionally peculiar ways of looking at the world. Many of us learn our fields in an "all other things equal" context. At times we have been trained to screen out those aspects of the world that conflict with our discipline's way of looking at things. If we are to develop effective solutions to complex problems, we need a better-integrated view of the problems.

An interdisciplinary team begins, like the multidisciplinary team, with a rich diversity of expertise. It differs, however, in its intentional effort to integrate the various disciplines in a way that better recognizes the whole of the big picture for planning purposes. An interdisciplinary team crosses traditional boundaries between academic disciplines or schools of thought and weaves more holistic viewpoints of planning issues. Interdisciplinary solutions tend to be more responsive to social needs.

The experts in an interdisciplinary team come to understand the language and basic concepts of the other disciplines and their perspectives. This enables engineers to appreciate and consider the viewpoints of economists and environmentalists. This can lead to more economical and environment-friendly designs. Likewise, economists understand the importance of nonmonetized values reflected in other disciplines that may lead to cost-effectiveness and incremental cost analysis of quantifiable but nonmonetized plan effects.

A transdisciplinary team not only crosses traditional boundaries among disciplines, it erases those boundaries and integrates knowledge at the edges of disciplines. A transdisciplinary approach bridges many disciplines at once and develops a synergy among disciplines that creates new knowledge and ways of looking at problems and their solutions that transcend the abilities of traditionally bounded disciplines. As new needs and new disciplines emerge, disciplinary approaches that bridge and accommodate these changes are increasingly valuable. The most exciting disciplines now are those that integrate the traditional ones.

Litman (2011) points out that planners work at the intersection of many disciplines and need basic knowledge of a lot of subjects. Planning is an ideal field for people with diverse interests. Among the skills planners need, Litman includes the ability to:

- Accurately, critically, and objectively evaluate problems.
- Collect and analyze data.
- Apply general concepts to specific situations.
- Manage complex processes.

Transdisciplinary knowledge is greater than the sum of all the disciplines that comprise it. Local water issues are analogous to a multidisciplinary approach to planning. A systems approach represents the interdisciplinary step forward, and the watershed approach, when done well, is analogous to a transdisciplinary team approach to planning.

- Communicate complex issues with many types of people.
- Listen respectfully.

Planners rarely, if ever, work alone. A group of people is not a team unless it meets some essential criteria. Manktelow and Broadbeck (2009) propose seven defining factors for a team. In order for a group of people to be a team, its members must:

1. Share a common goal and approach.
2. Be small in number.
3. Work together.
4. Be mutually accountable for delivering the goals.
5. Possess complementary skills and abilities and work interdependently.
6. Interact in ways that have a direct bearing on results.
7. Include integration with other team members as a personal responsibility.

In a group, members work individually on distinct tasks that match their skills. In a team, members work jointly on tasks that require them to integrate their complementary skills and talents. The work product of a group equals the sum of the individual efforts, $1 + 1 = 2$. A team's work exceeds the sum of the individual efforts, $1 + 1 > 2$. In a group, the leader sets the goals, defines the approach, and sets the agenda. Goals, approaches, and agendas are determined jointly by the members of a team. Group members are individually accountable for their work. Team members are also individually accountable, but there is also a mutual accountability of the team to the project (Manktelow and Broadbeck 2009).

2.4 WHAT DOES A TEAM LOOK LIKE?

A team is made up of people. That happy-go-lucky biologist once served as an environmental advocate and is a painter. You'd probably never guess that your quiet unassuming urban planner leads a rock band and manages the books for his homeowners' association. Your team isn't just a group of technical experts but a simmering stew of human skills, ideas, values, and aspirations. A wise planner will find out what they are and put them to good use.

Every individual brings a unique set of personal qualities to the team:

- Expertise—the technical background people began learning in school and now practice professionally.
- Talent—skills and abilities not necessarily related to technical expertise, such as leadership, writing, and speaking.
- Affiliation—the groups people are formally associated with and represent, including employers, political, religious, fraternal, and other groups they belong to.
- Personal values—what we each believe is right and wrong, good and bad, the answer or not the answer.
- Personality—the essence of what makes each of us individuals. Are we introverts or extroverts, logical or intuitive, and just how do we approach each day?

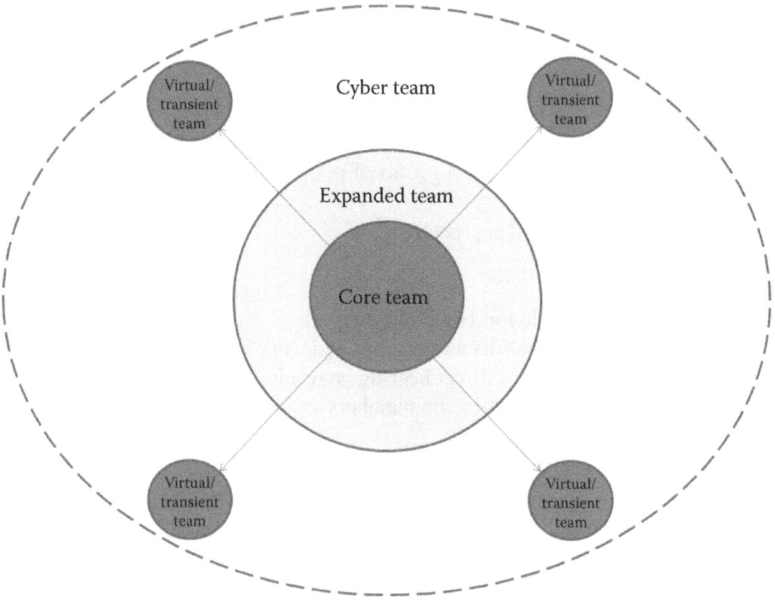

Figure 2.1 Structure of an idealized planning team.

If planning is an analytical and deliberative process, the two most critical elements of a successful team are to have access to the required expertise and the various perspectives about the planning issues on the team. The former is more common than the latter in my experience, because the essential core of a planning team is often housed in a single organization where diversity of perspective may not be rewarded or even appreciated. So let's begin with the team's structure.

There is usually a relatively small core of people who will do the vast majority of the analytical, administrative, and planning work. Figure 2.1 shows a stylized and idealized team structure. A planning team typically has a team leader/planning manager/project manager, or some similarly named individual whose job is to manage and oversee the planning process. It also includes content experts germane to the nature of the study. Water resource plans will require experts in hydraulics and hydrology, and wildlife plans will need biologists. Every planning study will have its required experts.

What kinds of experts are likely to be found in the core and expanded team of a natural resources planning investigation? The possibilities include

Architects
Archeologists
Biologists
Civil engineers

Communications experts
Cost estimators
Hydraulic engineers
Hydrologists
Economists
Epidemiologists
Environmentalists
Geographers
Geologists
Model builders
Planners (yes, they exist!)
Political scientists
Psychologists
Statisticians
Sociologists
Toxicologists
Wildlife managers

The core team is distinguished by its ongoing involvement in all matters of interest to the planning effort. They are the people responsible for the various analytical endeavors; they manage the budget, the schedule, and the conduct of the work. Often the required experts are found in a support team of "additional" personnel. This may include additional staffs from the planning organization, consultants, contractors, peer involvement and others who comprise the "expanded team." These are people with a more or less regular involvement with the planning effort who are not constantly engaged with the process. They contribute to the more labor-intensive tasks of the planning process.

To illustrate the distinction between the core and expanded teams, consider the economic analysis for a natural resource planning effort. A single more experienced economist may be a member of the core team. Her job is to advise the team on economic matters and to oversee the conduct of the economic analysis. In a more modest planning effort, she may do all the economics work herself, taking care to accomplish the work on time and within budget. In a larger investigation, she may contract the economics work to a consulting firm or she may supervise one or more support staff to accomplish the work. The contractor or the support staff would be a part of the expanded team. These team members are usually less concerned with the overall effort, including the administrative details of the planning effort and, thus, would not be as involved with the planning effort.

The expanded team is usually distinguished by the fact they are involved throughout much of the planning process. The virtual/transient team comprises individuals and groups who regularly plug in to and unplug from the planning investigation. They may work at a distance from the rest of the team. Virtual/transient team members may include decision makers and elected leaders who follow the investigation and provide input and feedback at selected critical points in the process. They also include experts who become involved in a limited way. This may include such things as special analysis, consultation, peer review, expert elicitations, and the like. Members of the public may periodically function as virtual team members, providing useful information and feedback to the rest of planning team at critical junctures. Numerous stakeholders and NGOs can function as transient team members who become intensely involved and interested in planning activities for short, sometimes intense, periods of time.

Figure 2.1 also shows a much larger cyber team. Virtual planning from a remote location enables the core planning team to open a window on cyberspace to involve interested parties from anywhere in the world. Discussion boards, listservs, wiki spaces, webinars, Twitter, and other means of communication and collaboration make it possible for planners to enlist help from virtually anywhere in the world. Problems can be identified and solved collaboratively in teams that far exceed the size of anything we have previously envisioned.

If it is not by now obvious, the team in Figure 2.1 is, indeed, a conceptual ideal. The core team is constantly involved with the planning process, the expanded team is regularly involved, the virtual team is sometimes involved, and the cyber team is involved on an opportunistic as-needed basis. Teams will vary just as planning efforts vary. The team planning a community walking trail will look very different from the team assembled to stop the great Gulf oil leak of 2010, and that will look different from a team assembled to conduct a congressionally authorized water resource study.

Each team will have an identity of its own. This identity stems in part from its structure or configuration, its purpose, its interrelationship with the larger culture, and the nature and qualities of its individuals. Every team, however, can be expected to pass through a series of stages before it becomes a good team.

2.5 STAGES OF A TEAM

Tuckman and Jensen (1977) define the five stages of a team's life as shown in Figure 2.2.* The first stage provides the team with structure and clarity of purpose. When a team first forms, it is a group of individuals on the way to becoming a team. There can be a lot of ambiguity, uncertainty, and anxiety. The members may be strangers to one another, and no knows whom to trust or even if the leader is trustworthy.

In the forming stage, the team leader, who may or may not be the study manager, should help the group focus on acceptance and trust. It is too early to try to dive deeply into the planning process. Trust building is crucial during this first stage. The team leader should invest time in free and open discussion that enables people to get to know each other. Spending time together, establishing clear goals, expectations, and rules for how the team will work while allowing time for asking questions, sharing information, and being sensitive to team members' needs is important.

* The discussion of the Tuckman and Jensen model that follows is based on a melding of the original works of Tuckman and Jensen (1977), Manktelow and Brodbeck (2009), and the Light Bryan Consulting Group (2011).

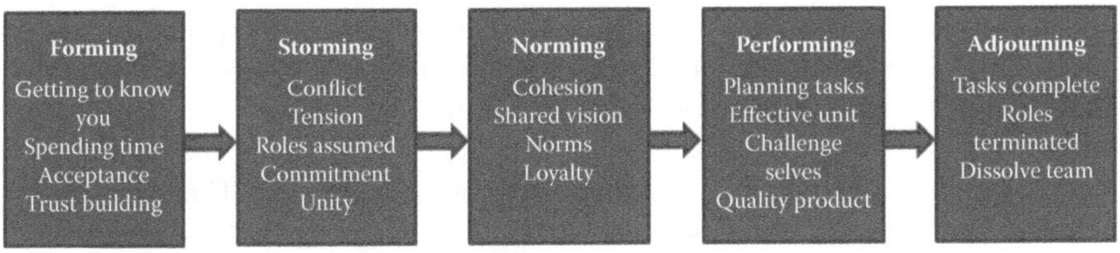

Figure 2.2 Tuckman and Jensen's model of team stages.

The storming stage is characterized by conflict. The wise team leader will focus his team-building efforts on team commitment and unity. As the many individuals begin to work together, conflict and tension are bound to arise. Team members have different roles and responsibilities that lead to different perspectives. It is common for the members to question the direction of the study, the leader's capability, and their teammates' abilities. As people become more comfortable within the team structure, they are more inclined to express their thoughts and feelings and to exhibit their true personalities. This is when the team will (or won't) learn to deal with adversity and conflict. This is a critical stage for the team. Teams can be destroyed at this stage. Team leaders must realize this conflict is not bad; people are just becoming comfortable enough to be themselves and take sides on issues. The team leader's priority now is facilitating effective conflict management. Getting to yes, finding win–win solutions, and actively avoiding "groupthink" is what the job requires.

The resistance that arises in the storming stage is hopefully overcome in the norming stage. This is the time for the team to come together and build cohesion, differentiate roles, and share a vision. The team begins to develop its norms, and members begin to conform to the norms. If the norm is overspending, underachieving, slacking off, and missing milestones, this is not a good thing!

In the norming stage, the initial uncertainty is gone, anxiety is reduced, and conflicts have been dealt with. Individual roles, mutual support, and commitment to the team are explored in this stage of the team's development. The more people work together, the more they begin to behave similarly. Loyalty to the team forms during this stage. Individuals' commitment to the team is expressed through cooperation, sharing information, ignoring disagreements, meeting obligations, and conforming to the team's standards of performance and behavior. This is when many of the team members' roles, described in Section 2.7, develop. In addition to technical roles, people begin to assume team function roles. Teams tend to get stronger or weaker at this point.

The best team leaders will notice who fills each type of role most naturally and then help these people build and use their skills to keep the team on track and working well together. Giving people feedback is especially important during norming. Leaders who can invoke a deeper meaning and purpose for the team at this stage will have the most successful teams. In-group feelings and cohesiveness develop and new standards evolve to support the new roles people adopt.

In the performing stage, group energy is channeled into the planning process tasks. This is the time for accomplishing things and raising the overall team performance. Now the team is working as an effective unit. The conflicts and challenges of the previous stages have been worked out. The lack of trust is gone, the shared vision is strong, and this is the most productive stage of the team's life. The team leader's job is to find ways to help the team continue to challenge themselves. That famous last kick by the team to finish things in a quality manner is simply the reward for surviving the earlier stages of the team's life.

The final stage of the team's development is adjourning and dissolving the team. All the tasks are complete, the roles are terminated, responsibilities are discharged, and the team returns to a bunch of individuals who are no longer strangers.

2.6 TEAM PROCESSES

Hire good planners, then get the right mix of people and skills on the team. Manktelow and Broadbeck (2009) suggest these skills can be broken down into three categories. First, your team must have the technical and functional knowledge required to address the problems of concern. Next, they must possess the teamwork skills necessary for an efficiently functioning team. These include both interpersonal and self-management skills. Examples of these include communication, problem solving, conflict resolution, collaborative, time management, goal setting, performance management, coordination, and other skills. Finally, a team needs an adaptive attitude. These are traits that enable a team to adapt, evolve, and remain proactive under changing conditions and uncertainty.

Manktelow and Broadbeck argue that an effective team must follow the these processes:

- Open communication—listening and sharing; do people share information, do they hear each other?
- Participation—building involvement and ownership; does everyone pull their weight?
- Consensus building—shaping a common purpose and establishing a common approach, setting aside differences, and finding points of agreement.
- Goal setting—shared and clear goals for planning content (small "p" planning) and procedures (capital "P" Planning).
- Accountability—individual and mutual accountability, schedule, budget, and other agreements are honored.
- Task orientation—clarifying roles, tasks, and assignments; what are the team's task-related skills? Economics, hydrology and hydraulics, hazardous, toxic, and radioactive waste (HTRW), and the like require expertise—do people know what to do, and how do we fill gaps in expertise?
- Civilized disagreement—hearing and honoring different viewpoints, seeking different perspectives.
- Style diversity—bring different experiences and styles to bear upon the combined team effort; no one personality or style dominates the team's work.

- Support for innovation—providing an environment that is conducive to innovation, breaking free of old ways of doing things.
- Collective learning—helping team members to learn from and along with each other, moving from multidisciplinary to interdisciplinary to a transdisciplinary team.
- Reflexivity—engage in constant reflection and review; adaptive learning is part of the team's modus operandi.

2.7 TEAM MEMBER ROLES

Good teamwork is essential to good planning. Think about the recurring and most important roles people play in a team, independent of their areas of expertise. There needs to be a leader of course, and there must be followers, too. What other roles might be useful for good teamwork? Over the years, I have noted a number of roles that recur on the more successful planning teams.

Doers make things happen. They accomplish things. They push the team forward and pull it along when necessary by the progress they make. Doers are needed to get things done. Duh!

Skeptics know consensus is a poor substitute for evidence. A major role of the skeptic is to slow down and when necessary to break up snowballing groupthink ideas. The skeptic regularly asks, how do we know it is so? What is our evidence? Skeptics ground planning in facts.

Visionaries see things that others do not, and they see in ways others do not. Visionaries are those mythical people who think outside the box and who can direct their creativity into the task before them. Visionaries see new possibilities and call them into being; they can imagine circumstances, situations, and things that do not currently exist.

Realists like to make sure they know what is in the box and to assure they have done their best in-the-box thinking before they are ready to move outside the box. Firmly grounded in the limitations of reality, they constrain the process to what is doable. Realists keep planning from spinning off into flights of fancy.

Thinkers are reflective, curious, and sometimes dogged. They can come back to an idea a week later and build on it because of the critical thinking they have done since it was introduced. Thinkers keep the process moving forward.

Reactors respond instinctively and at times emotionally. They blow hot and cold on ideas, sometimes on the same idea. They spur discussion by their incomplete enthusiasm. Reactors move the process.

Researchers remove uncertainty. They fill in the gaps in our knowledge and separate what we know from what we do not know. They seek knowledge. Researchers are planning's best weapon against uncertainty.

Writers communicate clearly with widespread audiences as well as with posterity. They are storytellers who understand that understanding is more important than the numbers. They write for their readers. Writers are a vital communication link.

Speakers communicate well with others in real time and face-to-face. They can marshal competence and expertise, empathy and caring, honesty and openness, or commitment and dedication as the circumstances require. Speakers are the face of the planning effort.

Technology wizards are the miracle workers who translate ideas into tangible results. They make the amazing maps, build the software, master the websites, and do astonishing things with software you've used for years but did not know it could do. They make digital things go. They understand the wires and the best ones can speak your language. Technology wizards make everyone more productive.

Bean counters and number checkers save the team's butt and its face. They are interested not only in getting the right science but in getting the science right. They make sure the models are verified, they check the calculations, and they read the quantitative stuff others give them. They keep track of dollars and balance budgets. They find your mistakes. Bean counters are the team's quality control people.

Peacemakers know that when emotions rise data do not matter. They are good at seeing all sides of a disagreement, and they can dispassionately resolve conflicts within and without the team. Blessed be the peacemakers.

What about the leader? This may be the most critical role of all. Every team needs a leader, and a team leader can emerge from virtually anywhere in the team. Normally, the study manager, by virtue of his experience, demonstrated ability, and responsibilities, will be the most logical candidate to be the team leader. That is not always going to be the case, however. In fact, many teams have been crippled when the manager is not the leader and no other leader emerges from the team.

Fisher (1993) says the best team leaders have certain recognizable behavior competencies. These competencies encompass some of the roles described above and also include living example, coach, business analyzer, barrier buster, facilitator, and customer advocate. These competencies or roles describe behaviors that would be valuable characteristics of any planning manager. When they describe someone other than the study manager, that person may well be the real team leader.

The leader unleashes energy and enthusiasm by creating a vision that others find inspiring and motivating. The living example serves as a role model for others by "walking the talk" and demonstrating the desired behaviors of team members and leaders. The coach teaches others and helps them to develop their potential, maintains an appropriate authority balance, and ensures accountability in others. The business analyzer understands the big picture and is able to translate changes in the business environment to opportunities for the organization. The barrier buster opens doors and runs interference for the team, challenges the status quo, and breaks down artificial barriers to the team's performance. The facilitator brings together the necessary tools, information, and resources for the team to get the job done, and facilitates group efforts. The customer advocate develops and maintains close customer ties, articulates customer needs, and keeps priorities in focus with the desires and expectations of the customers.

The leader on a planning team may change from time to time. As the study begins, the person most familiar with the people and places of the study area may be the team leader. At certain points during the study, team leadership may migrate to other shoulders based on the technical expertise of the team member. As schedules tighten, money runs short, and the frequency and importance of contact with the

A GOOD TEAM MEMBER

1. Attends team meetings regularly.
2. Participates in team brainstorming.
3. Works effectively as a team member by
   - Sharing communication
   - Negotiating
   - Facilitating
   - Participating
   - Cooperating
   - Trusting
   - Working toward and accepting consensus
   - Functioning as a teacher and learner
   - Valuing and using leadership skills
   - Using conflict-resolution skills
4. Makes original contributions to team issues; builds upon others' contributions.
5. Volunteers to handle action items or to participate in new teams.
6. Actively participates in establishing the team's purpose, direction, strategy, or goals.
7. Positively questions and challenges others; utilizes conflicting views in a constructive manner.
8. Acts to create and promote team cohesiveness.
9. Offers to relieve a team member's heavy workload.
10. Considers the impact on external relationships when influencing team outcomes.

Source: Adapted from DuBrin, A. J. 1995. *The Breakthrough Team Player: Becoming the M.V.P. on Your Workplace Team*. New York: AMACOM.

public increases, the study manager is more likely to resume the role of team leader. It is less important who leads the team than it is that it be lead.

2.8 WHAT DOES A TEAM DO?

Teams work together and do planning. To accomplish these two simple goals, however, some very basic needs must be met. Rees (1991) identified 10 essentials of teamwork that are expanded on below.

1. Common goals. You've already seen this many times in this chapter! A team needs a reason to exist. Commonly held goals rationalize the team's existence. Developing a clear, concise set of common goals that are completely understood and accepted by each team member is essential to the success of a team, and therefore to planning. Unarticulated or unclear goals are likely to be misinterpreted by team members; this virtually assures the team will suffer internal tensions, arguments, and cross-purposes. This is never more critical than when the team comprises members

of different organizations. Make sure it is always clear who is responsible for what. Avoid overlapping authority. Rather than have team members share responsibility for a single task, divide the task into distinctive parts and give each person more complete control in one of those parts.

2. Leadership. A team needs members who can lead when necessary, and that means multiple leaders. In addition to the skills described in the preceding section, the leaders motivate the team; coordinate its work; communicate clearly, consistently, and constantly; know how to get everyone involved; and get everyone's best work.
3. Interaction and involvement of all members. No one sits on the sidelines, and everyone plays the entire game. Everyone on the team must be a doer to some extent. Synergy and group spirit cannot be built on the backs of a few contributing members. It is essential to seek the active input and participation of all core team members.
4. Maintenance of individual self-esteem. Teams comprise individuals. Each individual's contribution must be heard, valued, and acknowledged. Nothing kills a team faster than a dominant personality, not to be confused with leadership!
5. Open communication. Team members must be and feel free to speak their minds openly in all team meetings as well as in informal communications. Every team member should be free to speak substantively to any other team member. Every team member should have free and personal access to the team leader and the study manager, if they are different people. Channeling all communication through a single source is a formula for team disaster. Regular contact and communication is essential to the team's success. It fosters an atmosphere of collaboration, provides opportunities for useful input and constructive feedback, and empowers team members to make joint decisions.

CHARACTERISTICS OF A POOR TEAM

1. Physical separation prevents members from meeting frequently.
2. The Team is not given adequate resources to do the job.
3. There is no recognition of team effort.
4. There is a lack of recognition by the organization or its leaders that a team exists.
5. Members do not participate in setting goals.
6. Goals are unclear.
7. Goals are not communicated.
8. Everyone is doing his own thing without attention to team goals.
9. Responsibilities are poorly defined.
10. No clear leader is identified.
11. There is buck-passing of responsibility.
12. Members engage in power plays for authority and control.
13. Members refuse to recognize their interdependence and act as if they were independent.
14. Decisions are always a crisis situation.
15. Decision making is dominated by one person.
16. Communications are one way—from the top down—and channeled through the leader.
17. Minor points are debated endlessly.

18. Meetings are unproductive with the issues unresolved.
19. Meetings cover trivia versus significant issues.
20. Actions are taken without planning.
21. Members work individually and ignore each other.
22. Members are late for meetings or do not attend.
23. Members are unwilling to be identified with the team.
24. There is covert conflict between members.
25. There are severe personality conflicts.
26. Relationships are competitive.
27. Members are defensive.

Source: Yoe, C. and K. Orth. 1996. *Planning Manual. IWR Report 96-R-21*. Alexandria, VA: Institute for Water Resources.

6. Power within the group to make decisions. Involve the whole team in deciding issues that require team consensus and commitment. The team goal, which will differ from planning objectives, is for each team member to feel ownership in the decision or idea. Teamwork needs to focus on the things the team has the power to influence. Teams should not spend time on things that will not get approved, that is unproductive, demoralizing, and a motivation killer. If the team has no power to make decisions, there is no real basis for the actual work of the team.
7. Attention to both process and content. Producing substantive content is the basic task of any team. Process pays attention to how work is structured, the general rules of working and playing well together, and how people get along. Without good process, the chances of good content drop markedly.
8. Mutual trust. Trust and openness are essential elements of good teamwork. They are often slow to develop naturally and once lost they are difficult to regain. Building trust through openness should be a priority for every team. Leaders can help build trust by spending one-on-one time with team members in an atmosphere of honesty and openness. Loyalty to individuals is an important aspect of trust building. The team-building exercises found in books and sitcoms are actually a useful way to build mutual trust. Team members must be afforded the opportunity to spend time together in situations that encourage open communication.
9. Respect for differences. The team must be a safe place for differences in people and in opinions. Every team member must be able to disagree without fear of being punished or ostracized. This respect must extend to personal differences as well as to professional differences among organizations, disciplines, and the like.
10. Constructive conflict resolution. Problems are going to come up. Conflict is going to happen. When it does, it must be addressed in a healthy way. Make sure there are no blocked lines of communications. Keep everyone fully informed at all times. Recognize interpersonal issues early. Deal with them. Do not expect them to blow over. Do not let them blow up. Be generous with positive feedback but be fair.

These essential elements will not just happen on their own, they must be built in place. If they are, your team has an excellent chance of doing some excellent work. If the team masters the planning process described in the following chapters, their chances of success are greatly enhanced.

Two other important aspects of planning teamwork are vitally important to mention here. The first has to do with spending time together; the second has to do with process, both team and planning process. First among these is that teams must spend time together, preferably face-to-face in the same place and at the same time. Technology makes it possible for teams to spend time together asynchronously or to meet in real time from many locations, and while we love and applaud that innovation there is no substitute for sitting in the same room together. If you cannot, then spending time together by any means possible working on tasks of mutual interest is the single most important thing a team can do.

Mutual interests do include matters of budget, schedule, and reporting recent activities, but this information can often be transmitted in more efficient ways, like email, memoranda, and such. Some anecdotal evidence gathered in my years of teaching planning to practitioners suggested that most meeting time was spent dispensing information with little to no time spent actually doing planning. Team together time is scarce in the best circumstance, so it should be spent solving problems and making progress on substantive planning issues of mutual interest. Because this is only Chapter 2 and you may have no idea what those things might be, it includes things like identifying problems, articulating planning objectives, identifying key uncertainties, projecting scenarios jointly, formulating alternative plans, evaluating those plans, tweaking those plans, comparing the plans, planning public involvement, and making decisions. The team can share ideas, brainstorm, coordinate their assumptions, argue for and against specific decision criteria, and collaborate with one another.

The second important item to mention is that process is important. There is a planning process, and it is the focus of this book; you'll see more on that later. There is also a teamwork process that is independent of the planning process, and it must be understood and built. Assigning people to a project, giving them some budget and a schedule, is not team building. That is the equivalent of dumping the contents of a "some assembly required" piece of furniture onto the living room floor. All the parts are there, but it is going to take a plan and some time to get this thing together. Be forewarned, you may only get one chance at some teamwork essentials, so measure twice and cut once.

2.9 TAKE AWAYS

Lesson 1: People are the most important part of planning, and people plan in teams.

Lesson 2: The core team is the heart and soul of a team but it is far from its only vital part.

Lesson 3: Good teams form, storm, norm, perform, and then adjourn.

Lesson 4: Teams are better than groups, multidisciplinary teams are better than teams, interdisciplinary teams are better than multidisciplinary teams, and transdisciplinary teams are better than interdisciplinary teams.

Lesson 5: There are many important team member roles; team leader may be the most important one.

REFERENCES

DuBrin, A. J. 1995. *The Breakthrough Team Player: Becoming the M.V.P. on Your Workplace Team*. New York: AMACOM.

Fisher, K. 1993. *Leading Self-Directed Work Teams, a Guide to Developing New Team Leadership Skills*. New York: McGraw-Hill, Inc.

Light Bryan Consulting Group. 2011. *Tuckerman's Stages of Group Development*. Accessed November 6, 2011, http://www.lightbryan.com/explain/tuckman.html.

Litman, T. 2011. *Planning Principles and Practices*. Victoria, BC, Canada: Victoria Transport Policy Institute.

Manktelow, J. and F. Broadbeck. 2009. *Team Tools: Build and Lead an Exceptional Team*. Mind Tools Limited. http://www.mindtools.com.

Rees, F. 1991. *How to Lead Work Teams*. New York: John Wiley & Sons, Inc.

Tuckman, B. W. and M. A. C. Jensen. 1977. Stages of small group development revisited. *Group and Organizational Studies* 2:419–427.

Yoe, C. and K. Orth. 1996. *Planning Manual. IWR Report 96-R-21*. Alexandria, VA: Institute for Water Resources.

CHAPTER 3

The Planning Process

3.1 INTRODUCTION

Planning begins with a chaotic mess and ends with the choice of a specific solution. There is no one right way to do planning. There are an infinite variety of paths from mess to solution. Planning, in best practice, is a creative and structured problem-solving process. This chapter introduces a process structure.

A process defines the sequence of tasks necessary to achieve an objective. It delivers something of value to people both inside and outside the process. The essential elements of a process are tasks to be performed, a sequence for their performance, resources to perform the tasks, an environment within which the process will operate, and a purpose to fulfill. Planning is the purpose of the planning process. The tasks of the planning process, once defined and put into a logical sequence, become the planning steps. Planning requires resources, the most important of which are human resources, whom you now know work in teams. The environment in which the process operates depends on who is using it and in what context. The planning process introduced in this chapter provides the organizing principle for the seven chapters that follow, the most important part of this book.

This chapter begins with a discussion of a good process and then it offers examples of a few representative planning processes so you know the process described in this book is not the only one out there. The process presented is an adaptation of the six-step planning process for water resources found in the U.S. Water Resources Council's *Economic and Environmental Principles and Guidelines for Water and Related Land Resources Implementation Studies* (1983), and it is introduced in Section 3.5.

3.2 WHAT MAKES A GOOD PROCESS?

Let's begin by considering the qualities we'd like a planning process to have. Davis (2009), writing about a business process, provides an excellent set of characteristics to build upon. He describes a good process as one that is effective, efficient, relevant, valid, usable, used, reused, managed, and measured. Here are some thoughts about how his structure might look adapted to planning.

Complete. The planning process must be complete. A complete process is predictable. The process must do what it is supposed to; in the case of planning that means it leads to a plan that will meet objectives, avoid constraints and in so doing will solve problems and realize opportunities.

Effective. An effective process meets the needs of its constituents. It is simple, flexible, and scalable. An effective process adds value for planners and those affected by plans by meeting the needs of the public.

Efficient. An efficient process uses all its resources to best effect. It avoids waste, unneeded tasks and steps, unnecessary iterations, and other wasteful practices. It collects only the data needed and uses the data collected. Planning is not research; it is applied science and does not do analysis to satisfy curiosities. It focuses on what is important and what will have a material bearing on decision making.

GET THE INFORMATION YOU NEED

I can recall dozens if not hundreds of hours spent early in my planning career collecting data to emulate the report content of my office's most recently completed planning document. Undoubtedly, that document itself was based on the most recent successful document before it, ad infinitum. The tendency to take the last successful report and add to it has, over time, lead to bloated and inefficient reports.

This wasted effort included such things as wind roses for fluvial flood studies, the age of housing stock, geologic histories of the Piedmont and other regions, and other esoterica that was not considered for a second in any decision-making process. Efficient planning collects all the data really needed and only the data needed to make good decisions.

One measure of efficiency is to reuse common processes. As planners iterate a task or step, it often improves the efficiency of the process. If an effective process meets the needs of the customer, an efficient planning process meets the needs of the planning sponsors.

Relevant. The relevant planning process has a significant and demonstrable effect on planning. It is not loaded with tasks and steps that contribute little or nothing to the intent and practice of planning. Administrative details, although essential to organizational operation, are not essential to planning. A relevant process is not overburdened or constricted by administrative details, policies, and procedures.

Valid. A valid planning process "gets it right" often enough to be valuable to planners and their constituents. The process needs to be verified and validated against the requirements of planners and their constituents. Does it meet the professional needs of planners? Does it meet the personal and societal needs of their constituents? If it does, it is a valid process.

Usable. The process has to be realistic to be usable. A realistic process is one that people can understand and execute with reasonable effort. It is clear enough in its principles that everyone understands them yet flexible enough in its detail that everyone can use it. A usable process can be easily updated as its users learn by doing. A person unfamiliar with the process should be able to understand it reasonably quickly. Its steps are tasks that are repeatable and predictable in terms of their outputs and outcomes. If people follow such a process, they should be able to do good planning.

Used. It is not enough for a process to be usable; it must also be used. The ultimate test of a good process is that people use it. For people to use it, they have to be aware of it and it must make a valued difference in their work. It needs to be better than what they were doing before. The planning process describes what people should do, and to the extent it changes what people do it is a "change management" effort. It must get buy-in, planners must be trained in its use, barriers to its use must be removed, and using the process must be rewarding and rewarded. The planning process needs the right environment and the right structures in place to be used. A top-down organization or a political body may lack the environment and structure needed for good planning, for example. Many organizations can point to a process on paper, but using the process is an entirely different matter.

Reused. The process must be reusable. Fortunately, the process is not consumed in the course of planning! However, if there is no effectiveness, efficiency, relevance, and validity, the process will not be used again. It has to work in all sorts of situations. A good planning process standardizes the approach taken to natural resource

IF LUKE SKYWALKER HAD BEEN A PLANNER

Obi-Wan: Use the process, Luke. Let go.
Darth: The process is strong.
Obi-Wan: Trust me.

problems, whether they are multi-year multimillion-dollar studies or ill-formed problems that walk through your door at 4 PM on a Friday afternoon demanding immediate attention.

Managed. The best planning process is one that is managed and improved. The six-step planning process of the P&G was a good planning process. The process in this book builds and, hopefully, improves on it, based on lessons learned using it for over a quarter of a century. It is not enough to use the process slavishly; it needs to be continuously improved as well. As you will learn, the process described here has steps and the steps have tasks and many of the tasks may involve their own mini-processes. So there are many parts of the planning process that can be tuned and polished over time. Planning, in particular, must stay aligned with the needs of society, and the needs of society continuously change. Societal improvement is the reward for a good process.

Measured. "You can't manage it if you can't measure it," might be the mantra of early twenty-first century managers everywhere. The planning process must not only be measurable, but it must be measured from time-to-time. Finding credible measures of a successful planning process should not be a problem. A process that leads to the desired outcomes is a reasonable measure of a successful process. This notion of measuring goes beyond the kinds of monitoring we've come to expect for plans with an adaptive management component. It includes measuring the outcomes of every plan to assure that the devised solutions are working. When they are not, the process may merit revisiting.

Some of these qualities of a good planning process are inherent in the process itself. Others are incumbent upon the users of the process. I'd like to add three generic qualities to Davis' list. A good process should be strategic, scalable, and organic.

Strategic. Planning is deliberate about affecting change. It strategically chooses a route to a different and more desirable future than would exist without planning. From the very outset it defines the problems to be solved and the opportunities for gain as well as the objectives to be achieved. Planning offers analytically derived causative explanations for its views of the future. It carefully explores the boundaries of knowledge and strategically chooses a plan of action intended to help society shape and achieve a desired future state.

PLANNER'S CHALLENGE

Litman (2009) identifies a fascinating challenge for planners involved in a forward-looking, change-oriented process. To facilitate change, successful plans must overcome entrenched values, interests, and practices. This is often easier said than done. Our species does not often do change well, and planners can encounter resistance from people who assume familiar = normal = good.

By nature, we tend to look back to the past rather than forward to the future. The successful planner needs to be able to overcome the inertia of the past to help communities create more promising futures.

Scalable. Planning is a perfectly scalable process. It fits any time frame and any budget. Planning can be done with the available resources, no matter what they are. You can iterate through the process in 15 minutes with no money, or you can iterate it many times over years with millions of dollars. The steps and the basic process are the same. The most significant difference is the nature and extent of the uncertainties with which you must deal in smaller-scale iterations of the process. Uncertainty is much greater when a problem walks through your door at 4 PM on a Friday and you only have an hour to react than it is when you have spent years studying a problem. Bear in mind your planning process works for problems of any scale.

Organic. Planning must live and breathe. It cannot become mechanized by the policy, guidance, and procedures of public agencies. It is a dynamic and creative process that cannot be reduced to a series of procedural obligations. The National Environmental Policy Act (NEPA) process,* for example, should be treated as a series of strategic opportunities to improve the quality of the planning process rather than a series of procedural hurdles. Federal agencies involved in natural resources planning often introduce stringent procedural guidelines and bureaucratic requirements that stifle the organic nature of planning. The more bureaucratic and mechanized the process, the less collaborative and creative it becomes. The goal of planning is not to execute mandated planning phases; it is to create a new future (Lachapelle et al. 2003). Nonetheless, planning must obey the law. It should comply with all applicable federal laws, executive orders, treaties, and regulations as well as all state and local laws and regulations that are binding on the study, its participants, and constituents. Planning must be an ethical practice that adheres to all relevant professional codes of conduct.

3.3 STEPS AND ITERATIONS

Before considering a specific process, it's important to understand the ubiquitous role of steps and iterations in planning. There needs to be a way to solve problems that will deliver you to an answer. Planning does that with a systematic approach that consists of the following elements:

- Planning process—defines a systematic sequence or series of coordinated tasks and activities to achieve a desired outcome, and these tasks are organized into steps.
- Planning practices—define the most efficient or best way to perform a task through principles and standards or the development of customary and habitual behaviors.
- Planning procedures—define a specific course of action to achieve a desired result.

3.3.1 Steps

The structure of a systematic approach gives planners a predictable and reliable way to proceed through the chaos of the most ill-formed problem situations to

* The NEPA process consists of an evaluation of the environmental effects of a federal undertaking, including its alternatives. There are three levels of analysis: categorical exclusion determination, preparation of an Environmental Assessment/Finding of No Significant Impact (EA/FONSI), and preparation of an Environmental Impact Statement (EIS).

WHERE IT STARTS NOBODY KNOWS

The planning process can begin at any step. Following Hurricane Katrina's devastation in Louisiana, many people were heard saying some version of "We need a Category 5 hurricane wall." That is a process that is beginning with a solution. That is an ugly way to begin a planning process, but the process begins wherever it begins.

a recommended course of action. Planning is an analytic–deliberative process for solving complex problems. That means there are cycles of careful analysis followed by deliberation that ultimately lead to a recommended course of action.

A good process requires clearly defined steps that define how planning is done in your organization. The steps represent a set of related and progressive activities. The steps are in order, and each step builds on the previous steps and provides inputs to the following steps. Somewhat paradoxically, planning can begin at any step. The steps are repeated as often as necessary, and every step must be done at least once. Some iterations of a step may take years, others occur almost instantaneously. Planning always ends on the last step.

There is no one planning process that will suit all resource planning needs, but every planning effort requires a specific planning process that is known to, understood by, and used by the planning team. The best processes will have a proven track record, quite simply they will work. Several processes are introduced beginning in Section 3.4.

3.3.2 Iterations

Perhaps the most important thing to understand about planning is that is iterative by nature and design. Although the process is described in a linear fashion, it is practiced in an iterative one. Both the analysis and the decision making are iterative. As evidence is gathered and uncertainty is reduced, our understanding of problems, objectives, potential solutions, and their effects are all improved. This in turn directs and influences the remaining analysis. Planning deliberations are also iterative. Many increasingly discerning and discriminating decisions are made throughout the planning process.

Some iterations can be anticipated in a study. Problems are understood and analyzed in an iterative fashion. It is rare for a study to begin with a complete understanding of the problems and opportunities. More often, that understanding comes in fits of analysis and starts of insight that lead to new cycles of analysis and so on. Cost estimates are usually based on relatively little design detail when planners are identifying and screening measures as they begin to formulate plans. Design and corresponding cost details increase as the final array of viable solutions is identified. The greatest level of cost detail is usually reserved for the recommended solution.

Thus, multiple cost estimates may be prepared for the same solution through the course of an iterated planning process. The attributes of the plans themselves are often tweaked many times between their initial formulation and their final disposition. There are iterations of the steps and iterations of tasks and activities inside the steps. Planning is iterations within iterations.

There are also iterations of the entire planning process. The life of a project, that is, the implemented manifestation of a plan, has three distinct implementation stages: preimplementation, implementation, and postimplementation. In some fields of practice—public works infrastructure, for example—the planning process is iterated during each of these stages. This book focuses on the planning done during the preimplementation stage of a project's lifecycle. This is when the bulk of the planning is done.

Preimplementation planning may, itself, consist of more than a single iteration of the planning process. The administrative process that directs planning will vary from one context to another, and it's impossible to do justice to all the variations on the limitless administrative themes. In natural resources planning, it is not unusual to have an initial stage of planning, sometimes called a reconnaissance stage, to establish the likely existence of a viable project. This "recon stage" is often conducted with a limited budget, less detail, more uncertainty, and a shortened time frame. This stage is then followed by a more detailed planning study, sometimes called a feasibility study. This is the stage during which the planning process is most intensive.

Either of the recon or feasibility studies may languish on the shelves for years for any number of reasons. At times, the delay may necessitate a "do-over" of the study. A good example of this would be the so-called "308 Reports" prepared during the 1920s through the 1940s, some of which were revived and updated decades later. Iterations may be days or decades apart.

Following the feasibility stage or its equivalent, a decision is made whether to move to the implementation stage of the project or not. Simpler plans may move from feasibility planning directly to construction. More complex plans will often undergo another round of planning, sometimes called advanced engineering and design (AE&D), project engineering and design (PED), or something similar. By this point, the recommended plan is generally identified and that iteration of the planning process is oriented toward refining the plan. The level of design detail available in early planning iterations is often quite low. These details are provided during this AE&D planning.

Significantly, as planning moves through the implementation stages, uncertainty is constantly being reduced. As this happens, surprises are sometimes encountered and

308 REPORTS

In response to an authorization in the River and Harbor Act of 1925, the U.S. Army Corps of Engineers completed comprehensive multipurpose planning studies for some 200 rivers in what came to be known as the "308 Reports." Although the plans did not set forth recommendations, they did include specific plans of improvements and projects.

iterative adjustments are made. An unanticipated HTRW site may be discovered in the planning area, for example. In other situations, new opportunities to improve a plan may appear. These are all considered part of the implementation iterations of planning.

Another implementation planning iteration may occur when construction of the physical features of a plan begins. As implementation moves forward and remaining uncertainties about actual conditions are resolved, it is sometimes necessary to amend or "tweak" the plan to make it fit actual conditions on the ground. The same tweaking may occur as changes in the ways of doing things (as opposed to constructed components of a solution) are implemented. These iterations comprise the implementation stage of a project.

Once a project is implemented, it moves into its postimplementation life cycle stage. Changes in social values may necessitate changes in the way a project is operated or maintained. These changes should be made in another iteration of the planning process. A common example is a reservoir storage reallocation study. These studies consider changing the proportions of reservoir storage dedicated to different project purposes like hydropower, water supply, flood risk management, and the like. The Comprehensive Everglades Restoration Project (CERP) under construction in Florida is, in a sense, a postimplementation study of the Central and Southern Florida Project authorized by Congress in 1948. Changing values and environmental conditions have necessitated a new round of studies and a reiteration of a planning process and decision from long ago.

The best planning processes will provide for monitoring the outcomes of an implemented project. Periodically, the monitored data are evaluated to assure that desired outcomes are being obtained. When they are not, it may be advisable to iterate the process again and modify the project and its operation. The final postimplementation iteration of the planning process may occur when a functioning project is deauthorized, decommissioned, or removed. Planning always ends at the last step, but that is often just the beginning of the next iteration of planning.

In an iterative process, significant time may pass between some iterations.* Iterations can also simultaneously overlap in such a way that you may be working

> Concerns like public safety, risk analysis, environmental justice, social vulnerability, social equity, heritage preservation, energy independence, homeland security, nonmotorized transportation, public health, and sustainability have all become legitimate resource planning issues in recent years. A few decades ago, no one was concerned about such things. In a truly iterative decision-making process, no decision is ever final; it is conditional on what was known at the time it was made. When conditions change, decisions may warrant revisiting. Changing social values alone are sufficient reasons to welcome an iterative planning process.

* As with the time that passes between a reconnaissance stage and a feasibility stage or the time that passed before the CERP study was undertaken.

SEC. 1508.13 FINDING OF NO SIGNIFICANT IMPACT

"Finding of no significant impact" means a document by a federal agency briefly presenting the reasons why an action, not otherwise excluded (Sec. 1508.4), will not have a significant effect on the human environment and for which an environmental impact statement therefore will not be prepared. It shall include the environmental assessment or a summary of it and shall note any other environmental documents related to it (Sec. 1501.7(a)(5)). If the assessment is included, the finding need not repeat any of the discussion in the assessment but may incorporate it by reference.

Source: Council on Environmental Quality. 1978. "Regulations for Implementing NEPA." *Code of Federal Regulations Title 40 Parts 1500–1508.* Accessed June 14, 2012, http://ceq.hss.doe.gov/nepa/regs/ceq/toc_ceq.htm.

on several steps at once. Other times the work you are doing may contribute to multiple steps. An iterative process is a reasonable adaptation to a world of uncertainty. Planning needs to be adaptive to new science, information, resource uses, technology, trends, conditions, and other circumstances of importance. Planning tends to evolve over time with new issues, changing social values, reductions in some uncertainties, and appearances of new ones. Planning's iterative nature is one of its greatest strengths. Don't worry about what step you are on or which iteration it is. Let the process flow and when your study blows up at some point, as many studies do, remember your process and return to it.

3.3.3 The Scoping Iteration

A good planning process will include a scoping activity early in the planning study. This is often the first formal iteration of the problem identification process. Scoping frames problems and opportunities in their planning context and provides as much information as possible to guide the subsequent investigation. Scoping assesses and defines the work that needs to be accomplished to complete your study. The NEPA process (Council on Environmental Quality 1978) defines scoping as follows: "Scoping is an early and open process for (1) actively bringing the public into the decision-making process, (2) determining the scope of issues to be addressed, and (3) identifying the major issues related to a proposed action (40 CFR 1501.7)."

3.4 EXAMPLES OF PLANNING PROCESSES

Before presenting the planning process used in this text, it is worth a moment to stress there are many planning processes available.

3.4.1 The P&G

Let's begin with the intellectual parents of the process I'll introduce in the next section. It's the six-step planning process found in the P&G (1983). Figure 3.1 is the invention of the U.S. Army Corps of Engineers and one I used for many years. The P&G description of these steps follows.

Specification of the Problems and Opportunities Associated with the Federal Objective and Specific State and Local Concerns

(a) The desire to alleviate problems and realize opportunities should be specified for the planning area in terms of the Federal objective and specific State and local concerns. The problems and opportunities should be defined so that their definition does not dictate a narrow range of alternatives.

(b) The problems and opportunities should be defined in such a way that meaningful levels of achievement can be identified. This will facilitate the formulation of alternative plans in cases in which there may be financial, environmental, technical, legislative, or administrative constraints on the total alleviation of a problem or realization of an opportunity.

(c) The problems and opportunities should be stated for both current and future conditions. Desired conditions for the future should be explicitly stated.

(d) The problems and opportunities should reflect the specific effects that are desired by groups and individuals as well as the problems and opportunities declared to be in the national interest by the Congress or the Executive Branch. This identification and detailing of problems and opportunities is the process of making explicit the range of preferences and desires of those affected by resource development. It should be understood that the initial expressions of problems and opportunities may be modified during the planning process.

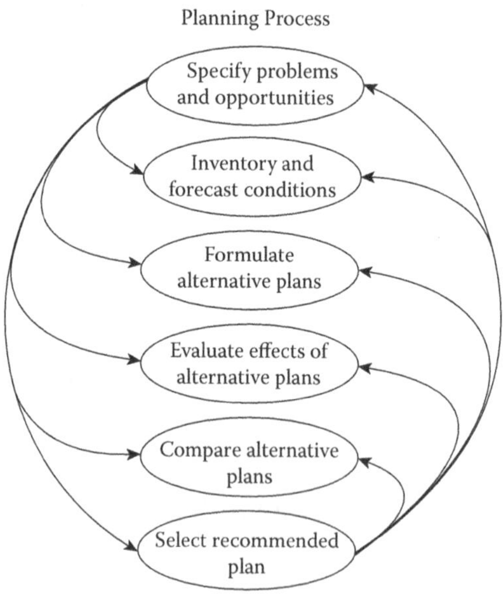

Figure 3.1 Planning process of the U.S. Army Corps of Engineers based on the P&G.

Inventory and Forecast of Water and Related Land Resource Conditions

The potential for alleviating problems and realizing opportunities is determined during inventorying and forecasting. The inventory and forecast of resource conditions should be related to the problems and opportunities previously identified.

Formulation of Alternative Plans

Alternative plans are to be formulated in a systematic manner to insure that all reasonable alternatives are evaluated. Usually, a number of alternative plans are identified early in the planning process and become more refined through additional development and through subsequent iterations. Additional alternative plans may be introduced at any time.

Evaluation of Effects

(a) General. The evaluation of the effects of each alternative plan consists of assessment and appraisal.
(b) Assessment. Assessment is the process of measuring or estimating the effects of an alternative plan. Assessment determines the difference between without-plan and with-plan conditions for each of the categories of effects.
(c) Appraisal.
 (1) Appraisal is the process of assigning social values to the technical information gathered as part of the assessment process.
 (2) Since technical data concerning benefits and costs in the NED account are expressed in monetary units, the NED account already contains a weighting of effects; therefore, appraisal is applicable only to the EQ, RED, and OSE evaluations.
(d) Displays. The results of the evaluation should be displayed according to the directions provided in Section VIII—Displays.

Comparison of Alternative Plans

(a) The comparison of plans focuses on the differences among the alternative plans as determined in the evaluation phase.
(b) The differences should be organized on the basis of the effects in the four accounts or on a combination of the NED account and another appropriate format for other significant effects.

Plan Selection

After consideration of the various alternative plans, their effects, and public comments, a plan is selected following the general guidance in Section X—Plan Selection.

3.4.2 U.S. Forest Service, Pacific Southwest Region

A Forest Service process, found on the U.S. Department of Agriculture Forest Service website (2012), consists of the following steps:

1. Identification of purpose and need
2. Planning criteria
3. Inventory data and information collection
4. Analysis of the management situation

5. Formulation of alternatives
6. Estimated effects of alternatives
7. Evaluation of alternatives
8. Preferred alternative recommendation
9. Plan approval
10. Monitoring and evaluation

Steps 1 through 8 are consistent with the P&G process. Steps 9 and 10 extend it beyond the P&G process. Where the process ends is perhaps the greatest divergence among planning processes. Some end at selection of the best plan, others continue through implementation, and some continue through a monitoring, evaluation, and modification stage. Some will single out tasks, subsumed in the steps of one process, and make them independent steps. None really stray far from the quick start process in the following text box.

3.4.3 Victoria Transport Policy Institute

Litman (2009) offers the following 10 steps for a generic planning process:

1. Establish the basic planning framework, including scope, stakeholders, schedule, and so on.
2. Invite stakeholder input to share ideas and concerns.
3. Create a vision, goal, or problem statement.
4. Develop a list of possible *options* (also called *solutions*) using various information resources (brainstorming, publications, websites, experts, etc.).
5. Evaluate and prioritize options from best to worst.

QUICK START

I bought a new wireless printer/scanner/fax machine/coffee maker/diaper changer recently and was dismayed to find a 200-page user's manual in four languages, one of which I hoped to speak. I was saved by the quick start card for those who wanted to jump right in and start using the printer. I can always go back to the manual when problems arise or when I want to understand the details.

(*Source:* Ken Orth, planner and friend)

Here is your quick start guide to the planning process:

1. Write down the problem you are trying to solve.
2. Make a list of things that will change if you solve this problem.
3. Think up as many solutions for the problem as you can.
4. Write down the pros and cons of the different solutions.
5. Pick the best solution and explain why you chose it.
6. Document the results of the previous steps in a narrative story.

6. Create a *plan*, which identifies who does what, when, and how. This may include a long-term *strategic plan* and short-term *action plans*. It may include contingency options that are only implemented if warranted by future conditions.
7. Gather baseline data (data collected before the plan is implemented).
8. Implement policies and programs.
9. Evaluate program (gather data after the program is implemented to determine whether it is achieving objectives as expected).
10. Revise plan as appropriate.

3.4.4 National Association of Recreation Resource Planners

The NARRP is preparing a book on best planning practices a chapter at a time. Haas (2010) in the second chapter describes the following planning process:

1. Identification of issues, concerns, opportunities, and threats
2. Establishment of planning and decision criteria
3. Inventory of resources and best available science
4. Formulation of alternatives
5. Evaluation of alternatives
6. Selection of preferred alternative
7. Implementation
8. Monitoring and adaptation

3.4.5 National Environmental Policy Act

The National Environmental Policy Act of 1969 (commonly called NEPA) is generally known for its reporting requirements, particularly the requirements on federal agencies to prepare Environmental Assessments and Environmental Impact Statements for many of their actions. But at the heart of these documents is a planning process that is best revealed in the recommended format for an impact statement (CEQ1978, 40 CFR 1502.10):

a. Cover sheet
b. Summary
c. Purpose of and need for action
d. Alternatives including proposed action
e. Affected environment
f. Environmental consequences
g. List of preparers
h. List of agencies, organizations, and persons to whom copies of the statement are sent
i. Index
j. Appendices

Here, and especially in parts (d), (e), (f), and (g), you see concepts and language very similar to the planning tasks included in almost every other planning process. Where other planning processes presented tasks like formulation, evaluation, and selection, the impact statement guidance presents the same concepts as report titles. If you write a NEPA document, you will by default have followed a basic planning process. Any number of additional planning processes can be identified. The purpose in showing you a few is simple; understand there is no one right way to do planning.

3.5 FIVE BY TWO

The 1983 P&G planning process is a good model. As time has passed, social complexity has grown and the pace of change has accelerated in many dimensions. Uncertainty, risk, and evidence-based decision making have become more prominent considerations. Stakeholders and public involvement are finally being taken more seriously. That has necessitated some basic changes to that model.

I'd like to draw a line between preimplementation planning and implementation planning and postimplementation planning. The planning process introduced here is the preimplementation planning process. It terminates with the selection or recommendation of a plan or solution. Implementing the plan is part of implementation planning. Monitoring, evaluating, and modifying a plan are parts of postimplementation planning.

The basic preimplementation planning process presented in this book is shown in Figure 3.2. It has five steps and two ongoing processes. The arrows suggest the iterative nature of the process. Although the predominate flow is from top to bottom, the process may at times reverse or skip steps in either direction. The five steps feed the two processes, which in turn feed the steps. The entire process is itself iterative.

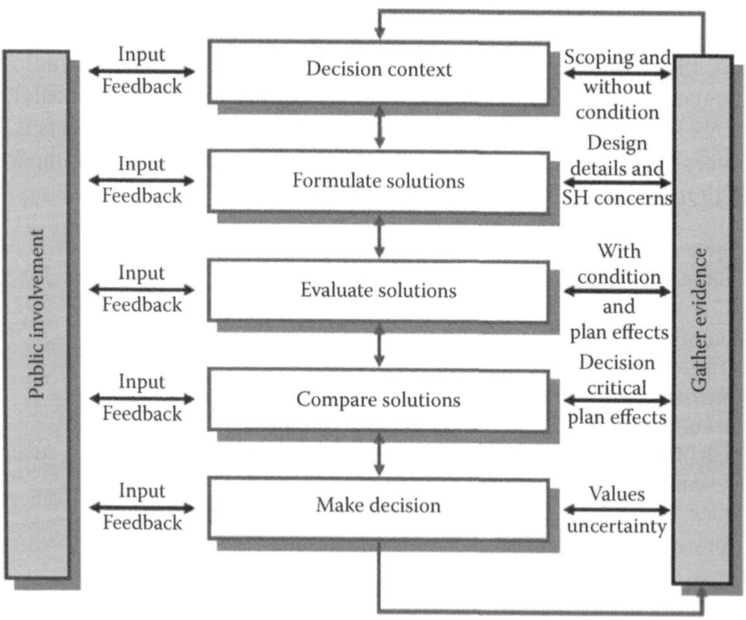

Figure 3.2 The "Five By Two" planning process.

An initial reconnaissance iteration may be followed by a feasibility iteration or a design iteration. In any one of these three iterations there can be multiple iterations. For example, a study may begin with a quick iteration of the entire planning process (see Chapter 15), followed by another complete iteration using the readily available data, and culminating in a final iteration of all the steps based on the best available data developed during the planning process.

The arrows connecting steps to the processes suggest some of the dominant activities at each step. The paragraphs that follow introduce each of the planning steps and the principle public involvement and evidence gathering tasks in each. Chapters 4 through 10 develop the steps and processes in detail.

The primary purpose of the first planning step is to identify the problems you will try to solve, the opportunities you will try to realize, and the planning objectives and constraints that frame and define a successful solution. This is the best time to identify the decision criteria that will be used throughout the study. An initial assessment of what you know and do not know is made, and information gathering strategies are laid out. Public involvement begins here with opportunities for input to the decision context as well as feedback as these Step 1 outputs are completed. Evidence sufficient for scoping the decision context is gathered in this step. This will include preparing a without condition scenario and sufficient analysis to verify the existence of the problems and opportunities identified.

The primary purpose of the second planning step is to identify as many different potential ways to achieve the objectives and avoid the constraints, thereby solving problems and realizing opportunities, as possible. The plan formulation step has three essential tasks, which include

1. Identifying measures that meet the planning objectives
2. Combining these measures to build plans
3. Reformulating the plans as necessary

Public input and feedback are critical to the formulation of viable alternative solutions. The public is an essential source of information and evidence about community concerns that must be considered in the formulation of solutions. In addition to gathering evidence on these social preferences, the team will require substantial information required to design solutions.

The primary purpose of the third planning step is to qualify plans for comparison in the fourth step by identifying viable solutions from among the many potential solutions formulated. The most important effects of the viable plans are identified in this process. Public involvement seeks feedback on the evaluation results. Evaluation includes the following tasks:

1. First evaluation iteration—is it a plan yet?
2. Prepare one or more with condition scenarios for each formulated solution.
3. Conduct the necessary analysis within the with condition.
4. Compare with and without condition scenarios.
5. Assess, that is, describe quantitatively where possible, all important differences between the two conditions.

6. Appraise, that is, judge, the solution's effects.
7. Qualify the plan for further consideration, reformulate it, or drop it from further consideration.

Accomplishing these tasks requires evidence-intensive analysis of the effects of the formulated plans.

The primary purpose of the fourth planning step is to compare viable solutions to each other for the purpose of highlighting the differences among them that will make a difference to decision makers. Public involvement in this step is likely to be dominated by feedback; it is largely an analytical step. The comparison step includes these tasks:

- Identify the differences.
- Compare the differences.
- Contrast the differences.
- Identify trade-offs among the differences.
- Display the differences with uncertainty.

This analytical step will identify decision-critical plan effects.

The primary purpose of the fifth planning step is to weigh the differences among solution effects and select the best solution for achieving the objectives and avoiding constraints, thereby solving problems and realizing opportunities. The tasks in this step include the following:

- Confirm or select the decision criteria from Step 1.
- Create a decision matrix.
- Identify the decision-making process or style.
- Weigh the criteria.
- Apply decision model, if one is used.
- Order the alternatives.
- Consider the uncertainty.
- Choose a plan of action—including no action—taking uncertainty into account.
- Document the basis for the decision.

Public input and feedback on the decision-making process are essential pieces of the public involvement program. Evidence about the social values and preferences important to decision making must be gathered and considered. It is critically important that all significant remaining uncertainties be presented and understood by decision makers.

3.6 HERE'S HOW IT WORKS

Planning is rarely done in a linear sequence of five discrete steps. More often it is simultaneous, random, and repeating. The process does not always begin at Step 1, but it always ends at Step 5. There may be times when your planning process borders on or steps over into chaos; a planning process brings order from chaos. When your planning process slips into chaos, that is the best time to pull out your "Five By Two" checklist and ask, have we defined the decision context, do we have the data we need, and so on. That is when it is important to return to the process and proceed again, returning each time chaos threatens. Eventually, each step must be completed and in a good process they will be.

If you are fortunate enough to become involved in a planning study from the beginning, please know that not many applications are orderly in the way they go about following the process. The process is never perfect. One step bleeds into another; you jump over some steps and double back on others. That is all okay. Just keep returning to the process as often as you need to and you will reach the end.

Sometimes, a plan begins with the solution. That solution may come from a previous study, from an elected official, from the public, the media, an interest group, or even a random individual. Someone will plop down a detailed plan for a path along the old canal, they will have a plan for preserving sea grasses and hard bottoms in the inlet, or Congress may tell the U.S. Army Corps of Engineers to provide category 5 hurricane protection for some coastal state. Plans begin where and how they begin, and that is not always at Step 1. Nonetheless, establishing the decision context is an essential early step in any planning process. Problems must be identified or confirmed, objectives need to be clearly established, and the many tasks you'll read about in the chapters ahead need to be accomplished at some point. Sometimes you will begin at the beginning, sometimes somewhere in the middle, and occasionally you'll begin at the end of the process, with a solution someone has already chosen for you. It's all good.

Planning investigations can be kicked off by data. The ducks are disappearing at Chelsea Marsh. Sightings of the American panther are becoming rarer. Dissolved oxygen is falling in Edith Grove Lake. Is sea level rising or falling? The polar icecap appears to be melting. Why? What is the problem? What is to be done about it? Data collection and analysis are often considered the meat and potatoes of the planning process. It is what analysts and specialists were trained to do. Aerial photography, foundation explorations, surveys, data collection, model runs, and engineering, environmental, cultural, economic, and other analyses are the things that take so much time and consume so much of the budget. While these are undeniably the greater part of the planning process, evidence gathering is also the least planning-oriented part of that process. Thus, it is shown as a process that begins in Step 1, continues through each of the planning steps and does not end until a decision is made in Step 5.

Evidence gathering usually begins in a scoping process. Chances are you began with some but not enough evidence about the problems and opportunities or even the solutions that motivate your study. Certainly, the study will require more gathering of evidence. Normally, evidence is gathered throughout the study, often right up to—and even beyond—the choice of a recommended solution.

Often the work is phased in a very logical and natural way. Evidence is gathered to identify or verify problems and opportunities, and a without condition scenario is defined. All this and more happens in Step 1. As plans are formulated, new evidence has to be gathered. For example, the costs of the solution eventually selected will be estimated. Early in the study, conceptual costs may be used, and these are very crude cost estimates based on some rather general assumptions and past experience. As the study progresses and more evidence becomes available, cost estimates are refined and improved. Foundation explorations, HTRW analyses, and utility surveys are just a few examples of work done later in a study that must precede the cost estimate.

You may find substantial portions of your study schedule are spent waiting for data. Some surveys and data-gathering efforts can only be conducted at certain times of the year. You might require data of a certain duration or frequency of observation. You may have to wait for special equipment, contractors, or conditions to become available. Some things cannot be done until after other things have been completed. A delay in one sequenced work unit can throw your entire schedule off. Don't underestimate the time you will spend waiting for others to provide you with items (data, permits, signed letters, access to property, approvals, comments, and such) you need to proceed. You will spend time waiting for others to review your evidence as well. Sometimes they will reject your evidence and tell you to start over. All this work comes under the general heading of evidence, that is, the data, information, and wisdom required for planning. It happens throughout the planning process.

Data are one of the most common causes of rebooting your study and beginning again. Natural resource problems are often more complex than they first seem. You may think the ducks are disappearing because of a lack of hunting restrictions. What begins as a wildlife management study may veer off into a water quality or ecosystem restoration study when your evidence gathering reveals the true cause for the ducks' demise is quite different. As evidence is gathered, uncertainty is reduced and your understanding of problems, opportunities, and their solutions can change dramatically. A growing evidence base can send your planning process back to a previous step faster than anything.

As you read through the chapters that follow, you will see a well-structured and very rational process. In Step 2 you formulate solutions, in Step 3 you evaluate them, and in Step 4 you compare them. It is nice when it works that way. The problem is it does not always work that way. Formulation, evaluation, and comparison tend to get mashed up all together at times. As soon as you begin to identify potential solutions, people will begin giving you good reasons why something will not work or why some other solution will work better. You may find yourself evaluating and comparing even before you've formulated your first plan. These middle steps really do tend to run together in practice. If the steps begin to bleed together, that is okay, let them bleed.

The two ongoing processes of evidence gathering and public involvement are included in the planning process to help you get it right the first time. It is always better to do it right than to do it over. On your way to a decision, it is essential that you involve a lot of people. Public involvement in the process is essential for some very basic reasons. Among them (Bryson and Carroll 2002) are so that you will:

- To Meet regulations and requirements.
- To Adhere to democratic principles—people have the right to influence what affects them.
- To Improve the process of finding problems that can and should be solved.

- To Enhance future problem-solving capacity.
- Arrive at better, more substantive decisions and outcomes as a consequence of
  - More information.
  - More perspectives.
  - Increased mutual understanding.
  - Free consultants.

The planning steps provide a path for an orderly process; they provide essential checkpoints for a messier one. It's great when you are planning and can say, okay, today we'll begin Step 3 and then continue to do so until you begin Step 4. When your process is not quite so orderly, it is important to take stock of the process from time to time. This means asking, how have we done with the first step? Is it complete? Do we have a without condition scenario? Should we revisit any aspect of it? No? Okay, how are we doing with our formulation of solutions? Are all the essential data available or underway? Use the steps as a checklist if your process is a messier one.

In summary, a process is a set of progressive activities called steps each with its own tasks and mini-processes. The steps are in an order, but the process can begin at any step. Each step builds on the previous step and provides inputs to the following step. Every step is at least done once, but the steps are repeated (iterated) as necessary. Some steps take years, and some occur in the blink of an eye. The process always ends with the last step.

3.7 RISK-INFORMED PLANNING

What does not appear in the step names and the brief description of tasks is the fact that planning ought to be a risk-informed planning (RIP) process. RIP is the confluence of risk analysis,* depicted in Figure 3.3, and planning. Think of RIP as planning with a keen focus on addressing significant uncertainties in the planning process in an intentional and honest way. RIP requires planners to identify and convey the significance of uncertainty in their investigations, and decision makers take that uncertainty explicitly into account in their decision-making processes. RIP develops and uses risk information to aid decisions made under uncertainty.

Planning is making decisions for the future, and the future is fundamentally uncertain. Planning, then, is a form of decision making under uncertainty. It always has been and always will be. When planners take care to separate what they know in a planning investigation from what they do not know and then choose to be intentional and purposeful in the ways they address the things they do not know, this is RIP.

Risk is the chance of an undesirable future outcome. What usually creates the "chance" is a lack of information. We lack information because there are facts we do not know and because the universe is inherently variable. Risk analysis is an emerging paradigm for decision making under uncertainty. It comprises the three tasks in risk analysis, that is, risk management, risk assessment, and risk communication, as shown in Figure 3.3.

* This topic is revisited in Chapter 12.

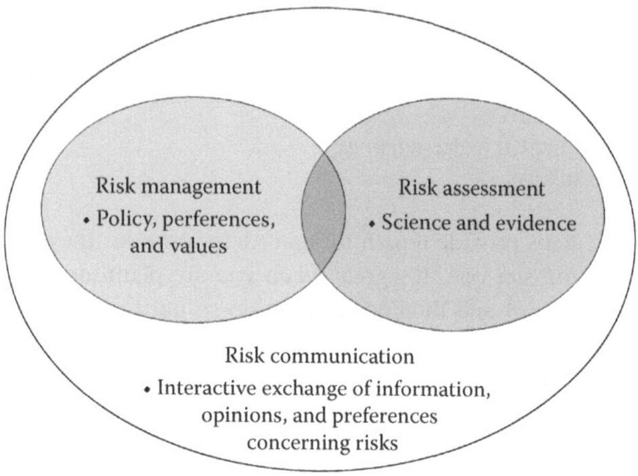

Figure 3.3 Three tasks of risk analysis.

3.8 WHAT DO YOU DO WITH THE PROCESS?

What do you *do* with the planning process? Use it. But, if you have 2 years for your study and you start scheduling time for each step, for example, months 1 to 3 for decision context, then you're doing it wrong. It doesn't work that way, and it is much messier than that. You might not even start with Step 1.

Use the planning process as a blueprint. You're building a plan, and it is going to need at least one of each of these steps. It helps if the entire team knows what you're building and, in general, how it gets built.

Use it as a map. If you're stuck figuring out what's next, check the process. Where are you? What have you done so far? What is the next set of tasks to do and questions to answer? Maps show you alternative routes to the place you want to go, and so does the planning process. Check your map, know where you are, see the ways to get where you want to go and choose one. Don't wander off the map!

Use it as a meeting agenda. In one of your first team meetings, walk through the planning process for your study. In a two-hour meeting, you can complete each step in about 20 minutes or so. That way everybody has experienced the flow of the process. It's a good way to begin. If you iterate through the process each time the team meets, summarizing what is done, focusing on what needs doing now, and anticipating what remains to be done, everyone will know and appreciate the process. The step you emphasize will change as you progress, but keep at it until the study is done. Meetings based on this process will guarantee you have time to actually plan together.

Use it as a touchstone. When all hell breaks loose and chaos reigns, ask yourself, "Where are we in the planning process?" Is this confrontation about a future condition (Steps 3 and 5); did the new city council endorse a totally new plan (Step 2 or maybe Step 5); is the National Environmental Policy Act demanding more studies (evidence gathering and maybe Steps 3 and 4)? Figure out where you are in the process, and it will help you understand the context of what's happening. It will help you understand what happened to get you into this mess, and where the mess ultimately needs to go.

Use it to tell your story. Don't use the process steps as a report outline, but the five steps and two processes make great storyboards to help you tell your story. What happened in each step? Why did it happen that way? How did it turn out? There's your story, go tell it.

Use it to measure your progress. How do you know when you're done planning? When you've completed the last step for the last time, and you've completed all the previous steps at least once. Then you're done.

3.9 TAKE AWAYS

Lesson 1: There is no one right planning process.
Lesson 2: All processes have steps and iterations.
Lesson 3: The "Five By Two" process is the focus of this text.
Lesson 4: A good process will intentionally address key uncertainties, that is, it is risk informed.
Lesson 5: Trust the process.

REFERENCES

Bryson, J. M. and A. R. Carroll. 2002. *The What, Why, Who, How, When and Where of Public Participation*. Minneapolis, MN: Hubert H. Humphrey Institute of Public Affairs and the University of Minnesota Extension Service.

Council on Environmental Quality. 1978. "Regulations for Implementing NEPA." *Code of Federal Regulations Title 40 Parts 1500–1508*. Accessed June 14, 2012, http://ceq.hss.doe.gov/nepa/regs/ceq/toc_ceq.htm.

Davis, R. 2009. What makes a good process? BPTrends. Accessed February 6, 2012, http://www.bptrends.com/publicationfiles/FIVE11-09-ART-Whatmakesagoodprocess-BPTrends.pdf.

Haas, G. E. 2010. *Chapter Two: Principles of Recreation Resource Planning*. Accessed February 6, 2012, http://74.91.226.186/Best_practices_book/Chapter_2_Recreation_Resource_Planning_Principles.pdf.

Lachapelle, P. R., S. F. McCool, and M. E. Patterson. 2003. Barriers to effective natural resource planning in a "messy" world. *Society and Natural Resources*, 16:473–490.

Litman, T. 2009. *Planning Principles and Practices*. Victoria, BC, Canada: Victoria Transport Policy Institute. Accessed February 6, 2012, http://www.vtpi.org/planning.pdf.

U.S. Department of Agriculture Forest Service. 2012. Pacific Southwest Region. Accessed February 20, 2012, http://www.fs.fed.us/r5/projects/planning/process.html.

U.S. Water Resources Council. March 10, 1983. *Economic and Environmental Principles and Guidelines for Water and Related Land Resources Implementation Studies*. Washington, DC: U.S. Government Printing Office.

CHAPTER 4

Establish the Decision Context

The primary purpose of this first planning step is to identify the problems you will try to solve, the opportunities you will try to realize, and the planning objectives and constraints that frame and define a successful solution.

4.1 INTRODUCTION

Here's the short story. If you don't get the problems you're dealing with right, nothing else you do in the study is going to matter much. At best, you'll find a good solution to the wrong problem. You've got to understand the context of the decisions you're going to be making. That means you need to understand all the problems you are trying to solve and that not everything is a problem. You may be trying to make some good things better, so it is important to understand the opportunities for gain you're pursuing. Solving problems and realizing opportunities are the reasons for your planning study.

Part of your job is to consider what the future of your study area is going to look like if no action is taken as a result of your study. This is the without condition scenario. If that future is desirable, there may be no reason to pursue your study further. If you do decide a planning study is needed, it is important to define what a successful planning effort would look like. What are the things you are hoping to do and to avoid doing? These outcomes of a planning study are the objectives and constraints of your decision context.

How will you make key decisions? You need to think about that now to make sure you estimate and measure the right things in your study. Even if you are not sure what all the decision criteria will be, you need to take a crack at identifying them at the outset of your effort. Some studies are familiar, and some may even be routine. When you do familiar work, everyone knows what is expected of them and they can swing into action on day one because they all have done this many times before. Other times, some or all of your problems, opportunities, objectives, and constraints may be new to you. In these situations, it is important to know what kinds of information are necessary for decision making. This may include information that goes well beyond your decision criteria. Decision makers have an obligation to request the information they will need in new or unique planning situations.

No matter how much time and funds you have for your study, there are always going to be things you don't know. Uncertainty is ubiquitous in planning. It is essential that you understand the information you have and the information you need to get at the outset of a study. A great part of risk-informed planning (RIP) is about purposefully and intentionally reducing the uncertainty you face. Keeping track of the uncertainty that could affect decision making must begin in Step 1 as you define your decision context. Figure 4.1 summarizes the key tasks in this step of the planning process.

There are a few more things to know before leaping into this chapter. A lot of people consider scoping to be the first part of a planning study. Defining the decision context is more than scoping. Scoping is a term that has taken on a specific meaning for many people within the National Environmental Policy Act (NEPA) process (see the following text box). In that context, it is often used more narrowly to mean identification of resource concerns and the evaluation of impacts and effects of alternatives. This first step includes some but not all of that scoping process, it also includes a lot more than the NEPA scoping process. Planners establish the context of the decision problem in a much broader way than the NEPA view of scoping entails.

I'm going to suggest that a good decision context has at least the five tasks, as shown in Figure 4.2. The without condition can be considered to be rolled up into the

Step1: Establish decision context
Problems and opportunities
Without condition scenario
Objectives and constraints
Decision criteria
Information needs
Uncertainty

Figure 4.1 Decision context topics in this chapter.

NEPA SCOPING

NEPA is the national charter for the protection and restoration of the environment. NEPA includes a requirement to prepare a detailed statement, either an environmental assessment (EA) and finding of no significant impact (FONSI), or an environmental impact statement (EIS) of major federal actions significantly affecting the quality of the human environment and record of decision. The Council on Environmental Quality (CEQ) regulations direct federal agencies that have made a decision to prepare an EIS to engage in a public scoping process. This process determines the scope of the EIS and is intended to ensure that problems are identified early and properly studied. It assures that issues of little significance do not consume time and effort. The scoping process should identify the environmental issues and alternatives to be examined in the EIS while eliminating nonsignificant issues.

Source: Derived from Guidance Regarding NEPA regulations. 1983. 40 CFR Part 1500, accessed June 14, 2012, http://ceq.hss.doe.gov/nepa/regs/1983/1983guid.htm.

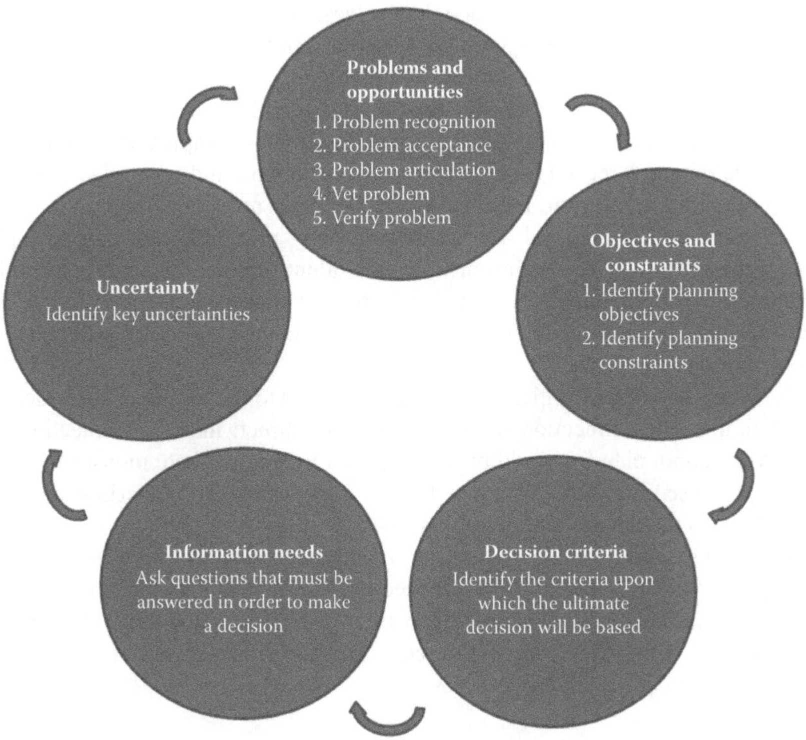

Figure 4.2 Decision context cycle for natural resources planning.

problems and opportunities and objectives and constraints work. Notice the cyclical nature of these tasks. Almost every part of the planning process is iterative and the decision context has a powerful feedback loop. As problems are better understood, the objectives and information needs are refined and change. Uncertainty, usually great at the outset, is subsequently reduced, the problems are often understood differently, and the cycle repeats.

Planners need to begin with a clear view of the big picture, which is the focus of this chapter. The outputs of this first planning step are simple and essential. They are as follows:

1. A written problems and opportunities statement
2. A without condition scenario
3. A list of planning objectives and constraints
4. A list of decision criteria that will lead to the choice of a course of action
5. A list of unique questions to be answered in the investigation
6. A list of the most significant uncertainties

4.2 PROBLEM (AND OPPORTUNITY) IDENTIFICATION

Problems don't always announce themselves. More often than not they lurk in a chaotic mess of tangled and jumbled symptoms, potential causes, uncertainties, pressures, values, and shortcomings. Stakeholders don't often offer planners well-posed problem statements. We operate in a world of uncertainty. Cause and effect relationships are confused, our understanding of complex and even simple systems is incomplete, our models are rarely sufficient, and when there are data not all of them are relevant. Identifying the problems and opportunities in a mess is a creative and intentional act. There are many problems that can be pulled from a mess. Pulling the right ones out is a talent planners are well advised to develop. The opportunities we choose to pursue may be few from many. The framing of a decision problem begins with the identification of problems and opportunities.

Problems and opportunities ought to be connected to the resources for which you are planning. The connection need not always be direct; indirect connections are okay. You cannot plan for world peace or an end to political hate mongering, but if we restore these urban streams it may help property values, reduce crime, and other things beyond the natural resources themselves.

Increasing social complexity and the rapidly increasing pace of change in our world guarantee an endless supply of concerns. It's the planner's job to sort through these concerns and clarify which of them are important to your study. These important concerns, then, need to be structured as problems to be solved or opportunities to be attained via the planning process.*

* To limit the need to write sentences that include clumsy references to both problems and opportunities, I'll use the term problem identification. Understand this to include opportunities as well.

Einstein is quoted as having once said, "If I had one hour to save the world, I would spend 55 minutes defining the problem" (quoted in Poole 2012). This is the key to good planning as well as to saving the world. Defining the problem is the real focal point for establishing the decision context. One of the most common organizational failings is that as soon as a problem arises we're so eager to jump in and solve it that we spend very little time understanding and refining our understanding of it. Consequently, we're often unclear about a problem and we solve the "wrong" problem correctly.

It helps to have a problem identification process. I propose a five-task conceptual model with a toolbox full of methodologies for carrying out the tasks. Each task is developed in turn in Sections 4.2.1 through 4.2.5. The model tasks are the following:

1. Recognize the problem.
2. Accept the problem.
3. Articulate the problem.
4. Vet the problem.
5. Verify the problem.

4.2.1 Recognize the Problem

First, you have to become aware of a problem. Some problems burst through your door, find you, and follow you everywhere you go. Other problems are more subtle and do their best to avoid detection. Some are as familiar as the back of your hand while others you've never seen before and will never see again. Some problems stay only a brief while and others refuse to leave. Among these extremes are an infinite variety of ways for problems to be discovered. Although the precise manner in which a problem is recognized will vary from one organization to another, there are some useful generalizations that can be made about the origins of problems and how we initially become aware of them, likewise opportunities.

Explore the mess. Problem recognition is the simple act of recognizing that a problem exists. It happens in one of two broad ways. A problem finds you or you find a problem. Problems that find you are triggered by outside influences. Stakeholders bring you a problem or an event occurs that results in a problem. This is reactive or passive problem recognition. In contrast, there is proactive or intentional problem

TWO QUESTIONS

"What kinds of problems are you having here?" is a nice icebreaker when you're in this stage of the study. Don't worry about whether you have the authority to solve the problems or not yet. Be interested in every problem you hear about. Authorities and such can be sorted out later.

"Beyond problem solving, what kinds of things could we do to make things better here?" is another question that can turn up some useful information. Don't worry about evaluating ideas and information in the early going. Just collect them.

finding. This occurs when planners look actively and often strategically for the most important problem to solve. See the following text box for examples of things that can trigger recognition of a problem.

Some problems are pretty simple, and their solutions are almost automatic. When your tire is flat, you change it. These well-structured problems (Powell and Baker 2010) have similar characteristics that include the following:

- Objectives of the analysis that are clear
- Required assumptions that are obvious
- Necessary data that are available
- A well-understood logical structure

Well-structured problems are often amenable to relatively obvious solutions.

Planning, by contrast, usually faces ill-structured or wicked problems. These are planning problems where the objectives, assumptions, and structure are unclear and data may be sparse or missing. Lachapelle et al. (2003) describe wicked problems and messy situations as typified by multiple and competing goals. There is little scientific agreement on cause-and-effect relationships, time and resources are limited, there is a lack of information, and structural inequities in access to information and the distribution of political power are often additional barriers. These kinds of problems must be explored and understood before they can be solved. Recognizing an ill-structured problem is a creative, iterative, and open-ended process. Here are some examples of ill-structured problems:

- Should nonindigenous aquatic species be prevented from spreading to new waterways?
- Should coastal areas take action to respond to potential sea-level rise?
- Should government-owned land be leased for private development?
- How much of Springfield's open land should be preserved for potential use as detention areas?

PROBLEM TRIGGERS CAN BEGIN THE PROBLEM IDENTIFICATION PROCESS

There are several types of problem triggers:

- Authority trigger—Laws, resolutions, and similar authorities direct a study
- Crisis trigger—Real or perceived, media, public outcry, adverse comments, changing public values or awareness, decreased consumer confidence
- Science and technology trigger—New knowledge or technology, emerging health problem, improved detection, surveillance, or method
- Emerging or "on the horizon" trigger—Planned search, forecasting, scan risk landscape, natural and anthropogenic disasters and events, imports
- Strategic plan trigger—Strategic planning, social needs, opportunities, can be beyond the horizon, historical precedents

Source: Food and Drug Administration Center for Food Safety and Applied Nutrition. 2003. *CFSAN's risk management framework*. College Park: CFSAN.

- How should the city's waterfront be redeveloped?
- Should a wolf population be reestablished in this forest?
- How can we minimize adverse effects on forest ecosystems while reducing forest fuel loads?

This task requires public involvement. A strong effort must be made to identify and engage the appropriate stakeholders and publics. All phases of the problem identification process must be open, transparent, and accessible to stakeholders and the public. Public involvement, discussed at greater length in Chapter 10, can be an essential source of important information at this stage of a planning study. An issue can be important because science tells us it is important. It can be important because some group has taken some significant, formal, or official action to stress its importance. Or, it can be important because people think it is important. Evidence gathering, the topic of Chapter 5, is also essential to this task. The task is complete when you have explored the mess and produced satisfactory preliminary descriptions of the problems and opportunities of most interest to you.

4.2.2 Accept the Problem

What's in the study, and what's not? It's not enough to be aware of a problem; one must also decide to own it, too. Problem acceptance is a priority-setting step. It is deciding to act or not to act. Accepting a problem as one to be solved or an opportunity as one to be pursued is a significant organizational commitment and a critical piece of the scoping process.

Planners have to identify the study resources required to address the recognized problem(s) in a timely manner. Then they have to evaluate the adequacy of their available resources in the context of their program authorities, the organizational mission and vision, and public values. This obviously implies consideration of competing activities; we can't solve every problem, we can't pursue every opportunity. This task is complete when you decide whether you will address the problem or not and when you have decided which opportunities to pursue.

4.2.3 Articulate the Problem

When you've recognized the problem and decided to address it, it's time to write the problem down so that planners and others can understand it. Stakeholder and public input is essential to this task. This is when the problem is fully articulated and reduced to words. The specific manner in which a problem is defined begins to link it to possible solutions.

There is likely to be substantial uncertainty attending this problem articulation task in its earliest iterations. There may be copper in the sediment of your water control structure and snails may be absorbing it. Eagles eat the snails and absorb the copper as well. But is that a problem? Water quality may be degrading, the mottled duck may be disappearing, flooding may be getting worse, and no one may yet know if any of this is actually so, much less why it is so. As the uncertainty attending your problem articulation is reduced, the problems will become better understood. Some may be dropped from your list of problems and new problems may appear. It is not unusual to realize that the first articulation of a problem got it all wrong.

A problem can be articulated in a simple sentence. For example, "Franklin Creek causes flood damage in the industrial sector of Central City" is a reasonable problem statement. It is rare that the fullness of a problem, including its root causes, can be captured in a simple sentence. That sort of understanding requires a bit more information, but it's not required to articulate the problem to others.

At the end of this task, if you can't clearly and concisely finish the sentence, "The problem is…" then nothing that follows will be clear either. This task is done when you have completed a written problems and opportunities statement. That statement provides the rationale or reason for your planning investigation. It should be considered a conditional statement that will change as you begin to vet your problems statement, verify the problems, reduce uncertainty by gathering information, and come to better understand the problems and stakeholders' concerns. Date this piece of paper. Its contents will change. Planning is an iterative process and you can expect to revise and refine the problems and opportunities statement several times before it is finalized.

4.2.4 Vet the Problem

Now that the problem is written down it's time to try it out on others. What do they say about your problem statement? Vet your problems and opportunities with your stakeholders. Publish it appropriately. Make it public if your stakeholders include the public. Show them your best thinking and ask, did we get the problem(s) right? What's missing? What is here that should not be? Does the public have information about these problems and opportunities that would be helpful to share? Do they agree or disagree?

Stakeholders and the public should be involved in the problem identification process. The appropriate level of involvement will vary with the decision problem. Some problems will be identified for you by stakeholders; other times stakeholders will have to be made aware of the existence of a problem. The recognition and articulation tasks provide the public with input opportunities, and vetting the problem statement is a feedback opportunity. The message to stakeholders and the public is "We have listened to your concerns and this is how we see the problems and the opportunities in this investigation." You need others to evaluate and approve your articulation of the problem. Getting a buy-in to the problem definition is critical to the ultimate success of a study.

Revise the problem statement, if necessary. Always keep it up to date. Let people know how and why it changes as it changes. The vetting task is complete when you have sought, received, and considered feedback from stakeholders and the public and modified your problem statement accordingly.

4.2.5 Verify the Problem

The final task in problem identification is to verify the problem. This is done by objectively assessing the problems to assure they are as people say they are. The assessment may be qualitative or quantitative. The quantitative assessment of the problem conditions begins the evidence gathering process in earnest. Verification through assessment is an important part of the problem identification task. Until there is objective evidence that verifies the existence of a problem (or opportunity) much of your problems and opportunities statement may be speculative in nature. The evidence gathered while assessing the problem confirms or challenges the problem definition. Evidence that challenges your understanding of a problem launches a new iteration of the problem identification process.

Planning problems (and opportunities) are complex and there is usually more than one factor contributing to the problem. The ubiquitous nature of uncertainty often will mean the real sources of those contributing factors are not always clear. Until the many potential causes of a problem are understood, the prospects for successfully solving the problem are not especially good. The Addendum describes how an Ishikawa diagram can be used to identify problem causes and effects. These diagrams can be useful for several parts of the problem identification task.

4.2.6 Opportunities

There are opportunities in natural resources planning. Examples include

- Reintroducing and reestablishing wildlife populations in areas that have supported populations in the past
- Restoring ecosystem function in existing areas
- Maximizing old-growth forest for habitat purposes
- Increasing the abundance and availability of invertebrates for foraging shorebirds
- Improving water quality for wildlife and ecosystem function
- Using prescribed burns to maximize the probability of species persistence at wildlife refuges over the long term
- Reconnecting urban communities to their ecosystems

Until now we have focused on problem identification. Opportunities for gain should also be an essential part of the decision context. They are identified in the same manner as problems. Uncertainty will play an important role in identifying opportunities as well because the gain may or may not be realized. The tasks parallel those for problems:

1. Recognize the opportunity.
2. Accept the opportunity.
3. Articulate the opportunity.
4. Vet the opportunity.
5. Verify the opportunity.

Trusting the reader to understand the parallel between problems and opportunities in performing these tasks, we turn now to other comparisons of the two kinds of concerns. Table 4.1 identifies similarities between problems and opportunities. The similarities tend to describe how the structures of the two concerns resemble each other.

There are differences between the two as well. These are summarized in Table 4.2. In general, problems tend to describe situations that exist or will soon exist unless action is taken to prevent them. Opportunities, on the other hand, tend to describe conditions that do not exist or will not continue to exist unless action is taken.

Table 4.1 Similarities between Problems and Opportunities

Characteristic	Similarity
Number	Variable, few to many
How Stated	In practical, meaningful, operational terms in a single statement
Source	Developed from people, observation, analysis, and documentation
Specificity	Specific, narrow; essentially limited
Subject	Usually limited to a specific resource or asset
Location	Usually found in a particular place or locale (e.g., "study area")
Measurability	Moderate to high; usually measurable or easy to recognize change that would result in a "better" or "worse" condition
Ability to Achieve	High; problems can be solved, opportunities can be realized
"Ideal"	An "ideal" usually exists and can be identified The "ideal" is not the same as the existing condition The "ideal" is not the same as the long-term "without" condition

Table 4.2 Differences between Problems and Opportunities

Characteristic	Problem	Opportunity
Focus	Existing undesirable condition; a description of what is or might be	Future desirable condition; a description of what could or should be
Core Message	Negative—an objection	Positive-a desire
Occurrence	Past—usually occurred Existing—usually occurs Future "without action"—usually expected to occur	Past—usually didn't occur Existing—may or may not occur Future "without action"—may or may not be expected to occur
Relationship to Other Assets	Existing condition may adversely affect other assets	Existing condition does not affect other assets
Implicit Objectives of Action	Return to a past/ideal condition not considered objectionable Create a future condition that would not be objectionable	Create a future condition considered to be desirable Return to a previous condition considered to be desirable
Consequences of Doing Nothing	Usually direct, immediate, and adverse	Usually indirect and long-term due to benefits foregone

4.2.7 What Should the Problem and Opportunity Statement Look Like?

It is written down. It is short, a page or two at most. Each problem and opportunity is succinctly stated in a sentence or two. They are numbered for convenience. Eventually, once you assess your problems and opportunities, you may develop an expanded profile and description of each item. None of this brevity is intended to reduce you to sound bites and clichés. It is intended to focus the planning effort and to crystallize the attention of planners, stakeholders, and publics alike. This statement makes clear what will and will not be considered during the course of the investigation.

If a stranger asked you, "Why are you doing this study?" you could hand him this statement and he would understand after reading it. The statement enables you to say why you are conducting the planning investigation. The planning document can still weave a story and explain the problems and opportunities in as much detail as necessary. The problems and opportunities statement needs to be clear, concise, and complete. It is essentially the investigation's mission statement. An example is shown in Table 4.3.

This statement is direct and concise. It is also admittedly spare and you may feel a little uncomfortable with such parsimonious information. The next step up is to provide a profile of each problem or opportunity. This stops short of the full accounting of the assessment of the problems that may appear in a technical appendix, for example. The elements of a profile are summarized in Table 4.4. You will not be able to complete a profile during the first iteration of a new planning study. Much of the required information will not become available until the study progresses.

Notice the natural iterative nature of problem and opportunity definition that evolves from the approach presented in the template. The assessment of the problem, that is, the technical analysis that verifies and fleshes out the description, does not even begin until after the problem is initially identified. Its completion often helps to refine the very problem identification. Iterative loops on small parts of the planning process, as opposed to the entire process, are common in natural resource planning investigations. Each time some uncertainty is reduced, it is likely that some aspect of the process will be iterated to some extent.

Table 4.3 Examples of Simple Problem and Opportunity Statements

Franklin Creek Basin Problem Statement
The problems in the Franklin Creek Basin are as follows:

1. Loss of fish habitat in Franklin Creek due to urbanization
2. Flood damages in the industrial section of Central City
3. Stream bank erosion along Campus Park
4. Saltwater intrusion in the Franklin Bay estuary
5. Loss of coastal wetlands along the South Ditch section of Franklin Bay

Franklin Creek Basin Opportunity Statement
There are opportunities in the Franklin Creek Basin to

1. Increase wildlife habitat along Campus Park
2. Restore indigenous fish species in the upper basin
3. Provide increased recreational opportunities along the waterfront

Table 4.4 Sample Template for a Problem/Opportunity Profile

Problem/Opportunity Profile

1. Source. Who first identified the problem or opportunity? Examples: higher authority, media, stakeholders, conversation with elected officials, experts, field observations.
2. Public Concerns
   a. Advocate—Who is the spokesperson for the problem or opportunity? Identify specific groups, agencies, and individuals.
   b. Basis—What is the advocate's basis for the problem or opportunity? Examples: homeowners who have experienced flooding, state agency legally mandated to oversee wildlife resources.
   c. Background—In the advocate's view, what is the problem or opportunity, and what are the causes and effects?
   d. Other Stakeholders—Who else believes that the problem or opportunity does or does not exist? Why or why not? Identify specific groups, agencies, and individuals.
3. Technical Analysis
   a. Subject—State the problem or opportunity.
   b. Location—Describe the location of the problem or opportunity; map it if possible.
   c. Measurement—Identify one (or more) measurable indicator that is used to measure change in the problem or opportunity.
   d. Uncertainty—Identify the things you need to know that are not yet known.
   e. Conditions—Describe past, present, and future conditions related to the problem or opportunity:
      (i) Historic condition
      (ii) Existing condition
      (iii) Future "without project" condition
   f. Decision Criteria—Identify any standard, target, or other criteria that may be used to define the magnitude of a problem or opportunity. Examples: state water quality standards, design specifications, legislated targets, strategic plan goals, profits, costs, and so on.
   g. Causes and Effects—Describe causes and effects of the problem or opportunity.
   h. Conclusion—Does the technical analysis support the public concerns about the problem or opportunity? Why or why not? How do the historic, existing, and future conditions compare to any applicable decision criteria?
4. Problem/Opportunity Statement. Write a clear and brief description of the problem or opportunity as it is currently understood.
5. Problem/Opportunity Profile. Write an expanded narrative description of the problems and opportunities based on the information above.
6. Information.
   a. Sources—List sources of information about the problem or opportunity.
   b. Information needed—Briefly describe the types of additional studies needed to address the problem or opportunity.

Source: Yoe, Charles, and Kenneth Orth. 1996. *Planning Manual. IWR Report 96-R-21.* Alexandria: Institute for Water Resources.

How do you know you have a good problems and opportunities statement? When people understand it, it's good. If you show it to someone and they understand why you are doing your study, you have succeeded. How do you know if they understood it? You must vet it, publish it, communicate it, and seek feedback on it. How do you know when it is final? When people agree with it, when you no longer find a reason to change it, it is final.

Don't wait until you have a final problems and opportunities statement to begin using it. A good statement takes time to develop, but a useful statement is the one you have. There are two things to do with a useful problems and opportunities statement: one, you can make it a good one, and two, you can use it.

4.3 WITHOUT CONDITION SCENARIO

What will the future of the study area look like if no planning action is taken? Will the problems get worse or will they go away? Will new problems emerge? Will any of the opportunities for gain be realized of their own accord? What's going to happen if the planning team takes no intervening action? That view of the future if no action is taken as a result of your study is called the without condition scenario.

The "without condition scenario" is critically important to any planning study. If the without condition scenario describes a desirable future state of the study area, no further action by the study team is necessary. If that scenario is less than desirable, the without condition becomes vitally important to the formulation, evaluation, and comparison of alternative plans. No single element of the planning process is more critical to Steps 2 and 3, of the planning process than the most likely future conditions that will exist in the study area if no action is taken as a result of your study.

This task has two distinct parts. Part one is to develop a most likely scenario for your study area absent any action due to your study. Part two is to assume the conditions of that future scenario and then to analyze problem and opportunity conditions in accordance with the assumptions of that future scenario. For a simple example imagine an effort to restore ecosystem values along a small urban stream. A without condition scenario may assume a future with continuing degradation of water quality, increased trash dumping in the floodplain, and rising criminal activity in the wooded areas of the stream. That narrative description is the scenario. Then analysts will work with that basic storyline to do realistic analysis of variables and relationships critical to the problems and potential opportunities that exist in that watershed.

For example, imagine that evidence is gathered and models are run that estimate species abundance may have a 50% or greater chance of decreasing by more than 20%. Water quality studies may forecast quality degradation beyond the point where the stream can support life in the stream. Hydrologic and hydraulic analysis may show that flood problems are exacerbated by the dumping. These and other effects are not determined in the scenario preparation. They are the product of the analyses the study team does within the basic without condition scenario. More is said about scenarios in Chapter 5.

For now, think of a scenario as a synopsis or description of some condition or situation of interest to us, specifically future conditions in your study area without any action as a result of your study. Scenarios are sometimes factual, sometimes speculative, and sometimes a mix of the two. Aspects of the future that can be inventoried and described with certainty contribute to the factual dimensions of a scenario. Uncertain aspects of the future may be little more than informed speculation. The without condition scenario is usually a mix of fact and speculation about the future. It is a story about what might be. In preparing this scenario and doing analysis within it, we look into our crystal ball and try to describe the most important aspects of life in the study area over the next several decades.

In scenario development we paint several pictures of what the future of the study area might look like. When it is reasonable to do so, we select one of these alternative future conditions as the most likely future condition. Although it may be the most likely future scenario, it is not necessarily the only possible future condition. Even so, it does become our baseline picture of the future. When we consider how our alternative solutions will alter the future, we are always comparing alternate future conditions, with different plans in place, to this one without condition scenario analysis.

If the uncertainty about the future makes it difficult to choose only one future scenario as most likely, there are two primary options. The simpler option is to conduct a sensitivity analysis of specific plan effects for other realizations of the future. The other option is scenario planning, which relies on multiple without condition scenarios. Scenario planning is discussed in Chapter 13.

4.4 OBJECTIVES AND CONSTRAINTS

What would you like to see happen as a result of your plan? These are your objectives. What do you want to be sure does not happen as a result of your plan? These are your constraints. Objectives are based on the problems and opportunities. Constraints are not.

What would a successful plan look like? Never mind how or if you can achieve it just yet but, given the problems and opportunities you have identified, how will you know when you have succeeded? Your planning objectives and constraints define success.

Objectives and constraints are a statement of the community's values. They are a public declaration of the desired outcomes of the planning investigation. They must be precise, practical, and measureable so that they can guide the actions of the planning team.

An objective is a statement of the intended purposes of the planning process; it says what an alternative plan should try to achieve. They state in broad and general terms what planners intend to do about the problems and opportunities they face. Done well, objectives reflect the most important values in the decision process. If flooding is one of the problems, protecting lives and reducing property damages might be objectives. An objective does not say how those things can or should be done; only that it is an objective to do so.

Objectives do not identify specific solutions to problems or specific means by which an opportunity will be attained. They say what is desired. Don't confuse

OBJECTIVE AND CONSTRAINT VERBS

Eliminate, reduce, minimize/maximize, enhance, harmonize, identify, define, describe, increase/decrease, raise/lower, strengthen/weaken, avoid, adapt, blend, reconcile, coordinate, affirm, diminish, weaken, promote, encourage, complement, strengthen, ensure, control, do not.

objectives with strategies. An objective is a clear statement of a desired end that subsequently formulated plans are intended to accomplish. Objectives reflect the most important social values in the decision-making process. They identify the things planners are trying to do.

Sometimes there are important things we're trying not to do. We'll call these things constraints. Study-specific planning constraints are statements of things unique to a specific planning study that alternative plans should avoid doing. Examples of study-specific constraints might include the following:

- Do not induce saltwater intrusion to freshwater aquifers.
- No loss of flood protection from an existing levee system.
- No increase in shoreline erosion related to navigation.
- Avoid regional loss of jobs or income.
- Do not adversely impact the southern sea otter.

Where do these objectives and constraints come from? Values! They reflect what is important to people. You can find values in what concerns "the public," experts, and our institutions (law, regulations, guidance, policy, organizational missions). People are more likely to attain objectives they set than they are to attain objectives set for them. Involve stakeholders and the public in this process, especially those whose cooperation is going to be needed to implement the plan and make it successful.

A written planning objectives and constraints statement is the desired output of this task. Writing them down makes them specific and knowable. Consider the statement conditional and subject to change as you iterate your way through the planning process. This is the critical document that reflects key social values in a planning investigation. It's a good guide to choosing your decision criteria. Stakeholder input is essential for identifying objectives and constraints. Like the problems and opportunities statement, this statement should be published and vetted as appropriate to the decision problem's context.

The similarities between objectives and constraints are identified in Table 4.5. A good objective/constraint is specific, that is, it is clear and free from ambiguity. It is flexible and can be adapted to new or changing requirements. You need to be able to measure and document its achievement by some objective means. Good objectives are attainable; they can be reached by your alternative solutions. Ideally, they would

Table 4.5 Similarities between Objectives and Constraints

Characteristic	Similarities
Number	Variable; few to many
How Stated	In practical, meaningful, operational terms and in a single statement
Specificity	Specific; narrow; essentially limited
Subject	Usually limited to a single subject, resource, or issue
Measurability	Moderate to high; usually measurable or easy to recognize achievement
Acceptability	Low to high; may be conflicts; consensus may be lacking
Changeability	Variable; low to high
Achievability	High; can be achieved, in whole or part, by a single solution

Table 4.6 Differences between Objectives and Constraints

Characteristic	Objective	Constraint
Source	Developed from problems and opportunities	Given (e.g., some legal design constraints); or developed based on area-specific conditions (i.e., public views, resource limitations)
Specificity	Dated; time phased; can be or are intended to be achieved or phased within a particular time frame	Variable; may be dated and time phased or intended for the long term
Location	Applicable to a particular place or locale ("study area"); regional or local	Variable, depending on the subject being constrained
Message	Do good	Don't do bad

be congruent, that is, in harmony with the other objectives. This may not always be achievable and trade-offs may be necessary. If objectives reflect social values, they should be acceptable, meaning they are welcome or pleasing to key stakeholders.

In contrast to what good objectives/constraints are, we can consider what they are not. Good objectives are not absolute targets, that is, they do not specify a particular level of achievement. Neither do they include specific measures for solving problems, nor should they prescribe a specific course of action. They are neither political or governmental goals nor are they planning tasks. Completing an EIS may be a required task, but it is not an objective. As mentioned in the "Planning Constraints are Different" text box, planning constraints do not include resource constraints such as time, money, and expertise. Some differences between objectives and constraints are shown in Table 4.6.

ROLE OF PLANNING OBJECTIVES

Planning objectives play a central role in the planning process. They are an essential output of the public involvement process in collaborative planning, and they represent the desired outcomes of the planning process. In best practice, planning objectives will play an essential role in plan evaluation, comparison, and selection.

Multiple stakeholders and complex problems lead to multiple planning objectives. Plan formulation purposefully strives to meet these multiple objectives. The best plan is identified based on its relative achievement of these many, and sometimes conflicting, objectives.

PLANNING CONSTRAINTS ARE DIFFERENT

Two distinctly different categories of constraints can be identified. They should not be confused. First, there are resource constraints. These include limits to our knowledge, expertise, experience, ability, data, information, money, and time. Although these constraints are real and may limit the scope of a study in significant ways, they are not the constraints of interest to us here. Our concern is with planning constraints that restrict plan formulation and the range of solutions that are available to us.

Planning constraints can be divided into universal constraints and study-specific constraints. Universal planning constraints are the legal and policy constraints that need to be included in every planning study. They may vary from study type to study type, but for a given type of study, there are some predictable constraints. For example, you don't formulate plans that intentionally and adversely affect threatened or endangered species. Others are defined by the laws and regulations of the federal government and the applicable laws and regulations of the state and local governments.

4.5 DECISION CRITERIA

Be explicit about the criteria that will be used to make key planning decisions at the earliest possible point in time. You'll need explicit criteria to judge what's good and what's bad, what's beneficial and what's adverse, what's desirable and what's undesirable. When you evaluate plans, how will you know if a plan makes things better or worse? Which solutions are good enough to be contenders, and which should be dropped? You'll need screening criteria to eliminate solutions from further consideration. How do solutions stack up against each other and what's important to decision makers? Ranking criteria may be used to order solutions from best to worst. Selection criteria are needed upon which to base decisions.

Planning is an analytic-deliberative decision-making process. It is analysis followed by deliberation in many cycles of decision making, large and small. The ultimate decision in the planning process is choosing a solution. That is when the planning team or decision makers, if the team does not make that choice, say this is how we would like to meet our objectives and avoid our constraints and by doing so, we will solve our problems and attain our opportunities.

PLANNING DECISIONS

Planning culminates in the selection of a recommended course of action, that is, choosing "the plan." To get to that point, however, it may be necessary to literally make dozens of different decisions. Each of these decisions will require criteria. The criteria used to support scoping, evaluation, comparison, and selection decisions are the focus of this discussion. These decision activities should be based on explicit decision criteria and the best available scientific evidence. They need to be evaluated and reviewed as appropriate on the basis of new data in later planning iterations, and based on a clear consideration of uncertainty and variability.

There are many decisions before this one, however, and decisions require information. That information does not appear by accident. Knowing what information to gather is an essential part of the evidence gathering and analytical aspects of planning. One of the most useful things to do in this first planning step is to identify the criteria that will be used to make the most important decisions in the planning process as accurately, completely, and early in the process as possible. The idea of decision criteria is parsed into the following "key decision" categories:

- Scoping criteria
- Screening criteria
- Evaluation criteria
- Comparison criteria
- Selection criteria

The scoping process sets boundaries. Scoping criteria may include considerations such as policy, geography, politics, authorities, and other such broad concerns in addition to all that is entailed in defining the nature and extent of the decision context. Screening criteria are used to select some from many. They encompass the broadest category of criteria and include criteria used to screen problems and opportunities in or out, criteria to measures that will be useful for formulating plans and dozens of other routine decisions. Scoping and screening criteria should be identified.

For example, the scoping criteria used to define a study area might be the language of a congressional authorization, the defined area of a planning authority, or political boundaries. The criteria used to screen the potential problems to be accepted might include an authority to address the problem and the consensual support of constituent stakeholders.

Evaluation criteria are used in planning Step 3 to evaluate plans; comparison criteria are used in Step 4 and selection criteria in Step 5. These criteria may be identical; they may overlap to varying extents or they may differ from step to step. What they all have in common is they should be identified here in Step 1. In practice, they are made explicit in the documentation far less often than they should be.

Some planning programs are well structured by legislation, policy, and other authorities. They often deal with recurring types of problems. In those instances, policy may have established explicit decision criteria or experience may indicate the criteria that are going to be most useful to decision makers. In other words, for some recurring planning problems the decision criteria may be well established by policy and practice.

The U.S. Army Corps of Engineers Civil Works Program provides a good example of recurring problems. They have been authorized by Congress and the president to provide the following for the nation through water resource projects:

- Coastal protection
- Disaster preparedness and response
- Environmental protection and restoration
- Flood risk management

- Hydropower
- Navigable waters
- Recreational opportunities
- Regulatory oversight
- Water supply

The specific decision criteria for these projects vary from purpose to purpose. From 1983 until the time of this writing, the *Principles and Guidelines for Water and Related Land Resources Planning* have dictated that net national economic development benefits will be the primary decision criterion for selecting a solution. In other instances, decision criteria will be developed anew for each planning investigation. If analysts are going to gather the best evidence and do the best analysis, they need to know the criteria upon which decisions will be made as early in the process as possible. This is a critical part of the decision context.

Any stakeholder, person, or organization that will be party to the final decision-making process should be asked, what information do you absolutely require to make a decision about how to meet these objectives and solve these problems? What information will you base your decision on? These are candidate decision criteria. They need to be identified early in the process to assure that the data collection and analytic steps that follow produce the information that is needed to evaluate, compare, and select plans.

As a practical matter, the uncertainties may be too large early in the planning process to identify all the specific criteria that will be used to choose the solution. On the other hand, once the problems and opportunities statement and the objectives and constraints statement are completed, it should be easier to anticipate the decision criteria used for the ultimate decision. In any event, planners, with input from appropriate decision makers, should identify the criteria that are most likely to be used for decision making throughout the planning process.

A GOOD CRITERION IS

Appropriate to the decision being made
Supportive of decision making
Explicit
Measurable
Based on evidence
Quantitative
Specific
Discriminating
Reliable
Honest about uncertainty
Reproducible
Verifiable
Clear about thresholds
Documented

4.5.1 What Is a Criterion?

Criteria are the attributes, variables, and values associated with a decision problem that are important to decision makers. A criterion is something the decision makers care about and something that can influence the decision makers' choice. You should expect the decision criteria for a study to reflect the planning objectives and constraints.

Plan effects are not the same as criteria. There are many things people care about and will want to see that will not play an explicit role in the decision making. It is common practice in some studies to prepare a "list of plan effects" that summarizes the many and varied effects of the plan. This can be valuable and useful information, but most of these effects or impacts are not criteria. Decisions are not made based on all these effects.

A criterion is important to the decision makers' thought processes. It is important to know that a plan does not violate the Endangered Species Act. That is likely to be a screening or evaluation criterion. It is something that is important to consider at some point in the process, but it is not likely to be the basis for a final decision. Once a plan "passes" the qualifying test of the evaluation step, those "test questions" or criteria may no longer be specifically important for subsequent decision making. Later-stage decision making often proceeds on the basis that all surviving plans are acceptable based on the prior screening criteria. Thus, if all plans are in compliance with the Endangered Species Act, such knowledge does not assist the decision maker at all in distinguishing a preference for one plan over another. Consequently, although this was an important screening/evaluation criterion it is not a useful decision criterion for the final selection of a solution; it no longer discriminates.

In a similar vein, it may be very useful to anticipate and identify the effects of a plan. It may be helpful or even important to know how much noise will be made during construction or how traffic flows will be disrupted. But if the decision maker is not going to consider such effects in the decision-making process, these are planning effects, not selection criteria.

A study can generate information on dozens or even hundreds of variables and impacts during its course. Most are included in the study because someone cares about these effects. Some are identified as criteria out of habit. They were criteria in a previous study, so they must be criteria in this study. Others may be items from checklists, such as those used for the NEPA process. Still others function simply as filler, chosen because analysts wanted to see their favorite effect included among the

PLAN EFFECTS ≠ DECISION CRITERIA

A plan effect is something that happens because of a plan. It is a measurable impact. A criterion is a plan effect or factor extrinsic to the plan that is explicitly considered in one or more decision-making processes during a planning investigation.

criteria upon which a decision was based. These may all be important components of a study, but they are not decision criteria unless they are used to make a decision.

Criteria are those things that actually influence the decision-making process. There may be many different sets of criteria used throughout the planning process. This task in Step 1 focuses on those criteria that are important to the planning team in their scoping, evaluation, comparison, and selection decisions, as well as those things important to the decision makers who will make the final plan selection.

Identify the criteria you'll need to choose the solution to the greatest extent possible at this point in your study. This may require identifying and meeting with the decision makers early in the planning process. Depending on their awareness of and involvement in the planning process, they may need considerable coaching to understand the importance of the team knowing what is important to them.

If you can identify the comparison, evaluation, and selection criteria at this point in time, it is wise to do so. They should be identified for the public and stakeholders as early in the process as possible. This aids the transparency of the process. If you tell me before the numbers are available what you will be considering for the purpose of decision making, I sense a more transparent and honest process than I do if a decision magically descends from the hill after months or even years of study.

You are not explaining the trade-offs that may have to be made among these criteria or how they will be made, at this point. You are simply trying to say that by knowing the likely decision criteria, you can gather the necessary evidence and do the appropriate analyses to assure the desired information is available. If decision makers are going to want to know the impact of a plan on regional jobs and income or the ecological or water footprints of the alternative solutions, analysts need to know that in advance. What you want to avoid most is a process that relies on producing a potpourri of plan effects from which decision makers cherry-pick a few upon which to base a decision. Know what is important to the decision maker and then provide it.

The output of this step is a list of potential decision criteria needed to identify a solution. Ideally, this includes evaluation, comparison, and selection criteria. The list should be communicated to the public to see who may have information to contribute and to assure that everyone understands the basis for the decision as early in the process as possible.

4.6 INFORMATION NEEDS

Ideally, once the problems and opportunities and objectives and constraints statements are drafted, decision makers will examine them and ask themselves, what information do I need to know to meet these objectives and avoid these constraints? What do I need to know in order to solve these problems and attain these opportunities? Then they would request this information in a series of written questions. That is the desired output from this Step 1 task—a written list of questions, which when answered will provide decision makers with the information they need to select a solution. This question set will be much broader in scope than the identification of decision criteria.

There may already be a template for conducting studies for recurring problems. When the Corps of Engineers undertakes a flood risk management planning study everyone swings into action knowing from decades of experience most of what will be required to successfully conclude the investigation. There will be project costs and a benefit–cost ratio using national economic development as the perspective for benefits. To estimate these criteria, there must be hydraulic and hydrologic work of a rather familiar nature, foundation and real estate surveys, and much other predictable work. Likewise, transportation planning agencies usually know exactly what is needed to support decision making.

When Asian carp are headed upstream on the Illinois River toward Lake Michigan, however, there is no template. This is a new problem. Many planning investigations are unique, one-of-a-kind efforts. Everything about the investigation, but the process itself, is ad hoc and the decision makers' information needs must be carefully specified. Decision makers may ask, "Can the carp reach Lake Michigan? Is there a hydrologic pathway? Can they survive once there? Can they establish a breeding colony in the lake? Can they spread from there to other waterways? What harm will they cause if they do spread?" It is the decision makers' responsibility to request the information they will need to make decisions. They will often need help formulating those requests. Analysts may be in the best position to anticipate what information is likely to be most useful to decision makers.

What kinds of questions should decision makers be asking? The kind that gets them the information they will use for decision making. This presupposes they know what that information is. Some decision makers will know, others will need coaching. But they need not ask down-in-the-weeds detailed questions. They need not know a roughness coefficient is used in a hydraulic model nor do they need to know how many houses in the floodplain have basements.

In general, think of the decision questions as coming from four broad categories:

1. Problem/opportunity characterization
2. Objectives and constraints
3. Plan performance
4. Public involvement

Problem/opportunity characterization questions ask for information at an "effects of the problems/opportunities" level, not at a details level. How long can the mottled duck survive if we take no action? How many people could die if the levee fails? How many habitat units will be lost in the next decade if we do nothing?

> Decision makers should not ask things like...
> What does the mottled duck prefer to eat?
> When was the levee built?
> What is the suitability index for dissolved oxygen in the marsh?
> Decisions will not be made based on this information. Decision makers must focus on the information they truly need and will use for decision making.

The details will be there. An experienced staff will know all the details they must address to answer these more general questions. A less-experienced staff may require more questions or more-instructively detailed questions until they gain the experience needed to conduct the analyses necessary for a planning study.

It is also wise to ask questions that focus on objectives and constraints. How will you achieve your objectives? What do you need to know to protect the boreal forest? To reduce flood damages? How can you best avoid your constraints? What kinds of information do you need to have in order to formulate options that achieve goals and avoid constraints?

There are plan performance questions. What options do we have? Is there anything special we should be considering? How well will the options work? What are other people doing about this situation? What do you need to know to formulate and choose the best option? What do you need to know to monitor the effectiveness of your actions? What are the outcomes of interest, that is, what does success look like?

Public involvement questions focus on stakeholders and their perceptions. What are the third-rail issues for the public? What do your planning partners care deeply about? Who cares about the study and how much? The answer to each question on this list should bring decision makers one step closer to knowing the best course of action for your study area.

4.7 UNCERTAINTY

When planners begin a study, one of the first things they should do, figuratively, is to make a pile of the things they know and a pile of the things they do not know. At the start of an investigation, the pile of unknowns can be expected to be quite large, at times much larger than the pile of things they know. Let's call the pile of things we don't know uncertainty.* As planners begin to establish the decision context and then to gather evidence and do some analysis they systematically begin to learn things and reduce uncertainty in the study.

Uncertainty is the result of imperfect knowledge concerning the present or a future state of a system, event, situation, or (sub) population under consideration. Knowledge uncertainty is the uncertainty attributed to a lack of knowledge on the part of the observer. It is reducible in principle, although it may be costly to reduce or it may require significant time in advance of a decision to do so. Knowledge uncertainty arises from incomplete understanding of a system, modeling limitations, and/or limited data. Knowledge uncertainty is sometimes called epistemic uncertainty. It can be addressed by a variety of means that might include quantifying the ranges of uncertainty, applying factors of safety, scenario planning, adaptive management, or other techniques.

Natural variability is the uncertainty that stems from the inherent variability in the physical world. This "randomness" is assumed to be irreducible in principle.

* Uncertainty is a recurring theme throughout this text. The most focused discussion on this topic is found in Chapter 12.

Variability can occur over time, space, or among the elements of a population. For example, stream flow at a given location varies over time while soil properties vary over space. Natural variability is sometimes referred to as aleatory uncertainty. It cannot be altered by obtaining more information, but additional data may improve one's ability to characterize it. Natural variability is often dealt with by statistical or probabilistic methods.

When the term "uncertainty" is used in this text it should be understood to include both knowledge uncertainty and natural variability unless otherwise specified. Planning should be intentional about identifying and addressing significant uncertainties in its analytical tasks so that the effects of the most significant uncertainties can be characterized and taken into account by decision makers.

Much of the uncertainty at the outset of a planning investigation will be reduced or even eliminated as evidence is gathered and analyses are conducted. Thus, some amount of the initial uncertainty is simple ignorance of the facts and it will be reduced as evidence is gathered and analyzed. However, some of the uncertainty will persist. Anticipated sources of significant uncertainty should be identified as soon as possible so that they can be addressed intentionally throughout the course of the study. The output of this step is a written list of the key uncertainties at the outset of the study.

4.8 SIX PIECES OF PAPER

If you do this first step well, planners and all interested publics will understand quite precisely the decision context for the planning investigation. There will be six very useful pieces of paper available early in the study, be sure to keep them up to date, they are the following:

1. A problems and opportunities statement that explains why you are doing this planning study
2. A list of the planning objectives and constraints that state in broad terms what a successful planning solution will do
3. A narrative description of the without condition scenario
4. A list of decision criteria that identify what the planning team knows or believes are important to decision makers
5. A list of questions decision makers would like to have answered as the investigation proceeds
6. A list that singles out for attention those uncertainties that are likely to be most significant in the study

These six pieces of paper make an excellent newsletter or fledgling website for the planning study. Make sure everyone understands each one of them will change as the study progresses and evolves. These six items define the decision context and comprise key elements of the eventual documentation of the study. This information should be published as appropriate and circulated as widely as interest allows.

Armed with these six pieces of paper planners are ready to begin the creative work of planning.

4.9 TAKE AWAYS

Lesson 1: If you don't get the decision context right, then little that follows will matter. Understand the problems you are trying to solve and the opportunities you are trying to realize.

Lesson 2: Define what success looks like early in the process by saying what a successful solution will and won't do.

Lesson 3: Know what information decision making will require.

Lesson 4: Carefully separate what you know from what you don't know and be honest about what you don't know.

Lesson 5: It's not real until it's written down; get those six pieces of paper in your hands.

4.10 ADDENDUM: CAUSE AND EFFECT DIAGRAMS AND PROBLEM IDENTIFICATION TECHNIQUES

Some problems are routine and familiar. They require little effort to identify. There is little uncertainty about what the problems are or how to solve them. Other problems are more difficult. There is often a lack of information and much uncertainty about what the problem really is in these situations. Symptoms of problems are often confused with the causes of problems. The most difficult problems are complex, ambiguous, and accompanied by a lot of uncertainty. These require more effort to identify.

Several techniques for identifying problems are introduced in this addendum. Although the focus is on problem identification, they are equally adaptable for identifying opportunities. Successful planning requires the planning team, their stakeholders, and the public to have a common view of the problem(s) to be solved. When a diverse group of people, however large or small, is expected to come to a common understanding of anything, process is important. In my experience, many adults have little use for a formal process. We are all adults; certainly we can identify a problem, right? Maybe not, or at least not easily. A formal process can help a study team work more effectively and efficiently.

The techniques presented in this addendum can help people better understand complicated and difficult situations. Having a structured and methodical way to accomplish this important organizing and framing task keeps problems from seeming huge and overwhelming. The techniques also provide the laser-like focus needed for a successful problem identification process.

There are dozens of problem identification techniques in use and more are invented all the time. VanGundy (1988) provides an excellent starting point for anyone looking for ideas as do the resources of Mind Tools (2008–2010) a web-based management training site. Rather than attempting a review of the best of these techniques, the Addendum presents the techniques the author has found most useful over the years.

4.10.1 Ishikawa Diagram

Cause and effect or Ishikawa diagrams can be effective tools for detailing the cause and effect nature of your problem(s). A cause and effect diagram is shown in three parts in Figure 4.3.

Problem articulation begins when you write the recognized problem in the problem box shown on the right in the top of Figure 4.3. The straight line indicates the "cause" of this problem "effect." The next step, shown in the middle of Figure 4.3, is to identify all the relevant factors that contribute to this effect. These are broad areas in which causes

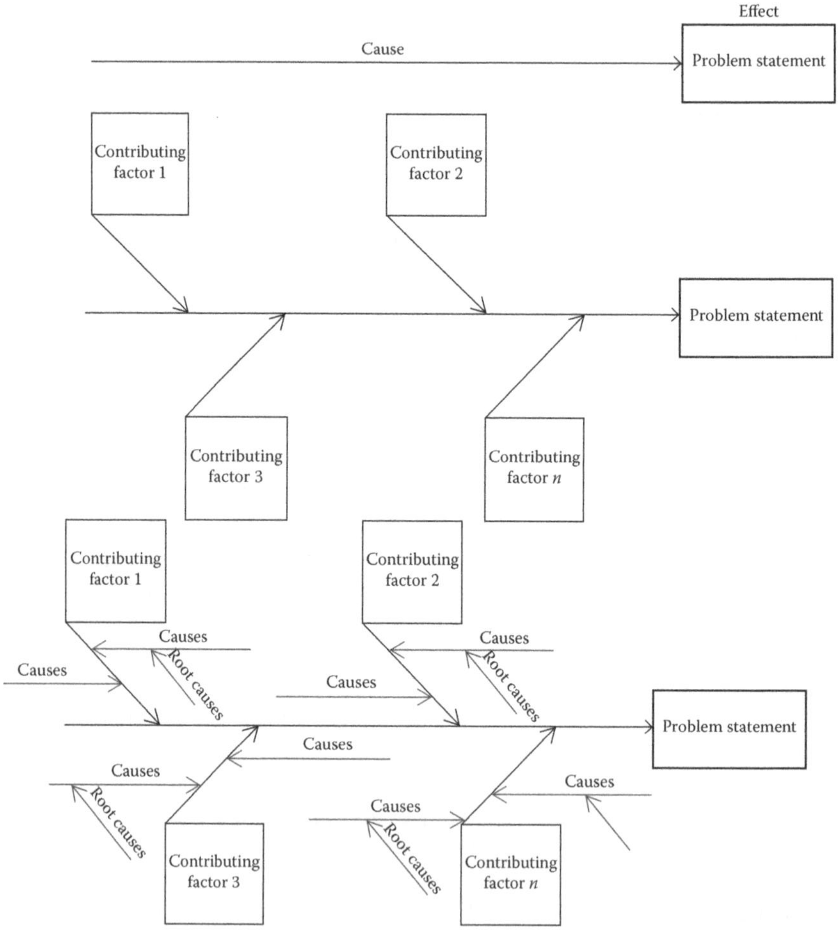

Figure 4.3 Evolution of a cause and effect diagram.

of this effect may lie. These factors can be brainstormed or decided in any manner suitable to the team. Mind Tools (2008–2010) has suggested using five Ms and an E to help identify these factors. These would be man, method, machine, material, measurement, and environment. Man refers to people-related factors. Method refers to process, procedure, and instructional and similar causes. Machine would include technology, capital, equipment, and the like, as well as their condition, capacity, and so on. Material refers to the resource inputs, their quality, quantity, and other relevant attributes. Measurement refers to the quality of data and instruments used to record it. The environment refers to the external environment, which would include the natural environment.

The bottom depiction of Figure 4.3 shows how specific causes under each factor cluster would be included in the diagram. There would be as many of these as the team can identify. For each factor identified, the team should ask what causes are contributing to the effect. These causes are related to the appropriate factor by drawing lines into the factor branch.

The remaining step in producing an Ishikawa diagram is to make sure you have gotten to any root causes that may exist. This can be done using the why-why approach. For any given cause of the effect, simply ask of that cause, why? Answer the question. Then ask why of that answer. Keep repeating this process until there are no more whys. Three to five levels of why ought to be enough to get you to a root cause. Write that on the diagram as shown. By the way, if the diagram begins to get too busy, just tear off a portion of it, expand its size, and work on it separately.

A sample cause and effect diagram for a flood risk management problem is shown in Figure 4.4. Once the diagram is completed it is time to analyze and use it. The first and primary use of this diagram is to help the team articulate a common understanding of the problem for itself and others. With the more complete view of the problem provided by the diagram, planners can write a more objective problem statement and write a more complete problem profile. You also have sufficient detail to support a rich narrative description of the problem if desired.

The diagram provides an effective jump off point for identifying measures that may eventually help you solve the problem. Branches with lots of details may be ripe sources of ideas. When a particular cause is found under several factor categories, there is a good chance this is a root cause. Root causes of problems are a good level at which to begin to consider solutions. This task is complete when you have evidence that confirms your articulation of the problem and its root causes.

4.10.2 Brainwriting

You can easily find a lot of information on this topic with a simple web search, so let me take you directly to what may be my favorite technique for getting a group started with problem identification. All it requires is a group of people, some index cards, and pens. I like it because it involves everyone and it is not easily dominated by a strong personality. More importantly, it has invariably produced useful results.

The example that follows is for articulating the problems for a problem statement. The assembled team members are split into groups of four or five. Each person is provided

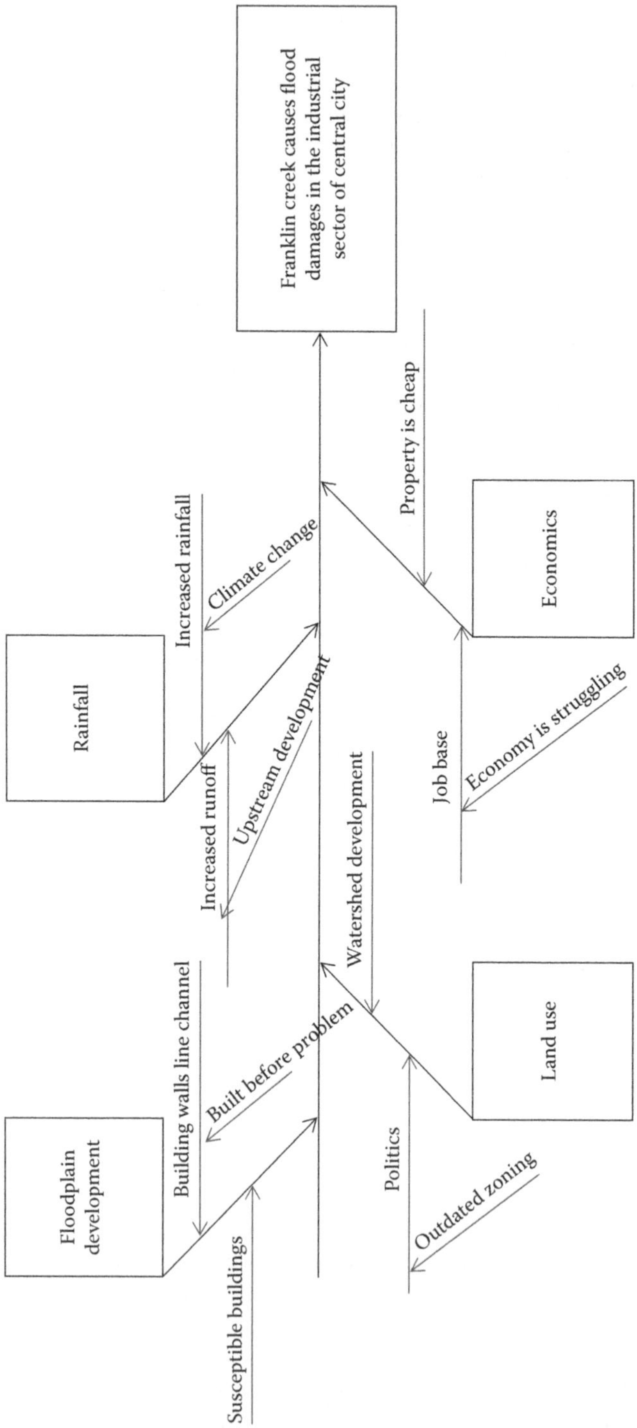

Figure 4.4 Cause and effect diagram for a flood risk management problem.

with an index card and is asked to write down the most important problem in no more than two sentences, beginning with the phrase, "The problem is."

The initial problem statement is passed to the left so that each person is looking at another person's problem statement. The task now is to improve on the problem statement you are reviewing. The writer is not to make it sound like their original problem statement; they are to wordsmith and improve the statement they receive. When the review is completed, the cards are once more passed to the left. This process continues until the writer receives his original statement back. Looking at the edits to their original, each person prepares a revised problem statement.

At this point the small group votes one of the four or five statements as its choice for the most important problem. This sometimes involves discussion and negotiation over the final wording. Each group reports back on its decision and a plenary discussion and refinement of the initial list of problems ensues. At the conclusion of the discussion it is normal to expect to have the top one to three problems articulated in an initial form. The entire process takes less than an hour.

In a situation where there are many problems rather than a few standout problems a different process may be used. A six-step technique I have had success with follows. It works best with groups of 4–20 people. A large pool is preferred.

1. Each individual silently writes down his ideas about the problems on a sheet of paper.
2. The papers are all placed in the center of the table (or some other pool).
3. Each individual takes a sheet of paper from the pool, reads the ideas on the new sheet and uses them to stimulate new ideas, which are added to the sheet.
4. Once any new ideas are added to the sheet, it is exchanged for a new sheet from the pool. Often there may be a rule that at least one new idea must be added to a sheet before it can be returned to the pool.
5. Each individual continues to write down ideas and exchange sheets for three or more iterations.
6. At the end of the process the ideas are read aloud for clarification and wordsmithing. The problems so identified represent the initial articulation of problems.

4.10.3 Brainstorming 3X Yeah

The Internet is lousy with brainstorming techniques. My favorite is one I learned from Ken Orth, a planner and friend. It is pretty close to many classical techniques. I like it because anyone can facilitate it. The process is called "3X Yeah," (as in the Beatles' "Yeah, yeah, yeah"). It works like this:

- Provide materials.
- Identify the question.
- Explain the process.
- Silent idea generation (brainwriting).
- Group idea generation (brainstorming).
- Preliminary evaluation.
- Award prizes.
- Follow-up.

You'll need a workspace with enough room to break out into small groups of six or less. Index cards and pencils are needed for private idea generation and a flip chart with markers (or some other communal form of recording, such as a laptop and projector) is needed for each group.

The question can be as simple as what are the problems we face in the XYZ planning study? The process is easy to explain. It has the following simple rules:

- No evaluation.
- No judgment.
- Quantity counts.
- Time is limited.
- Follow the process.
- Generation of ideas is separate from judgment of ideas.

The goal is to generate as many ideas about problems as you can. It's important not to censor one's thoughts. This means individuals must not evaluate or judge any ideas. Write down anything that comes to mind during the process. Capture ideas. Write it down, do not worry if it is a good or bad, complete or not. This is a time-limited process—let people know that. They will be committing an hour or less of their time. It's important to follow the process. Give it a chance to work. Let everyone know they will get a chance to evaluate the ideas when they are all done generating them.

To avoid the effects of group domination, strong personalities, and intimidating relationships the process begins with each individual silently generating a list of his—I think you have been using all male references? own ideas. Make sure everyone has several index cards or a piece of paper and something to write with. Ask them to write legibly because their cards will be collected. The facilitator begins the process by telling the group they have 3 minutes to list 10 problems the study faces. If you get more than 10, keep going but shoot for 10. Everyone is to work silently for the 3 minutes.

After 3 minutes have passed, ask if anyone got more than 10 ideas. Then ask if anyone got 10, 9, and so on until most of the group has raised their hands. If you'd like to inject a little fun into the process, toss the person with the most ideas a token prize of a candy bar or some other trinket that would be regarded as fun without crossing over into competitiveness or jealousy.

Congratulate everyone on their efforts and results and don't feel guilty if you lead them to believe they are done. Then tell everyone to draw a line beneath their last idea. Explain that the first time through everyone was getting the obvious stuff, the low-hanging fruit. Give them 3 more minutes and ask for 10 new ideas. When the time is up, repeat the query of your group, asking if anyone got more than 10. Before moving on, ask if anyone failed to get any new ideas. It will be rare that someone did not, be gentle with anyone who comes up blank. Take the process seriously, but have fun with it.

If some people were still generating five or more new ideas, it may be worth one more round of silent idea generation. This time give them 1 minute and ask for three ideas. You never want to ask people to generate their own ideas more than three times.

Use a random process of some sort (see the following text box) to form small groups for some oral brainstorming. Counting off is a simple way to break the chosen seating patterns and to form groups. An ideal group will be five or six people. The group's charge is to use their personal lists of ideas and as a group make the longest list of unique (nonrepeating) ideas they can. These ideas will be numbered and recorded on flip chart paper, a computer, or some other medium that others will be able to see. Start the process by asking everyone for one idea from their list. Going around the group until no new ideas are coming out is polite, but not much fun. An alternative is to encourage people to blurt out any new ideas they get from the ideas of others. The sole goal is to keep the ideas coming, without any judgment or evaluation of anyone else's ideas. If you want a rule for this session, it is no cross-talking! Offer the group with the most ideas a prize—ice cream on a stick, candy bar, an hour off with pay—to spur some friendly competition among the groups. Thirty minutes is usually sufficient for this task. If you need more time, take it.

When the group idea generation is completed, the team moves into its evaluation phase. Ask each group to select their best idea, their worst idea, and their wildest idea. For a variation and to get people to read the ideas of others, have the groups exchange lists and vote these "honors" to another group's list. Each group reports its selections. Collect the index cards and flip chart paper, award prizes if you use them, and call it a productive hour.

The critical question is what will you do with all this information? Without productive follow-up, you have just wasted an hour of everyone's time. What you have likely done, however, is to have generated a lot of ideas from a lot of people in less than an hour.

Compile the problem ideas and feed them back to all participants and interested others as swiftly as you can. It's time for convergent thinking. You want to go from the many ideas to weeding out the weaker ideas to get to the best ideas. This requires more group evaluation. Dot voting, discussed in Section 4.10.7, is one option that works transparently well.

FORMING GROUPS

Instead of counting off by 3s or 5s to form small groups try this. Ask someone for their favorite letter of the alphabet, TV show, city to visit, color, beer, or any random thing you can imagine. Write their response down for all to see. Then ask your next goofy question of another person. When you have as many answers as groups you want to form have them count off by the answers, for example, A, *Leave it to Beaver*, Berlin, Green. These become your group names and the exercise usually throws enough people that it can provide a few laughs along the way.

Warning. Never attempt humor if you are humor impaired.

4.10.4 Be a Reporter

When it comes time to identify problems, you can do a lot worse than to ask some who, what, when, why, where, and how questions. Who is causing the problem or suffering from the problem? What is causing the problem? What are the consequences of the problem? When does the problem occur? Why does it occur in this way? How did we become aware of the problem? Where does the problem occur? Where are the consequences visible?

Do not feel limited in the ways these questions can be posed. For example, some questions to ask of significant resources could include the following:

- What happened or will happen to the resource?
- What is or will be that event's impact?
- Who or what does that or will that impact affect?
- When did or will that impact happen?
- How did it or will the impact occur?
- Why did it or will it occur?
- What could be done about what happens, its impact, who or what it affects, when, how, or why it occurs?

Ask and answer these questions and you have a good start on identifying problems.

4.10.5 Mind Maps

Mind maps are growing in popularity and ease of use. These are diagrams used to represent ideas and their linkage to a central concept. They are extremely useful for visualizing, structuring, and classifying ideas. Mind maps are useful for analyzing, comprehending, synthesizing, recalling, and generating new ideas as well as for illustrating problems. Mind maps can be done by hand. To draw one:

- Write the key problem word or phrase in the middle of a blank sheet of paper. Draw a circle around it.
- Think of as many related subtopics as you can. Write them down and connect each of them to the center with a line.

PITFALLS IN PROBLEM DEFINITION TO AVOID

1. No focus—definition too vague or broad.
   Example: Lack of biodiversity in the watershed.
2. Focus is misdirected—definition is too narrow.
   Example: How can we improve conditions for the mottled duck?
3. Statement is assumption driven.
   Example: How can we stop harmful human disturbances?
4. Statement is solution driven.
   Example: Mallow Marsh needs a water control structure.

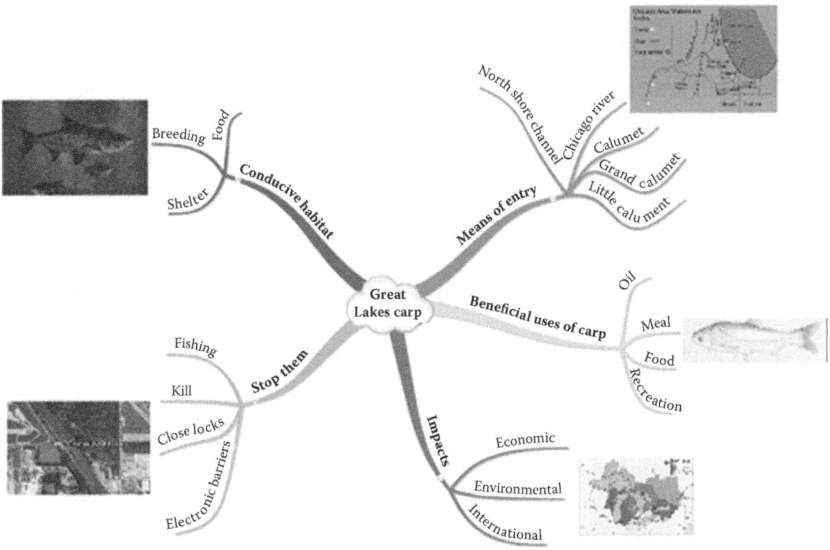

Figure 4.5 Sample mind map diagram for an aquatic nuisance species.

- Identify subtopics or facts for your subtopics and repeat the same process. Generate as many of these lower-level subtopics as you see fit. Connect each one to its corresponding topic.
- Be as visual as you can be. Use colors, stick figure drawings, sketches, and symbols generously.
- Limit your words. Keep your topics, facts, and phrases as short as possible. One word is good; a picture is better.
- Engage your brain, use variety. Different size print or script, colors, thickness, lengths, curvature, alignments, and so on should be used.

The simple illustration in Figure 4.5 shows how a mind map aids thinking about the problem of the Asian carp reaching Lake Michigan in the United States. There is concern that this invasive species may become established in the Great Lakes. Once a map like this is drawn it becomes easier to see the complexity of the problem and to develop a well-focused problem statement.

4.10.6 Restatement

Once a group has articulated a few problems, I like to use this technique to make sure everyone understand how different a problem can look to others. The idea is simple; restate a problem in as many ways as you can. Imagine it from the perspective of different stakeholders, then different professions and different individuals like Homer Simpson, Superman, and the mayor. How would the Environmental Protection Agency (EPA), the State Department of Natural Resources, recreationists, neighbors, scientists, uneducated people, politicians, and so on state the problem? Imagine how different magazines (e.g., *Mad*, *Popular Mechanics*, *Home and Garden*) would

OTHER RESTATEMENT TECHNIQUES

Instead of focusing on different groups or individuals, there are other ways to restate a problem, using different words or perspectives. Here are a few examples that have proven effective in my experience. Restating should take no more than 5 or 10 quality minutes if you start with a basic statement of a problem.

Initial Statement: Freshwater fish and invertebrate species are being eliminated.

1. Paraphrase: Restate the problem using different words without losing the original meaning.
   Example: How can we preserve and restore freshwater fish and invertebrates?
2. One-hundred-eighty degrees: Turn the problem on its head.
   Example: How can we eliminate freshwater fish and invertebrate species?
3. Broaden the focus: Restate the problem in a larger context.
   Example: How can we achieve greater biodiversity?
4. Redirect the focus: Boldly, consciously change the focus.
   Example: Saltwater fish species are on the increase.
5. Use the why-why approach mentioned above: Ask "why" of the initial problem statement. Then formulate a new problem statement based on the answer. Then ask "why" again. And again restate the problem based on the answer. Repeat this process a number of times until the essence of the "real" problem emerges.
   Why? Because salinity levels are changing in the marsh.
   Restatement: Salinity levels are changing in the marsh.
   Why? Because salt water has entered this formerly freshwater marsh.
   Restatement: Salinities in the freshwater marsh are increasing.
   Why? Because the navigation channel has introduced a saltwater wedge to the river that feeds and abuts the marsh.
   Restatement: Dredging has increased salinities in the marsh, eliminating freshwater fish and invertebrates.

describe the problem. The goal is to identify as many different ways of seeing and expressing the problem as possible. Write them down. Do you see the problem any differently? Do you understand it better? If so, it may be time to re-articulate it.

4.10.7 Dot Voting

Here is a process for finding out quickly how a large group feels about a lot of things. It's a good way to find out how your team feels about a long list of brainstormed problems. It can also be used with stakeholders or the public. It requires a room large enough for people to move about freely. Flip charts or flip-chart-sized paper affixed to the walls, along with markers for the flip charts will be needed. In addition, you'll need a supply of colored sticky dots.

Candidate problems should be written on the posted paper. You could, for example, use the group sheets from the 3X Yeah process described earlier. Give each person a supply of colored dots; five is a nice round number of dots. Read off the problem statements for the audience. Ask those participating to circulate around the room and vote for the problems they consider most important. They can distribute their votes in any way they like. They can give one dot to each of five different problems or they can vote all five to a single problem.

The accumulation of dots around a relative few problems statements can reveal the growing consensus among the participants about what is important. Conversely, the lack of a consensus will be visually clear as well. In my experience, this process usually results in the emergence of a few consensus problem statements.

REFERENCES

Food and Drug Administration Center for Food Safety and Applied Nutrition. 2003. *CFSAN's risk management framework*. College Park: CFSAN.

Guidance Regarding NEPA Regulations. 1983. 40 CFR Part 1500, accessed June 14, 2012, http://ceq.hss.doe.gov/nepa/regs/1983/1983guid.htm.

Lachapelle, P. R., S. F. McCool, and M. E. Patterson. 2003. Barriers to effective natural resource planning in a "messy" world. *Society and Natural Resources*, 16:473–490.

Mind Tools. 2008–2010. Accessed December 28, 2011, http://www.mindtools.com.

Mind Tools. 2008–2010. *Problem Solving*. Mind Tools Limited e-booklet.

Poole, C. 2012. One hour to save the world. *The Work Institute*. Accessed July 20, 2012, http://workinstitute.com/One-Hour-to-Save-the-World.

Powell, S. G., and K. R. Baker. 2007. *Management Science: The Art of Modeling with Spreadsheets*. 2nd ed. Hoboken, NJ: John Wiley & Sons.

VanGundy, A., Jr. 1988. *Techniques of Structured Problems*. New York: Van Nostrand Reinhold.

Yoe, C., and K. Orth. 1996. *Planning Manual. IWR Report 96-R-21*. Alexandria: Institute for Water Resources.

CHAPTER 5

Gather Evidence

The primary purpose of this ongoing process is to develop, gather, organize, compile, and analyze data to produce information and knowledge useful for decision making. This will include conducting analyses under a variety of scenarios describing the study area future with and without planning-related interventions.

5.1 INTRODUCTION

Here's the short story. You are probably going to be gathering evidence and doing analysis from day one until day last of your planning process. Get used to it. Expect it. This is an ongoing process. Some planning models have a step devoted to data collection or analysis. That's not the way it works. Analysis is found in every step of the planning process, it can't be isolated to a single step.

Planning needs to be based on the truth or at least the truth as best as we can discern it. That means assertions, beliefs, and consensus are not good enough; we must let the evidence speak. Everyone might believe the lead shot used by hunters is killing waterfowl, but it also has to be true, and we only know what is true through evidence. People can feel strongly and believe deeply in things that just are not so. Facts and truth matter, and evidence points to the truth. Gathering evidence is also the best way to reduce uncertainty in a planning study. Be forewarned, however, evidence is not always or often going to be enough to change minds. Beliefs, once formed, are highly resistant to facts.

Analysis is going to take up most of your time, money, and effort. That analysis should be based on the best available science. That means good data, good models, good analysis, and conclusions tied to all this evidence. Use the scientific method as often as you can. When you have a void in your understanding, formulate a hypothesis you can test and design strategies that will enable you to test your hypotheses.

Risk analysis has risen in popularity largely because it provides a framework for making decisions under uncertainty. Planners live with uncertainty, and planning can be enhanced by using risk-based analysis. It is the planning team's job to account for uncertainty in their analysis and to convey the significance of that uncertainty forward to decision makers. It is the decision makers' job to consider that uncertainty during their decision making.

This book is not going to tell you anything useful about how to do analytical stuff. That is just way too big a topic that involves far too many disciplines. So you won't learn how to discount money or calculate a habitat suitability index here. The nitty-gritty work of doing analytical work basically gets a hand wave in this book. Instead of analytical details, you'll find scenarios. Scenarios are the stories we tell about conditions of interest to us. In planning, these are most often future conditions. Scenarios provide the narrative logics used to frame the analyses planners do. The without condition scenario, introduced in the previous chapter, reappears in this chapter as the most important scenario used in a study. Several other scenarios are introduced in this chapter. One of them, the with condition scenario, will be prominent in Chapter 7 as well. Evidence is gathered to flesh out the details of scenarios. Much of the most important analysis done in a planning study involves the comparison of decision criteria under different scenarios. That is how planners figure out whether their plans make things better or worse as well as how much better or worse.

Here are a few more things to know before leaping into this chapter. Even though evidence gathering and analysis are ongoing, there are specific points in the planning process—like verifying problems and opportunities in Step 1—where significant new analytical activities are launched. Table 5.1 identifies planning tasks where evidence gathering activity tends to become rather focused. Three kinds of evidence gathering are identified in Table 5.1. The data column indicates a task where evidence has to actually be gathered and organized. The analysis column indicates analytical tasks using the available data. PI stands for public involvement, and it indicates the need to gather information mostly about values, beliefs, and other subjective kinds of

Table 5.1 Evidence Gathering Activities Throughout the Planning Process

Step	Evidence Gathering Focus	Data	Analysis	Public Involvement
1	Identify problems and opportunities	√		**√**
	Verify problems and opportunities	**√**	**√**	
	Without condition scenario	√	**√**	
	State objectives and constraints			**√**
	Assess without condition scenario	**√**	**√**	
	Decision criteria			√
	Uncertainty	√		
2	Measures	√		√
	Plans	√	√	√
3	First evaluation iteration	√	√	
	With condition scenario	**√**	**√**	
	Assess plan effects	**√**	**√**	
4	Compare		**√**	
5	Selection criteria	√	√	√
	Formal weighting process	√	√	√
	Order alternatives		√	

information from the public. A large bold check indicates more intensive effort and activity than a smaller check. Table 5.1 presents the author's opinion, not hard facts. Notice that evidence gathering activities take place in every step.

Consequently, discussions of evidence gathering, my euphemism for all the analytical work planners do, are sprinkled throughout the next five chapters. Figure 5.1 provides a simple guide to this chapter, whose major topics are evidence, analysis, and scenarios.

Evidence gathering process

Evidence

Analysis

- Best available acience
- Risk-based analysis
- Uncertainty
- Assess and verify problems and opportunities

Scenarios

- Common scenarios
- Types of scenario analysis
- Comparing scenarios

Figure 5.1 Guide to Chapter 5 contents.

5.2 EVIDENCE-BASED DECISION MAKING UNDER UNCERTAINTY

Planning analysis should be based, to the maximum extent possible, on facts and truth. An in vogue phrase used to capture this idea is "science-based." I prefer evidence-based because not all of the information used by planners will meet the standard of being considered science. Let me be clear, however: planners should strive to get the right science focused into a planning effort, and then they should assure that the science they use is right. Planning should be as science-based and as informed by good science as is possible in any given study. Know that evidence includes good science, but it also includes reliable data, professional judgment, expert opinion, and all kinds of social values, which may not avail themselves of the scientific method.

Planning should be grounded in reality so planners should constantly demand evidence as part of every analytical and deliberative task in the process. The point of this principle is that analytical parts of the planning process are to be analytical. They should not be based on default or summary judgments, what you did last time, consensus, what the boss believes, opinions, whim, or fancy. "What is the evidence?" is an important question to ask repeatedly throughout the planning process.

Gathering, organizing, analyzing, and presenting data are essential activities for every study. Ackoff (1989) describes the content of the human mind classified into five categories. They are *data*, which consist of symbols (e.g., numbers on a page). *Information* consists of data that have been processed in a way that makes it useful. Information provides answers to "who," "what," "where," and "when" questions. When information and data are applied, they provide *knowledge*, which answers "how" questions. This leads to *understanding*, that is, an appreciation of the "why." The final step is *wisdom*, an evaluated understanding. This is an apt classification for planning, which passes through all five of these classes of mental activity. At the base of all of it is the evidence. It is the duty of every professional to consider the best available science and expertise in the planning process. If you are going to gather data to make a decision you ought to use the best available data. The best

HOW DO WE KNOW WHAT WE KNOW?

Get used to asking for evidence. If a colleague says it is a good day ask her to give you three pieces of evidence to support that statement. It is important to be clear about the things we know and the things we don't know. It's also okay not to know things. When you are assuming and believing things, that information should not look like facts to anyone. Let the evidence speak. When there is none, make sure decision makers and stakeholders know that and how you are dealing with that uncertainty.

available data are those data that reduce the uncertainty to a point that planners and decision makers can feel reasonably comfortable with the decisions they are making. Risk-informed planning will explore the effect of the remaining uncertainty on the decision and its potential outcomes in order to fortify the decision maker's confidence level. Don't go overboard with the idea of best available. This need not mean all the available data. The best data for a decision will not always be the highest quality and most expensive and time consuming. The best data for some decisions will often be qualitative, it may even be crude. It may be entirely appropriate to screen plans early in the planning process using qualitative data even when quantitative estimates could be made available with more time and effort. In this instance planners ought to use the best available qualitative data. Thirty-six months of water quality data will be better than three months, but if decisions can be made on three months of data that is fine. Use the best available data for those three months. If you are evaluating the cost of trails and can do that simply based on the length of the trail you may not need any actual cost data to do so. Use the best available estimates of the length of the trail.

Planning should use the best science, data, models, and analysis available. To do anything less is irrational. If "evidence-based" is a planning principle, using the best available science is the standard. The only fly in this ointment is how one interprets "available." The best may not be available to you because of resource constraints or other limitations. There may be better data or models than you are using, but if you lack the time, money, or expertise to acquire and use them they are still, nonetheless, available.

Use the best scientific methods. Samples should be probability samples. Experiments require appropriate controls. Models should be verified and validated. Assumptions need to be investigated and verified with data or subjected to sensitivity analysis or peer review when they cannot be verified. Analytical methods should be appropriately chosen and rigorously applied. When using the data, knowledge, and results of others, the literature should be rigorously reviewed and cited. The best available science defines what planners know in an investigation. Uncertainty defines what they don't know.

WHAT IS THE EVIDENCE THAT...

...trout are disappearing?
...new growth forest is stunted?
...there is a flood problem?
...a bike trail is needed?
...the community opposes that?
...this sits on a seismic zone?
...that will cost too much?
...the super colony will return if water quality improves?

5.3 EVIDENCE

Evidence comprises the facts, science, data, information, and opinion used to understand a study area, its problems, opportunities, and potential solutions. Although evidence is gathered throughout the planning process in each of the steps, several significant evidence gathering analytical tasks occur early in the planning process.

Without a proper assessment of the problems and opportunities and a common future without condition scenario, it is not possible to formulate the most effective solutions. Thus, these tasks are accomplished early in the planning process. Formulating solutions requires substantial evidence gathering that will likely include inventories of numerous existing conditions. Once solutions are formulated, another rigorous iteration of evidence gathering and analysis follows as the efficacy of each solution is assessed in subsequent steps. All this analysis should be based on the best available evidence and the best available scientific approaches. Because of budgetary and schedule constraints, this is not always, or perhaps even often, the case. The result is planners must deal with the uncertainty that ultimately remains when evidence gathering is completed.

Evidence comes from many different types of sources. Some are going to be more valuable than others, and this will vary from study to study. It is useful to consider three levels of evidence or information: primary, secondary, and tertiary. Primary sources are the original works of research and raw data gathering. This includes such things as aerial photography, topographic and other surveys, interviews, observations, questionnaires, behavioral analysis, experiments, and any other organized effort to gather information that is not yet available in some form. Primary data may also be found in the administrative data of organizations: admissions, visitor counts, inventory records, process control charts, rain gage data, stream flow data, and all such routinely collected raw data. Primary sources of information may also be found in letters, videos, interviews, speeches, court decisions, transcripts, government and private databases, and so on.

Secondary sources are interpretations of primary data. Several secondary information types encountered in a planning study are summarized in Table 5.2, prepared by Swanson (2007). Tertiary information sources tend to be interpretations of secondary sources. These are generally found in bibliographies, indexes, and other finding aids like Internet search engines. All evidence is not of equal value; therefore, it is vitally important that planners carefully document their evidence.

5.3.1 Generic Information Needs

Information is evidence put to a useful purpose. Describing the scenarios presented later in Section 5.5 so you can analyze differences among them is the primary purpose of much of the data collection and analytical work. Several important generic types of information can be identified for any planning effort.

Table 5.2 Types of Secondary Information Sources

Information Type	Description	Sample Sources
Scholarly	Author: has some degree of authority in the field; typically has an academic post or is a researcher Audience: other experts in that field Purpose: advances a field a study by reporting new findings or ideas, increase author's authority and credentials in field	Where to find: Books, articles in scholarly journals, websites • *New England Journal of Medicine* • *The Explicator* • Research findings on authoritative websites such as PubMed • *Irony in the Short Stories of Edith Wharton*[a] by Charlee Sterling
Professional/trade	Author: member of a profession or trade but not necessarily a researcher Audience: members of a particular field or trade Purpose: informs, promotes, and generally strengthens the profession	Where to find: Trade magazines and journals, association websites • http://www.ala.org (American Library Association) • *Transport Topics* • RN
Government	Author: varies (could be government employee, elected official, or expert in a particular field) Audience: varies (could be public, elected official, or government agency) Purpose: generally created to run the government, inform decision makers, or inform the public	Where to find: Printed government publications and official government websites • Congressional record • *Supreme Court Reporter* • http://www.nih.gov (National Institutes of Health)
Facts, definitions, and statistics	Author: staff writers Audience: general public or researchers Purpose: presents facts, definitions, and statistics with little explanation or evaluation	Where to find: Books and websites, but not Wikipedia since it is edited by the public, not a staff. • http://www.worldalmanac.com (World Almanac) • Statistical abstracts • *American Heritage Dictionary*
Overviews	Author: staff writers, freelance writers, or scholars Audience: general public Purpose: provides overviews and background on a subject	Where to find: Encyclopedias, dictionaries, other books, websites • *Encyclopedia Britannica* • About.com
Entertainment/popular	Author: nonexpert in a field usually with a degree in journalism or training as a writer Audience: general public Purpose: presents information in an interesting manner that does not necessarily focus on depth of coverage	Where to find: Magazines, websites • *Rolling Stone* • *InStyle* • *Entertainment Weekly* • http://www.etonline.com (*Entertainment Tonight*)

(*Continued*)

Table 5.2 Types of Secondary Information Sources (*Continued*)

Information Type	Description	Sample Sources
News	News author: nonexpert in a field usually with a degree in journalism or training as a writer Audience: general public Purpose: reports current events in a timely fashion possibly influenced by editorial policy	Where to find: Newspapers, news magazines, news websites • *New York Times* • http://www.msnbc.com • *TIME*
Special interest/ opinion	Author: typically a nonexpert in a field, but could be an expert expressing his or her opinion Audience: general public/people subscribing to a particular point of view Purpose: advances a particular point of view or expresses an individual's point of view (the attribution of authority may heavily depend on the beliefs of the reader)	Where to find: Books, magazines, websites • *Cult of the Amateur*[b] by Andrew Keen • National Rifle Association website http://www.nra.org • American Cancer Society website http://www.cancer.org
Unsubstantiated or uncredited information	Author: unable to substantiate identity of the author or author's credentials do not carry authority Audience: general public or unable to determine Purpose: hobby or personal interest	• Where to find: Websites, info from friends • Personal websites • Blogs • Bogus websites like http://www.malepregnancy.com

Source: Based on Moraine Valley Community College Library/Learning Resources Center. 2012. Types of information. Accessed July 18, 2012, http://www.morainevalley.edu/library/handouts/types_of_information_0705.pdf.

[a] Charlee Sterling. *Irony in the Short Stories of Edith Wharton*. Lewiston, N.Y.: Edwin Mellen Press, 2005.

[b] Andrew Keen. *Cult of the Amateur: How Today's Internet is Killing Our Culture*. New York: Doubleday/Currency, 2007.

First, information is needed to identify, verify, and adequately describe the problems and opportunities of the study area. If your study is addressing aquatic nuisance species (ANS) you are going to have to identify the species and learn something about their life histories and life requisites. A forest management study will need information about species, ages, acreages, disturbance history, ability to regenerate, and the like. Flood risk management requires detailed hydrologic and hydraulic studies as well as property damage surveys that will be needed to describe the flood problem in a probabilistic manner. Subject matter experts know what kinds of information and analysis are necessary for their evidence gathering activities.

Second, you are going to need information to measure success toward meeting your objectives and avoiding your constraints. You'll also need to measure your decision criteria. If you seek solutions to a flood problem that reduce social vulnerability, you are going to have to be able to measure and talk about social vulnerability. If your aquatic nuisance species (ANS) study is to minimize impacts on existing commercial and recreational fisheries, you are going to need information to measure your progress toward those objectives.

Third, you are going to need information to estimate life cycle costs for your proposed solutions. Someone is always going to care about costs, even if you do not. As you will learn in Chapter 14 on economics, not all costs are financial. For example, if a plan limits the way people can access and use resources they have grown used to using, they will bear a cost. Although there is indeed a comprehensive framework within which all costs can be accounted for, not all planning studies adopt that framework. A state that is developing a state park system to increase tourism is not going to care if their program takes tourist dollars from neighboring states. From a national perspective, their gain may come at a cost to another state, and it may represent no net increase in national output. From the state perspective, every "stolen" dollar marks an increase in state output. Consequently, the definition of costs is often context dependent.

Fourth, information is needed to describe the more important effects of a plan. Some of these effects will include the objectives and constraints. There will be other impacts of interest, however, that must be described. Some of these may be required by law. You will probably be interested in endangered species, cultural resources, and a variety of environmental effects. There are also likely to be a number of impacts of special interest to your planning partners, stakeholders, and the general public that must be described.

The cardinal rule for information gathering is to get what is needed to make good decisions, not what is available or what is traditionally gathered. "Planner as storyteller" is a useful role to assume when thinking about the information you will need. What kind of information will you need to tell a convincing story about why you made the decisions you made? If you cannot tell a complete, logical, and easy-to-follow story about what you did, bolstered and supported with the information that makes your story credible and transparent, then you cannot plan successfully.

As the planning process develops, a story does unfold. That is the story you must be able to tell convincingly if stakeholders are to trust in the soundness of your decisions and the rationality of the process by which you arrived at them. Because you need to guess at the information you'll need at the end from the very beginning, it is inevitable that you'll gather some information that is ultimately not useful (then don't use it!). Likewise, you'll find yourself needing information you don't have (then go get it or address the uncertainty its absence causes in your analysis). Revising information needs as your understanding of the problems and opportunities evolves, as planning objectives are refined and the story develops is a constant in every planning study. Consequently, evidence gathering is more than a discrete step in planning; it is a continuous process within the planning process.

5.3.2 Four Parameters of Evidence

The four parameters of evidence gathering are quantity, quality, timing, and location (Yoe and Orth 1996). How much information is enough? How accurate and how representative must the data be? At what point does the collection process start, how long do you have to collect information, and for what period of time are data

AVOID THE WIND ROSE

Many of us learn to plan on the job. I did. When you're new and everyone is too busy looking to ask for help, the easiest way to survive is by imitation. That meant including a wind rose in the report. The last report had one, so this report must need one too. Never mind that the information may be superfluous. If you gather any evidence at all make sure you understand what it is evidence of and how you will use it in your investigation.

required? What geographic area is to be covered? The definitive answers to these questions may not be known until the end of the study. Like virtually everything else in the study process, evidence gathering is an iterative process.

Quantity. For now, let's keep it simple and divide data into two broad categories: physical data and socioeconomic data. Physical data describe the physical environment. Socioeconomic data are those that depend on the human element in the environment. For any given analytical or scenario building purpose, you need enough of both kinds of data to be reasonably certain you have the information you need to move forward in the decision process. Collect the data you need not all the data there are.

Could your understanding of the life history of *Branchiura sowerbyi* (an ANS) be more complete? Could you further refine your estimate of expected annual damages (EADs) for your flood risk study? The answers to questions like these are almost always going to be yes. In general, it is always going to be true that if you broke the study area into smaller sectors or reaches; if you had more stream flow data; used a larger wildlife sample; developed site-specific dose–response curves; and so on, your answers would probably be better. But, have you used reasonable data collection and analytical methods to obtain the information you need to feel reasonably comfortable that you now understand the magnitude of the ANS, flood, or other problems you seek to solve? If so, you have enough information and it is time to move on. If significant uncertainties remain in the "pile of things you don't know" then you need to address them, but there comes a time in every planning study when you have all the information you are going to get. Time is up. Pencils down. Eyes on your own paper.

Quality. The quality of data you work with often depends primarily on the stage and type of planning study you are doing. Cost estimates of your solutions in a reconnaissance-level study may be based on 10% of design detail. Feasibility study costs may be based on 30% of design detail. More advanced studies will have more design detail. You can make reasonable judgments about which solutions are less costly when all solutions are evaluated at the same level of detail. You do not need top quality data for every decision. Early iterations of the planning process do not always require the same data quality later iterations do. The notion of best available data is relative, not absolute, as used here. The best data for an early estimate of costs will not suffice when the work is going out for bids.

THE VALUE OF INFORMATION

How much uncertainty can you live with in your study? Which is the more horrible fate: paralysis by analysis or extinction by instinct? These choices reflect one of the more difficult decisions a planner faces in this evidence gathering process. How much information is enough?

Having more information may reduce your anxiety. Unless it changes your decision, it is not worth the cost of obtaining it, however. An important question to ask when considering what information to gather or how much more of it to get is, "Could this information affect your decision?" If the answer is no, do not get it. If the answer is yes, it's necessary to ask how likely it is to change your decision. If the possibility is remote, you may not need the information. Only get the information when it might affect your decisions.

Data should also be representative. If you're using sample data, it should be an unbiased sample from the population of interest. A short stream record taken during unusually wet years would not be representative of the long-term stream flows. Accuracy is a fundamental aspect of data quality. The accuracy of your data must be known and communicated to decision makers, stakeholders, and the public.

Timing. There are three dimensions to the timing characteristic of data collection. First is time as it relates to the planner, that is, how much time do you have to do the analysis? When can data collection begin and how long do you have to complete the study? These dates and periods will vary with the type and stage of study. Every question has the 1-hour answer, the 1-day answer, the 1-month answer, and the 1-year answer. Take the most time with the most important data.

The second time dimension relates to the data itself. The point in time during which data were collected may be important. Water temperature and dissolved oxygen measures taken in the summer differ from those in the winter. Economic surveys conducted during recessions will differ from those collected during economic booms. The timing of monetary values is important for the time value of money and the price level used to measure monetary values. A third aspect of data timing is the length of the data record. Duration of data is especially important for hydrologic studies and monitoring the health of ecosystems.

Location. The geographic area for which data are collected will usually conform to the study area. Normally data will not be collected for areas outside the affected area unless the outside data affect the study or are needed to provide perspective and context for the study area data.

When you gather evidence, pay attention to these four parameters. Gather all the information you need but no more. Insist on quality that is sufficient for the decision you're making at the iteration you are working on. Get temporally relevant evidence in a timely manner. Make sure your data are spatially relevant as well.

5.3.3 Appropriate Level of Detail

Decision making in planning is always going to entail a trade-off between the things we know and the things we don't know. Reducing uncertainty always has a cost. So planners sometimes agonize over the appropriate level of detail in a planning study. The answer is at once simple, elegant, and not terribly pragmatic; the level of detail shall be sufficient to make the decision at hand. Don't pay to gather evidence you do not need to make a good planning decision.

Decision making in a reconnaissance study can tolerate more uncertainty than decision making in a feasibility study. A 10% level of design detail for a cost estimate in a reconnaissance study may be enough to determine if there is sufficient interest in a plan to proceed to the next planning iteration. It is almost surely good enough to tell us if Plan A costs more than Plan B. On the other hand, advanced engineering and design or a design and specifications level of planning may require as close to 100% level of design detail as is possible.

Built into the notion of an acceptable level of detail is an implicit notion that the risks associated with not reducing the uncertainty further (say by going from a 10% level of design detail to a 30% or 70% level of detail) have been considered. This may or may not be true in any given instance. The standard for considering the appropriate level of detail is to develop a risk register. A detailed discussion of the risk register is found in Chapter 15.

The risk register is a structured approach to communicate the potential risks of using less information, that is, accepting more uncertainty, in a planning study. Planners identify specific technical or planning tasks, which if conducted with less than complete detail could affect decision making or future planning iterations in adverse ways. Each functional area of the team (e.g., geotechnical engineering, economics, hydraulics and hydrology, archeology, and the like) examines the potential risks associated with accepting more uncertainty in their areas of responsibility. Unacceptable risks must be managed.

5.4 ANALYSIS

Most of the evidence gathering begins early in a study and most of the analysis is done by the end of Step 3, when solutions are evaluated. But in an iterative process, who knows when Step 3 is done. Data gathering and analysis are expensive and time-consuming. Planners are expected to be familiar with the data resources for their areas of expertise, and all planners are expected to become familiar with the local data sources for an investigation. Observing and measuring aspects of the study area itself can be major sources of the physical and socioeconomic data that are analyzed by planners. Study area stakeholders and the public can usually be expected to be or to provide rich sources of evidence for much of the study's analytical efforts.

STEPS IN THE SCIENTIFIC PROCESS

1. A clear statement of objectives
2. A conceptual model, which is a framework for characterizing systems, making predictions, and testing hypotheses
3. A good experimental design and a standardized method for collecting data
4. Statistical rigor and sound logic for analysis and interpretation
5. Clear documentation of methods, results, and conclusions
6. Peer review

Source: Sullivan, P.J. et al. 2006. *Fisheries* 31(9). Accessed February 6, 2012. http://www.fws.gov/wafwo/fisheries/publications/fisheries3109.pdf.

There are three points to emphasize about analysis for planning. First, it should be based on the best available science. Second, it should be risk-based analysis. Third, it should assess the identified problems, verify that they are indeed problems, then qualitatively or quantitatively describe the effects of the problems and their solutions. These effects will include measurements of the objectives, constraints, decision criteria, legal requirements, and impacts of importance to interested parties. These analyses provide the basis and details for the without condition scenario, with conditions scenarios, and other scenarios of interest to planners. Each of these three points is taken up in turn below.

5.4.1 Best Available Science

Let's begin with an important caveat, which is that the role of the best available science is always going to be limited in the planning process. Science can help us measure changes in the natural universe; it helps us understand the complex natural systems with which our studies grapple. Science helps us predict how changes in conditions wrought by our solutions might affect the operation of these natural systems. Science is the foundation and the cornerstone for the analytical parts of planning. What science cannot do is provide a basis for choosing human goals and objectives for managing these systems. Objective setting is a value-based process. Science cannot produce objectives. That is the deliberative nature of planning that is guided by social values, not science. Science cannot answer, "How should we manage our natural resources?"

Thus, we seek the best available science in our planning analysis so that when the time comes to deliberate we can do so based on the truth of the situation. The role of science is also going to be limited by uncertainty. Uncertainty will always limit our ability to know the truth. Uncontested, value-free, universally applicable knowledge is what decision makers and the public like; it is not what science can usually deliver. It is rarely, if ever, what planners have. Evidence gathering is, in the end, knowledge development. Science minimizes subjectivity in the development of knowledge.

The U.S. Environmental Protection Agency (EPA; 2012) emphasizes the role of the best available science in its agency work. The Endangered Species Act requires that listing a species as threatened or endangered be based on the "best scientific and commercial data available." But, what constitutes the best available science? Despite its superficial appeal to all, it is not an unambiguous concept and planners, scientists, decision makers, stakeholders, and the public may have quite different ideas on what it is and how to get to it.

The American Fisheries Society and the Estuarine Research Federation established a committee to consider what determines the best available science and how to use it to formulate natural resource policies and procedures. Their efforts are summarized in Sullivan et al. (2006), the basis for much of the remainder of this discussion.

The best available science is based on the scientific process. It is absolutely essential than any primary evidence gathering planners be consistent with the scientific process. Planners are also responsible for assuring the data they do not gather but do use has been the result of a good scientific process. When exigencies require the use of data of uncertain origin, that information needs to be documented and included in the analysts' overall assessment of their confidence in the analysis upon which decisions will be based. The quality of the science is important piece of information to convey to decision makers. Frank communication of the limits of our knowledge is an essential element of the best science as well as a risk-based approach to analysis.

One could argue that failure to communicate the limitations of scientific knowledge and methods as a basis for decision making has led to the surge in interest in risk analysis, a topic taken up in the next section. The best available science is characterized more by the reliability of a transparent, repeatable scientific process than it is by the reliability of specific bits of knowledge it produces. A lot of the information in many natural resource planning investigations is gathered specifically for the study. Planners need to use good data-gathering processes and techniques.

The best science is verifiable, but in resource planning direct verification is not always possible. In its place, peer review is a useful tool for ascertaining the extent to which the available science is adequate for informing planning decisions. Peer review is a formal process conducted by active, knowledgeable experts in the field of interest. Peer review is becoming a more common aspect of much public decision making in and out of planning. Sullivan et al. (2006) describe a peer review that covers

- The validity of the methods used
- Whether the methods and study design adequately address the objectives
- Whether the results that are reported are adequate for interpretation
- Whether the results support the conclusions
- Whether the findings represent a significant advance in scientific knowledge

All science, even the best science, is embedded in uncertainty. Sample sources of uncertainty in resource planning include:

- Lack of basic biological information, exemplified through natural history or demographics
- Lack of information on functional relationships between populations and environmental factors

TRUST AND VERIFY

Stakeholder opinions and perspectives are often valuable sources off information. Marine pilots may be the only source of information on groundings and near misses in navigation channels. Hunters may speak of seeing cougar tracks or spotting a peregrine falcon in areas where these animals are believed to be extinct. Local folks have a lot of good information. But they also may understand what kinds of information will best help them get the features of a plan they are interested in, so as a planner, you have to ask for evidence. Who can corroborate what you are telling me? Do you have any photographs? How often does this happen? It is not an insult to ask for evidence if you do it consistently. It is a good scientific approach.

- Unpredictable events, such as the timing of floods and hurricanes
- High variability associated with key parameter estimates (Mangel et al. 1996)

Planners are expected to use the best available evidence, the best available models, and the best available methodologies in all of their analyses. If you are going to take a sample, make every effort to take a probability sample. Convenience samples or, worse, biased samples are not good practice. Do not assume things that can be proven true or false; always seek and present the evidence. Avoid shortcuts; use the scientific method. Never say there is no evidence of an effect if you have not made an effort to discover evidence of that effect.

5.4.1.1 The Risk Hypothesis

Risk, discussed at greater length in Chapter 12, is a measure of the probability and consequence of uncertain future events. Risk is often described by the simple equation:

$$Risk = Probability \times Consequence \tag{5.1}$$

Consider this expression a conceptual model that helps us think about risk rather than an equation that defines it. What this expression conveys is that both of these elements must be present for there to be a real risk. If an event of any consequence has no probability of occurrence there is no risk. Likewise, if there is no consequence or undesirable outcome, then there is no risk.

Assuming a general familiarity with the scientific process, the risk hypothesis is worthy of consideration as part of the best available science and as a segue to risk-based analysis. The risk hypothesis is useful when gathering evidence about risks. It begins by identifying the harm (or undesirable consequence) that could occur in a given situation. The next step is to produce a conceptual model that describes the sequence of events necessary to get to the harm. These are the links in the chain of events required to result in harm. Then, the required events are used to form testable risk hypotheses, that is, formulate hypotheses that say the necessary events do not occur. Planners can then test the risk hypotheses under conditions most likely to falsify them (i.e., show conditions do occur). This can guide the collection of evidence that enable planners to characterize risks.

If the potential for harm is not detected, testing under more realistic conditions is unnecessary. If potential harm is detected, planners have the option to evaluate under more realistic conditions, that is, to iterate the analysis at a greater level of detail or rigor. This sort of scientific approach builds confidence that your hazard poses a low risk by corroborating a risk hypothesis of no harm, not by falsifying a hypothesis of harm. Sufficient corroboration based on evidence is a judgment the hypothesis has withstood falsification by testing.

Consider the case of the risk of ANS entering a new waterway. The harm is economic and environmental damage to existing species. The probability of establishment of an ANS in a new basin is described by the following conceptual model:

$$P_{\text{pathway exists}} \times P_{\text{ANS is within pathway}} \times P_{\text{survives passage thru pathway}} \times P_{\text{colonizes new basin}} \times P_{\text{spreads through new basin}}$$

Each of these five events is necessary for an ANS to become established in a new basin. Establishment in a new basin is not sufficient for a risk. There must also be harmful consequences of establishment. Consequences are not explicitly covered in the model above, but they cannot occur unless all of these conditions are met.

The conceptual model describes the logical relationships among these events. *If* a pathway exists, *and if* the ANS is within the pathway, *and if* the ANS survives passage through the pathway, *and if* the ANS colonizes in the new waterway, *and if* the ANS spreads, *then* the ANS could cause harm *if* harmful consequences can be shown. The multiplicative form of the model reflects the fact that each probability is conditioned on all events preceding it. For example, $P_{\text{survives passage thru pathway}}$ is estimated based on the assumption that a pathway does exist and the ANS is within the pathway. If the probability of any one of these five events is zero or low, the overall risk of establishment is zero or low by nature of the multiplicative model.

The analysis proceeds by formulating testable hypotheses, the risk hypotheses, to guide the collection of data to aid the characterization of an ANS risk. The risk hypotheses follow, in order:

1. A viable hydrologic pathway does not exist.
2. The ANS is not within the pathway.
3. The ANS does not survive passage through the pathway.
4. The ANS cannot colonize in the new basin.
5. The ANS cannot spread through the new basin.

These are hypotheses of no harm from an ANS. If any one of them is corroborated no harm will occur. Analyses would be designed to collect data that refute these hypotheses. When the resulting evidence refutes all five hypotheses, a significant risk of establishment would be found to exist. At that point, it would be necessary to establish the existence of harm resulting from the establishment of an ANS in a new basin. That is when a consequence assessment would be undertaken. It is a wise use of resources to only conduct consequence assessments when the probability of establishment is found to be significant.

If the analysis done to refute any hypothesis corroborates the hypothesis (i.e., fails to reject it), then the likelihood of establishment may be considered low. Once a single element of the model is corroborated, it is no longer necessary to analyze subsequent elements of the model or the consequences of establishment. This science-based approach to evidence gathering limits the collection of unnecessary data. It also enables planners to test hypotheses rather than assert assumptions.

5.4.1.2 Forecasting

Planning requires forecasts. These include macrolevel forecasts of the future called scenarios (discussed at length later in Section 5.5) and individual microlevel forecasts of variables and specific conditions that are done within the forecasted scenario. Thus, forecasting takes place at two distinct levels.

Macrolevel forecasting, the first level of forecasting, is scenario forecasting. Scenario forecasts are integrated stories about future conditions that provide a framework for the individual quantity forecasts and analyses that comprise so much of the analysis in a planning study. Think of scenarios as the stories we tell, mainly about the future. They describe the future given a series of events that could happen. Several generic types of scenarios come up over and over again in resource planning. They are the historical condition, existing condition, without condition (do nothing or failure to act), and with condition (improvement) scenarios. Before and after conditions are variations of these themes. The existing condition scenario merely needs to be described; it is considered a matter of fact and does not require forecasting. Historical conditions are more accurately a hindcast than a forecast, but the techniques used to construct and integrate a hindcast are essentially the same as those used for forecasts.

Often the most critical forecast in a planning investigation is the forecast of what the study area will look like in the future without your solution in place. This without condition becomes the baseline future scenario against which all alternative solutions are compared. Each alternative solution brings with it its own scenario of future conditions if that solution is implemented. These futures are called "with condition" scenarios. There is no easy-to-follow or one-size-fits-all methodology for constructing a good scenario forecast. Chapter 7 describes how without and with condition scenarios are used to evaluate potential solutions.

The second kind of forecasting planners do is of individual quantities, phenomena, variables, resource conditions, demand/supply/balances, economic activity, environmental conditions, land use, other resource uses, and the like. Economists may forecast waterborne commerce or other measures of economic activity. Hydrologists may forecast stream flows in the future. Demographers may forecast land use and population change. Sociologists may forecast social trends. All of these forecasts are expected to use the best available practice appropriate to the forecasting situation at hand. These microlevel forecasts are framed by and based on the conditions described by the scenario forecasts. Thus, land use is forecast consistent with the without condition scenario and then it is forecast again consistent with the with condition scenario.

The variety of disciplines that will forecast individual quantities in a resource planning investigation are too varied to be treated in any depth here. The study of these forecasting techniques is best left to the texts that represent those disciplines. Quantitative analysis is numerical, qualitative analysis is nonnumerical. All other things equal, quantitative analysis of problems, solutions, and their effects is preferred to qualitative analysis of the same factors because it gives a more precise and complete description. When things can be measured, they should be. When things can be observed but not measured, qualitative analysis is appropriate. Decisions that can be made with qualitative data ought to be made with qualitative data.

5.4.1.3 Systems Approaches

Natural resources planning often deals with complex natural systems. Thus, a systems approach to planning is often appropriate. Systems theory traces its roots to the 1940s (Von Bertalanffy 1968). To understand the idea of a system, let's consider a living organism. If an animal is separated from its surroundings, it cannot survive without its life requisites of shelter, oxygen, water, and food. An organism is an open system; it cannot survive without its environment. Open systems are distinguished by their need to interact with other systems outside of themselves. The primary components of this interaction are inputs, which enter the system from the outside, and outputs, which leave the system for the environment. Thus, we must be able to distinguish the system (organism or other system) itself and its environment.

Bertalanffy explains that the output of a system usually results directly or indirectly from system inputs. A watershed provides a convenient example: the flow that comes out needs to have gotten in first. In most natural systems, the output is different from the input. The system is not a passive straw—it actively processes inputs. The transformation of inputs into outputs is called throughputs. The basic elements of a system are shown in Figure 5.2.

Consider the largest rectangle as the environment for the system shown at the top of Figure 5.2. Inputs to the system emerge from the environment and outputs from the system return to it. This environment is itself comprises other systems, each with its own corresponding environment. Figure 5.2 suggests both the complexity and the interconnectedness of these systems within a system.

The gray wolf (a system) interacts with its environment. The food, drink, and oxygen a gray wolf takes in (inputs), leave its body as urine, excrement, and carbon dioxide (outputs). Its environment, in turn, is made up of systems that produce the food, drink, and oxygen, each of which interacts with its own environment. Thus, we end up with a "systems within systems" approach as suggested by Figure 5.2.

The environment of each wolf includes other wolves. So each wolf is a system. If we consider the pack and how it interacts with its environment, we have yet another system. An awareness of the mutual interactions of the component systems and how they function as a whole is the "glue" of the systems approach to planning. It is

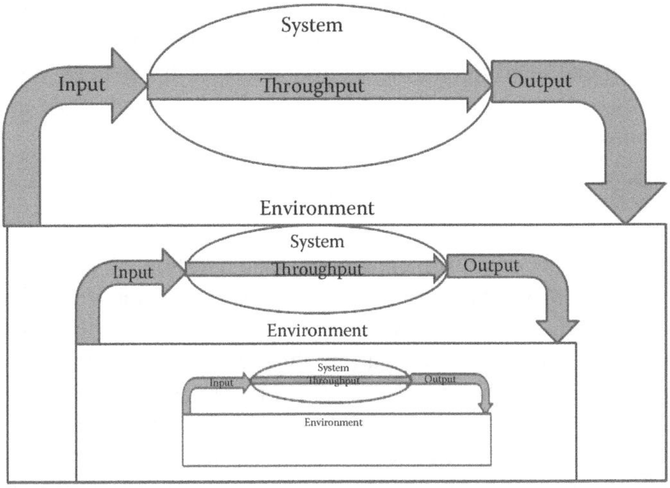

Figure 5.2 Systems within systems show how the environment of one system represents yet another system.

precisely because these systems interact that the whole of a planning environment is more than the sum of its parts. Natural resource planning should address the whole of a system as well as its parts. Principia Cybernetica Web (1992) suggests that with respect to the whole, the parts are seen as subsystems, while with respect to the parts, the whole is seen as a super system. When planners look at the super system as a whole, that is, they take a systems approach, they don't need to consider each part in detail. If it is sufficient to look at total inputs and total outputs; one need not worry about each subsystem. Thus, we can study a pack of wolves without worrying about each individual or we can examine river flows over time without examining each rainfall. In this black box approach, we do not see what happens between inputs and outputs. In a white box, we see the system's workings.

Although the white box approach to planning may be most appealing to us, the black box approach is often the best we can get. There are many processes in the natural universe that we simply cannot explain. We may have observed that if we put water in the right quantity and quality in the right place (input), we know that the ecosystem will react in a certain way (output) even if we cannot precisely explain the cause and effect relationships. Sometimes we lack knowledge of the details and other times we may prefer to ignore the details inside the box. In a reconnaissance study, for example, it matter less to know which mines or other sources contribute most to the acidity of a stream than to simply know the input of acid and its impacts.

The take away point for planners is this: Systems have a hierarchical structure. Systems exist at different levels. At the higher level, that is, the big picture of Figure 5.2, we get a more encompassing but more abstract view of the whole. The details are invisible to us. At the other extreme, the small picture of Figure 5.2, we see the details of the components of the interacting parts without understanding how they organize to form a whole. The systems approach integrates analysis (small picture details) and synthesis (big picture understanding) and has been widely used in engineering and the natural sciences. It has been very helpful in understanding and designing complex systems and is a valuable approach for natural resource planning.

5.4.2 Risk-Based Analysis

If there was no uncertainty, there would be no question about what will happen or how and when it will happen. It is because of uncertainty that risk-informed planning is needed. The key to risk-informed planning is for the entire planning team to be honest brokers of information. Planners come face-to-face with uncertainty in the evidence gathering and analysis process. There will always be things planners know with confidence (see Figure 5.3). These things provide the scientific basis for planning investigations. With luck that pile is large relative to the things we don't know. We hope there is nothing too important in the pile of things we don't know. Unfortunately, life does not often turn out like that, as you may have guessed.

In the early iterations of a planning process, the pile of things we don't know is often quite large. Evidence gathering is the primary means by which we reduce the uncertainty in a planning investigation. Inevitably, uncertainty will remain. If it is significant, that is, if it has the potential to affect decision making,

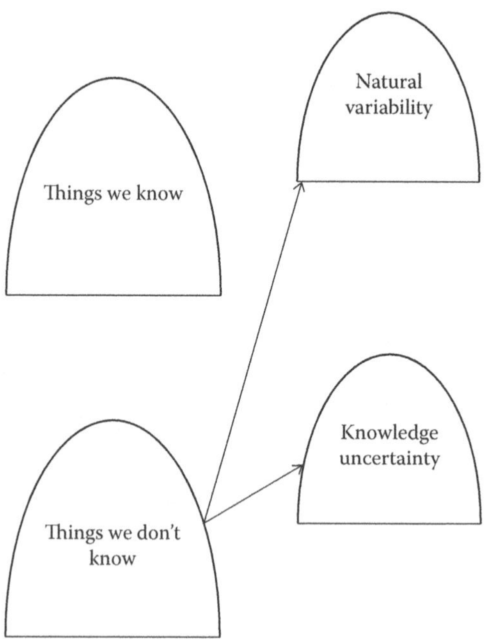

Figure 5.3 Separating what we know from what we don't know.

TWO KINDS OF RISK

Risk has been described as the chance of an undesirable outcome. That outcome could be a loss (fire, flood, disease, death, damage, financial setback, or any sort of hazard) or a potential gain that is not realized (water quality was not improved, the ecosystem was not restored, biodiversity was not realized, yields did not increase, cost savings were not realized, or any sort of opportunity). What usually creates the "chance" is a lack of information about events that have not yet occurred. We lack information because there are facts we do not know, the future is fundamentally uncertain, and because the universe is inherently variable. This is the uncertainty we've been speaking of.

risk-informed planning requires us to purposefully investigate and describe its potential impacts.

The best way to address significant uncertainties is through risk-based analysis. A risk-based analysis explicitly identifies the things that are not known and then carefully investigates their potential significance for decision making. Risk-based analysis is designed to deal intentionally and effectively with the uncertainty that is encountered throughout the planning process. Risk-informed planning is very clear about what is and is not known about the analytical results and their significance for making decisions.

One of the more straightforward ways to address uncertainty in planning is to learn to look at planning analysis in a risk context. Simply consider whatever you are working on at the moment in your planning analyses and ask yourself these four questions:

1. What can go wrong?
2. How can it happen?
3. What are the consequences?
4. How likely is it the consequences will occur?

When you ask and answer these questions about an issue, you are essentially doing risk assessment. More formally, risk assessment is a systematic process for describing the nature, likelihood, and magnitude of risks associated with some substance, situation, action, or event that includes consideration of relevant uncertainties. Risk assessment can be qualitative, quantitative, or a blend (semi-quantitative) of both. A simple risk assessment model is shown in Figure 5.4. The more formal steps capture the essence of the questions discussed in the numbered list above.

The risk assessor's toolbox can be partitioned for qualitative and quantitative tools. Quantitative tools lend themselves to numerical characterizations of the effects of uncertainty, while qualitative tools rely on narrative descriptions, categories, and the like to convey the effects of uncertainty.

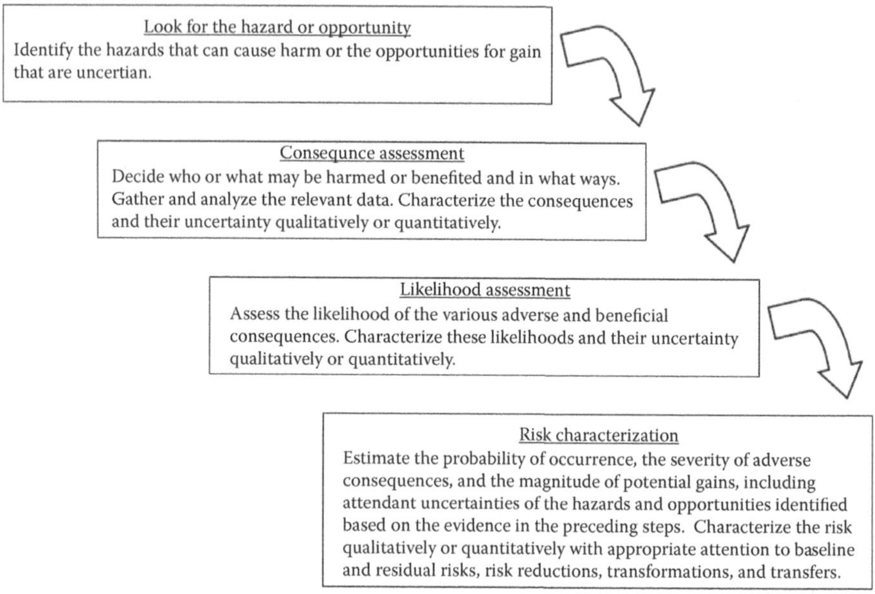

Figure 5.4 Four generic steps of risk assessment suitable for planning investigations.

Some qualitative risk assessment tools that may prove useful for natural resources planning include:

- Increase or decrease risk
- Risk narratives
- Evidence mapping
- Screening
- Ratings
- Rankings
- Enhanced criteria ranking
- Operational risk management (risk matrix)
- A generic process
- Scenario planning
- Multicriteria decision analysis

The risk assessor's quantitative tools include such things as:

- Safety assessment
- Scenario planning
- Scenario analysis
    - Deterministic scenario analysis
    - Probabilistic scenario analysis
- Sensitivity analysis
- Uncertainty analysis

- Modeling
- Simulation
- Vulnerability assessment

Descriptions of these methods are found readily in the literature; see Yoe (2012), for example.

5.4.3 Assess the Problems and Opportunities

We've already considered this task* as part of the problem identification task in the decision context step. Natural resource planning problems are wicked problems, that is, they are often ill-formed at the outset and poorly understood because of the unique interactions of the many complex physical systems involved. Assessing the problems is essential to their solution. Drawing an analogy of a problem or opportunity to a physical "thing," assessing the problem is like describing its properties. How big is this thing, how much does it weigh, what are its dimensions, how long does it last, does it move, how fast, where does it go? The dimension questions are contrived for the analogy, but I hope they make the point. Analysts must describe the problems and opportunities and the systems, resources, and activities that produce them thoroughly enough to facilitate solving problems and realizing opportunities. Along the way, it is inevitable that substantial uncertainty will be encountered. Be sure to say what you do not know about the problems and opportunities as well as what you know.

5.4.3.1 Verify the Problems and Opportunities

The focus of the science-based analysis at the outset of the planning investigation in Step 1 can be summarized by the simple question, "Is it so?" The problems were identified, but what is the evidence of the problems' existence? How do we know it is a problem?

This is the part of the planning process that should be the most scientific. Problems need to be described, and that requires scientific evidence. It is easy to posit and imagine problems, even easier to do so for opportunities; but the planning process demands evidence to verify that the problems and opportunities are indeed as they have been described. This is the major part of the initial analytical work, and it requires an inventory of current and perhaps past conditions in the study area. The methodologies and techniques required will vary from discipline to discipline, and planning draws on many different disciplines.

Research is frequently an important part of the planning process. It is not always the laboratory experiment, published research of the peer-reviewed journals but the primary research of field work that ground the truths the claims of others. A lot of the analytical works of planning are observing, counting, and measuring what is and then applying the best analytical techniques of the various scientific disciplines to

* It was called "verifying the problem" in Chapter 4.

verify the conditions that exist. This analysis is then used to describe the problems and opportunities in accurate and objective ways.

5.4.3.2 Impacts of Problems and Their Solutions

Planning analysis includes gathering, organizing, analyzing, presenting, and explaining data that will be used to measure planning objectives and constraints and the decision criteria that may be used throughout the planning process. Analyzing the impacts of plans that do not yet exist is speculative at best. Good analysis is more than simply extrapolating the future from past observations. It pays close attention to what we know and what we do not know, intentionally probing what is unknown and how important it is for analyzing uncertain future conditions.

As the analysis moves from inventories of past and present conditions to forecast the future conditions required to construct scenarios, the work becomes less certain. Estimating the effects of plans on the study area is a critical analytical task. This requires analysis of key variables and their relationships within the framework of one or more scenarios. Saritas and Smith (2010) suggest four variable classifications that are useful for planners to consider. They are tangible, technological, behavioral, and wild card variables. Identifying the most important variables from each of these categories for any given scenario is a useful way to begin an analysis of plan effects.

Tangible variables can be verified physically. They usually have a high degree of predetermination and, in principle, relatively easily reduced knowledge uncertainty, although the natural variability can be substantial. Examples include topography, land use, ground cover, stream flow, distances, sizes of things, and such. Some of these variables are dynamic (e.g., population data) while others are more static (e.g., the size and weight of physical quantities).

Technological variables describe how technologies work. How much fuel does a ship or barge of a given size use in an hour? How much water can be released from a structure with various-sized openings? How much hydroelectric energy can be generated from a turbine and so on? There is often a great deal of knowledge uncertainty about these variables on the part of planners. Technological variables often involve unfamiliar knowledge domains with a high degree of variability as well.

Behavioral variables describe the way people behave. They tend to be highly uncertain. Will people move into or out of a floodplain? How will people react to an energy price hike? Will manufacturers ship more or less by water? Will people be attracted to a restored area? Will farmers actually adopt the best available technology for phosphorous reduction just because we pass a law that says they must? Human behavior, always subject to natural variability, is one of the major sources of knowledge uncertainty in a planning study.

Wild card variables are sudden departures from the norm. Examples include hurricanes, undersea oil spills, revolutions, terrorist attacks, tsunamis, nuclear accidents, and other random events. These are the most uncertain of all variables, and they include many of the unknown unknowns.

The analysis done during Step 1 continues in the construction of future scenarios where verified and existing circumstances are forecast into the future to construct

YOU GET THE IDEA!

In natural resources, planning analysis may be done in one or more of these fields or others: agricultural sciences, anthropology, archaeology, architecture, area studies, biochemistry, biology, botany, business, civil and environmental engineering, computer science, cultural studies, political science, ecology, economics, fish and wildlife management, law, philosophy, endocrinology, environmental science and technology, epidemiology, floodplain management, food sciences, forestry and natural resource sciences, genetics and biotechnology, geography, geology, geomorphology, hydrology and hydraulics, land use planning and management, management and conservation, marine biology, mathematics, medicine, meteorology, microbiology, nanotechnology, operations research, organic chemistry, parks, recreation and tourism, policy and planning, psychology, social work, sociology, soil science, statistics, systems engineering, toxicology, veterinary medicine, watershed planning and management, wood and paper science.

a variety of alternative futures with planned solutions in place. The evaluation of alternative solutions in Step 3 is another major analytical iteration. The scenarios are forecast for conditions with alternative plans in place. Clearly, that analysis cannot be conducted until the alternative solutions are known.

5.5 SCENARIOS

Scenarios are the stories we tell about conditions found or anticipated in our planning investigations. They are narratives. Think of them as you might a newspaper article. Scenarios can be used to describe the present or the past. They are most often used to describe possible futures. Scenarios consist principally of a general description of some mix of resource conditions, assets, events, behaviors, and decisions. They be differentiated by the ways we address the uncertainty that attends the major threads of the story.

Scenarios are defined first, and then analyses and forecasts are done within the shell of the scenario to explore specific plan effects and phenomena under different scenario conditions. Planning scenarios are brought to life by forecasts of variables, models, and analysis that capture the relationships among variables. This stands in contrast to the mispractice of doing some analyses and bundling the results into something subsequently labeled a scenario. Scenarios are fleshed out by analysis. Analyses do not comprise scenarios.

When we speak of scenarios in the planning process we should always be thinking in the plural. No one scenario can "nail" the future. So even when the planning process uses a single scenario, it should be selected from a set of possible scenarios based on objective or at least explicit subjective criteria.

Scenarios are usually developed to facilitate evaluation and comparison of plan effects in order to help planners and decision makers choose the most desired future.

They also can be effective in bounding the uncertainty encountered in the planning process. A scenario is neither a forecast nor a prediction of what will happen. It is a plausible possibility of what the future could look like. Scenarios provide an indication of possibilities, not probabilities. Probabilities are not assigned to scenarios. A scenario is the unifying shell (or plotline) in which forecasts or projections may be made for specific variables and relationships. A forecast is considered a "best" prediction made by a particular person or with a particular technique or representation of current conditions for a given scenario. The credibility of a forecast depends on the forecaster or the forecasting technique as well as on the conditions defined by the scenario. Thus, a forecast or projection, unlike a prediction, is a more conditional statement. It describes what is likely to happen in the future if certain conditions (i.e., the scenario shell) develop. The boundary conditions used to frame a projection are found in the scenario. A scenario is based on assumptions about how key uncertainties will resolve themselves and how the future could consequently develop.

To illustrate the idea of a scenario, consider how planners might characterize a future water quality problem. The major drivers of this story might be whether the planning horizon is a relatively wet or relatively dry period and whether phosphorous loading from the agricultural sector is heavy or light. Figure 5.5 shows these two key uncertainties define four different future scenarios. Each of these futures will be compelled by a different logic and will tell a different story. The use of multiple scenarios like these is considered at more length in Chapter 13 on scenario planning.

Forecasting is not optional in planning. Be sure you understand the purpose of a forecast; it is not expected to accurately predict the future. It is an attempt to gather, analyze, and present the right information to reduce and bound uncertainty about the future so we can make better decisions in the present. A forecast does not have to be right to improve decision making. It is sufficient if it helps us better understand the decision or if it reduces or clarifies uncertainty. Thus, the scenarios of Figure 5.5 serve

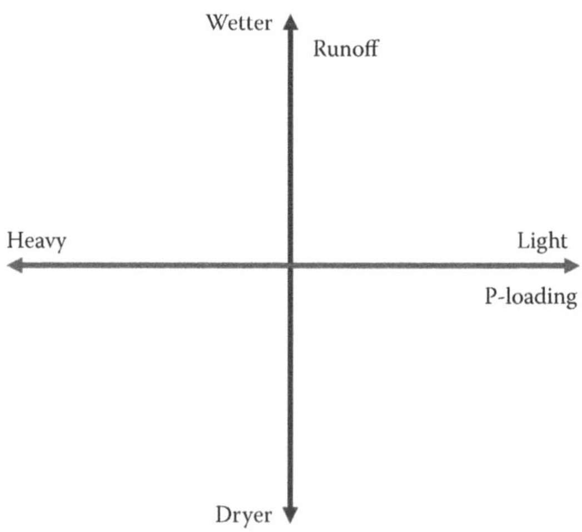

Figure 5.5 Four alternative future scenarios created by two uncertainty drivers.

planners best by asking them to consider the full range of possible futures. Then, when reasonable to do so, a single scenario is identified as the most likely alternative future.

The future is unknown and dealing with the unknown is always a challenge. Many aspects of the future are not specifically predictable. Recent examples include the end of oil, the dawn of ethanol, containerization, introduction of an invasive species (e.g., Asian carp), natural disasters (e.g., Hurricanes Katrina and Rita), the war on terror, and the rise of China. When the future is predictable, and it sometimes is in the short term, no one bothers with forecasts. We only forecast when the future is uncertain. So we need to begin by assuming our forecasts will be wrong more often than not.

Planners know they are not dealing with certainty when they make a forecast. Scenarios are used to describe a range of potential future conditions in which our forecasts are made. The forecasts are then used to inform the decisions we make in the present. So let's dispense with any notion that we can predict the most likely alternative future. It is not possible. We can have more or less accurate forecasts of variables, but we are not likely to ever successfully forecast the future. The purpose of forecasting is not to be right, it is to reduce uncertainty.

To reduce uncertainty and produce useful forecasts, scenarios and forecasts should be relevant, have a rationale, and be supported by research. To be relevant, planners must know exactly what aspects of the future are important to the planning context. A relevant forecast or scenario focuses on what people want and need to know. There should be a convincing rationale (logic chain) for the forecasted conditions or scenario to develop. The "terminally stupid locals" (see the following text box) rationale is not reasonable. There needs to be a compelling logic, a clear rationale for why we see things as we do. Decision makers and stakeholders will have to decide how much weight to place on the analysis and the forecast, and they can only do that if there is a rationale they can follow. Planners have to do their research. Evidence produces the information that reduces uncertainty. Planners must make a careful and concerted effort to gather and generate useful information. It is especially important to gather evidence in support of key variables in the decision process. With all of this in mind, let's consider some of the scenarios planners use most often.

If a scenario is an outline for a play, it stands to reason, then, to think in terms of well-established plot lines in order to consider the different types of scenarios

TERMINALLY STUPID LOCALS

There needs to be a clear rationale for our view of the future. Assumptions of convenience are not rational, they are just convenient. In my planning career, much of it with a federal agency, it was common for planners to justify their plans on the basis that they could solve a resource problem more cost effectively than the local government and other interests could. This requires one to maintain the belief that local folks are too stupid to figure out and use the cheaper, better, faster, fixes the federal agency had discovered. Not only must one present a rationale for one's view of the future, it must be rational.

planners use most commonly. Several types of scenarios come up over and over again in planning. They are existing/baseline, historical, without a plan, with a plan, and target/ideal scenarios.

Some scenario conditions are inventoried; others are forecast. Existing and historical conditions are generally inventoried or described as they are or were. Occasionally a target or ideal condition might be described based on a set of ideals or beliefs. A forecasted scenario describes the future or some other set of conditions that cannot be directly observed.

Inventoried scenarios are usually more certain than the forecasted ones. Because it is impossible to describe the future with certainty, best planning practice always produces multiple versions of the forecasted scenarios. The most likely one may then be singled out for special consideration or multiple scenarios may be used (see Chapter 13). It is not good planning practice to ever consider only one future scenario.

5.5.1 Existing and Baseline Condition Scenarios

The first story we might want to tell about the study area is the existing condition. This scenario describes the system we are interested in as it exists at the time of our investigation. It is the result of an inventory of the relevant variables and their relationships for the study area and the models we use to analyze it. Relevance is defined by the problems, opportunities, objectives, and constraints. The existing condition scenario should accurately reflect conditions as they exist, and it should be based on the best available evidence.

An alternative to the existing condition is the baseline condition. Although the two are sometimes identical, the baseline is sometimes the existing condition accelerated forward in time to a base year reference point (see the following text box). The base year is a slightly accelerated version of the existing condition, meaning a forecast of conditions a few years into the future. In a dynamic system where conditions are constantly changing the conditions at the time a study is done may be less relevant for decision making than the conditions that will exist when the project is operationally completed. A plan is operational when it is either completed or it is completed enough that it is producing the intended outputs to a significant extent. That time in the future is called the base year. Occasionally the baseline may refer to a point in the recent past.

BASE YEAR CONDITIONS

Suppose the study is conducted from 2010 to 2014, but the recommended plan will not be producing significant amounts of intended outputs until 2020. In this case, 2020 may be considered the base year. For many variables and relationships, there may be no significant difference from the existing conditions. When differences exist, they must be carefully forecasted. Otherwise, it is often sufficient to consider the existing condition and base year condition identical.

Consider the following hypothetical existing scenario for a flood problem in a small East Coast town.

> Elliott City was founded in the late eighteenth century and is home to several original buildings, many of which are built on or incorporate natural rock features in their architecture. Its business district has experienced a revival in recent years and consists of day tourist shops, taverns, and restaurants at the street level. It is home to 500 or so full-time and part-time jobs. Upper levels of the buildings that line Main Street are apartments, usually for shop owners and younger renters. There are only three side streets which climb steeply upward from the Main Street floodplain.
>
> There are seven structures on the National Historic Register and many more of local importance culturally, including unique old public buildings. Elliot City was the terminus for the famous race between a train, Ignatius Index, and a horse drawn cart. There are no longer any resources of environmental significance in the town proper. The town is located within one of the nation's wealthiest counties.
>
> The small town is subjected to flooding from two sources. Town Creek is a local drainage area of 2 square miles. It floods frequently at low elevations that disrupt traffic flow and affect a small number of houses. The town is also flooded by 38 square miles of the Patasco River drainage basin. This river caused major flooding seven times in the last century and once in this one. There have been incidents of coincident flooding from both sources. The Patasco River runs through a state park system both upstream and downstream of the town. The Town Creek watershed is as developed as it is ever likely to be. Upstream development is possible in the Patasco drainage area.

Note that there are few quantitative details in this scenario narrative. The details would be provided in the analysis of the flood problem. To estimate expected annual flood damages, for example, it will be necessary to inventory the elevations of the floodplain and to measure the cross sections of the waterways and their floodplains. A flow–frequency curve must be estimated, and an inventory of structures, their elevations, and their flood potential is necessary. A sophisticated hydroeconomic model will be used to combine much of these data in order to calculate EADs. Numerous sources of uncertainty will be encountered and addressed along the way.

The point to be taken here is that the scenario is the basic narrative description, as factual as possible but not detailed. The details follow in the analysis used to examine specific scenario conditions. Once a scenario is devised, planners move into that scenario and live within it, conducting the necessary technical analysis to produce the desired information. As the scenario changes, the analyses will change as well. It is by comparing the results of these analyses across different scenarios that significant impacts of solutions are identified.

5.5.2 Historical Condition Scenarios

The past is sometimes important to planners. It's not easy to understand the present without some knowledge of the past. In some planning investigations, it may be helpful or necessary to consider past conditions of the study area. This is especially true in restoration investigations that intend to restore conditions from some particular historical period. A good example of this is the Comprehensive

Everglades Restoration Project (CERP), which, in part, seeks to restore some water quantity and quality conditions of nineteenth-century Florida. There are many studies that seek to correct or mitigate adverse impacts due to some anthropogenic or, occasionally, natural disruption. These predisturbance conditions are examples of historical conditions.

It is not unusual to seek to restore conditions in the study area to some pre-event condition. People may want to restore their beach resources to prehurricane conditions or restore wildlife populations or plant abundance to pre-oil spill conditions. There are no hard and fast guidelines for choosing a nonspecific historic condition, other than that the condition must be relevant to the planning process. A town may want to restore stream vegetation to what it was like "before." Where before might mean nothing more specific than before it started to disappear. In such a case, the historic condition is nonspecific and simply refers to a reversal of a generally recognized negative trend in an ecosystem.

In the sense described here, historic conditions may function as a sort of target for planners to aim at in their planning efforts. Historic conditions do not always play a role in planning investigations.

5.5.3 Without a Plan Condition Scenario (the Without Condition)

Without a doubt, the single most important scenario in natural resources planning is the without a plan condition, shortened to the without condition. The without condition describes what the future is expected to look like without your plan. It is used as the basis for comparison for every planned solution the planning team formulates.

No one can say with certainty what the future of a study area will be like if the planning partnership takes no explicit action. The future is not knowable with certainty. Nonetheless, it is common planning practice to select one of the potential futures and identify it as the *most likely scenario* that will prevail in the absence of any specific intervention to solve problems or realize opportunities as a result of the planning process. Choosing a single scenario is often reasonable when the range of different scenarios (it may help to think "plotlines" here) is limited. In this situation, the future is not so much in doubt in broad terms as that there are some very specific conditions that are uncertain. Specific forecasts of these variables or conditions can often be adequately handled with sensitivity analysis or probabilistic analytical techniques within a single most likely without condition scenario. Nonetheless, that most likely scenario cannot be identified with any confidence if it is the only scenario considered. When multiple scenarios are considered, there may be instances when it becomes clear to planners that the fundamental direction of the study area's future is in doubt because of significant uncertainties. In these instances, scenario planning (see Chapter 13) is the best way to proceed.

There are always many possible future without conditions if nothing is done by the planning team. These conditions must be devised and explored before the most likely of them can be designated. Then, once the most likely scenario is identified, forecasts of important variables and relationships as well as model runs can be made within the framework of the scenario.

As mentioned earlier, this book does not focus on the techniques used to forecast a specific variable or value; they are too vast and varied and remain more properly the content of discipline-specific discussions. Instead, it focuses on the overall scenario, the story of the study area's future if no new action is taken. In the language of the National Environmental Policy Act (NEPA), this is equivalent to the no-action scenario. The central idea is to consider a variety of possible futures that include different possibilities for the important uncertainties in the system under study, rather than to try to focus on an accurate depiction of a single future. How does one do that?

Let's reconsider the Elliott City example. Imagine this is a flood risk management study done by the U.S. Army Corps of Engineers with the state. How would different scenarios be identified? We could begin by considering the most important tangible, technological, behavioral, and wild card variables in the study area. Alternatively, we could borrow some ideas from strategic planning and conduct a PEST and SWOT analysis of the study area. A PEST analysis considers emerging trends in the political, economic, social, and technological spheres. This looks at the study area in the context of its macroenvironment. A SWOT analysis considers the strengths and weaknesses of the study area relative to the emerging trends and uses matches and mismatches to identify opportunities and threats.

The simplest way to begin to describe the without condition may be to simply ask what is going to change here? What will the problems look like in a couple of decades? If the opportunities are not attained, how will things change? Imagine yourself a reporter 20 or so years in the future updating your existing scenario. Write that story.

Suppose the two primary uncertainties are the national economy and development in the Patasco River watershed. How will the economy behave in the years ahead? Will it be a discretionary income tourist-friendly world or will the developed world be mired in a global recession as it struggles to manage its debts? Locally, we want to know how much upstream development will take place. Just these few questions sketch plotlines from increased runoff with denser economic development and more property at risk of flooding to no change in the flood problem (i.e., no upstream watershed development) and a declining economy and damage base, with every possible permutation in between.

To illustrate how the scenarios might twist and change, what if an upstream reservoir had been authorized in the past but its eventual construction is uncertain due to funding priorities and opposition from Mallards Unlimited? There could also be a citizens group that is advocating that this very wealthy county takes its own actions to manage the watershed. One of its members is running for County Executive. The future without action by the planning team is suddenly more complex. There are many scenario plotlines to consider.

To illustrate the importance of these different futures, let's focus on estimating expected annual damages (EAD),* an important part of the analysis. The analysis

* Do not be concerned if you do not know this measure; it is an estimate of the flood damages that would occur in a floodplain, weighted by their probability of occurrence.

that needs to be done to estimate this value would be done under each of the scenario plotlines developed. Thus, scenarios with worsening runoff and denser development would yield higher estimates of EAD than would occur under other scenarios. For example, if the upstream dam is built the problem may revert to traffic-disrupting nuisance flooding with negligible EAD.

What is a planning team to do? Should they pick the most likely alternative future and do all the analyses assuming it is realized? Do they do a separate analysis for each of the plausible scenarios identified? In general, if the various scenarios are variations on a theme, one scenario will do. It is a simple matter to do a sensitivity analysis of the EAD estimates, for example, to characterize their potential range as a result of the uncertainty. However, when the key uncertainties result in very different plotlines for the future, it is important to use a set of scenarios as is done in scenario planning, described in Chapter 13.

It's time to summarize. It is important to understand that a scenario is not simply an extrapolation of recent trends. The one common element in every without condition scenario is that it is premised on the assumption that the entity doing the planning will take no new action to address the problems and opportunities identified in the study. In other words, the without project condition describes the study area's future without a plan implemented to solve the problem(s) at hand. Thus, the without condition represents the future without a response by the planning entity.

Once a scenario is devised, it is time for the planners to place themselves conceptually within that scenario and make the necessary forecasts of variables like EAD and their relationships and to conduct the important analyses of specific plan effects. When the scenarios do not vary drastically from one another, it is possible to identify one as most representative of the future without any additional action. Differences among the possible future values of specific variables can often be handled by alternative forecasts of a few key variables, that is, conducting a sensitivity analysis or doing probabilistic analysis. When the uncertainty is great and planners cannot legitimately identify one scenario as the most representative, then scenario planning may be the most effective approach to bounding the uncertain future.

The without condition is the one scenario that is used in the evaluation of every alternative plan. It is the one common element in all planning evaluation, and implicitly, in all comparison and selection steps. An error in the without condition will be reflected in the evaluation of every plan and it will carry through the decision-making process. Consequently, it is especially important to carefully develop a realistic, credible, and science-based without condition scenario in every planning study.

5.5.4 With a Plan Condition Scenario (the With Condition)

Each plan formulated by the planning team will have a different impact on the study area. Consequently, each plan will lead to a different set of conditions in the future. This forecasted scenario is called the with condition (sometimes called the with-project condition). There is a unique with condition for each unique plan formulated by the planning team. No two of them should be alike. Each with condition will be evaluated against the same without condition(s). The with condition is not any less important than the without condition, except for the fact that its impact is less pervasive in the planning process. It is used in the evaluation of only one plan.

Just like the without condition, multiple with condition scenarios need to be developed for each formulated plan. A specific solution could reduce future uncertainty to the point that it is ultimately possible to represent the future reasonably with a single representative scenario, that is, a most likely with condition. Planners would, then, make their specific forecasts and do their analysis of the plan's effects against the backdrop of this one most likely scenario. When the potential futures with a specific plan in place are markedly different because of uncertainty about the plan's efficacy and performance, multiple with conditions would have to be considered. This would once again steer the study effort toward scenario planning.

With conditions are actually devised in Step 3, the evaluation step of the planning process. Planners are looking for differences between without and with conditions that are important, that is, differences that make a difference to people in the study area and decision makers. This topic will be revisited in Chapter 7.

5.5.5 Target or Ideal Condition Scenario

In some instances, planners may be provided with a set of target conditions they have been directed to meet. These conditions may be prescribed by a higher authority or be offered formally or informally by stakeholders. A target identifies a set of desired outcomes. The scenario that describes these desired outcome levels is the target condition. When the target is realized, the desired outcome levels will be achieved. Thus, a target condition is a goal to be achieved. Target conditions are more likely to be identified for specific variables or conditions in the study area rather than as a plotline or logic for the future. Examples of targets might include a prescribed level of biodiversity, a water quality goal, a salinity level, a mitigation goal, a tolerable level of risk, or any other appropriate target.

If we generalize and extend the notion of a target, it becomes an ideal scenario. This is much more faithful to the idea of a scenario. An ideal condition scenario might be useful to develop before the planning team begins to formulate plans. Using the planning objectives and constraints and everything learned about the problems and opportunities during the analysis done early in planning, planners could devise an ideal scenario for the study area. Think of it as a utopian view of the study area with all objectives achieved, constraints avoided, problems solved, and opportunities attained. What does that look like? Describe it, and then write that newspaper story. It is the ideal condition. Planners then "move into this ideal condition" to devise and formulate plans that would come closest to making it a reality.

5.5.6 Before Condition Scenario

Some studies inventory and describe conditions that exist or existed before an action is taken. That could be conditions before the big flood, before the oil spill, or before any threshold event or period of time. Most often it simply refers to conditions before a specific action is taken, for example, before the refinery is built or before the herbicide is applied to the forest. The existing condition is a common choice for the before condition. The greater the uncertainty about the before conditions, the less it is like an inventory. When the uncertainty is great, multiple before scenarios should be considered before choosing a most likely or most representative one.

5.5.7 After Condition Scenario

Very similar to the before condition is its opposite, the after condition. If the actual conditions after an action is taken are described, this is an inventory. When the after condition is subject to significant uncertainty, multiple versions should be considered. When it describes a future condition that does not yet exist, for example, after the next big flood, or after sea level rises a meter or more, it is differentiated from the with condition scenario because it is triggered by an undesirable and usually uncontrollable event. In other respects, like whether one or several scenarios is appropriate to use, it is very much like the with condition scenario.

5.5.8 As-Planned Scenario

The "as-planned" scenario, sometimes called the surprise-free scenario, describes a future in which the system under consideration (e.g., a waterway, landform, ecosystem, park, or wildlife preserve) functions exactly as it is supposed to function. Or, in the case of manmade systems, it functions as it was designed to function.

The as-planned scenario is free of any failures. It is a scenario in which every feature of the natural or manmade system functions as it was planned to function. Scenario analysis, different from scenario planning, often begins by defining the success or as-planned scenario where all but one endpoint is a successful conclusion. Once the as-planned scenario has been defined, failure scenarios can be constructed and investigated.

Consider the simple illustration of an as-planned scenario in Figure 5.6. Three end states of the model define the as-planned scenario. There is one failure scenario in this example. If a flow is above flood stage, exceeds the levee, and inundates property, then the system has failed. Otherwise, the system functions as planned and flood damages are avoided. As-planned scenarios are more likely to be used in risk assessments and operation and maintenance investigations than in planning investigations.

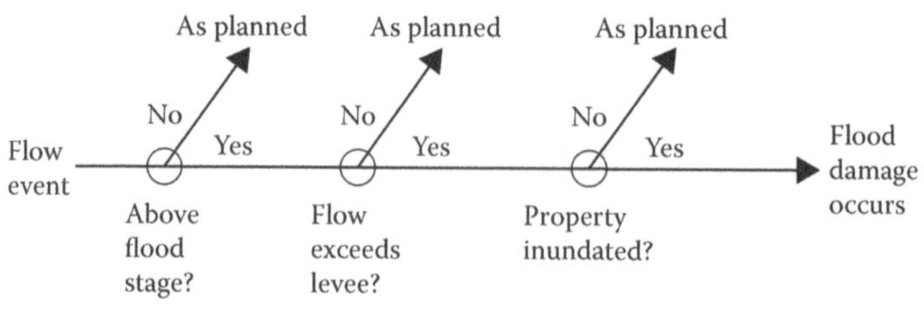

Figure 5.6 As-planned scenario for a flood event.

5.5.9 Failure and Worst-Case Scenarios

Scenarios that present alternatives to the as-planned scenario are called failure scenarios. They are developed to illustrate different failure modes. Any aspect of the as-planned scenario may be challenged in a failure scenario. One of the most common failure scenarios is the "worst-case" scenario.

Engineers and public health officials, to name just two professions, are trained and programmed with an irresistible drive to design systems conservatively. H.S. Lewis (Albert 1983) says of conservatism, "It is the accumulation of centuries of experience that the conditions of the real world are not always predictable and that it makes good sense to provide some margin of error for unforeseen events." This drive toward conservatism has led to the widespread propagation and use of worst-case scenarios. There is no real formal definition of a worst-case scenario. It is simply that future in which everything than can reasonably go wrong does go wrong.

If the worst-case scenario yields an acceptable or even a tolerable result, there is no need to manage the situation in a different manner. On the other hand, worst-case scenarios that result in unacceptable consequences often have lead managers to take precautions to preclude the worst-case scenario from occurring. To an extent, the National Dam Safety Program in the 1980s and 1990s was based on a worst-case analysis of the nation's federally owned and managed dams.

Despite its widespread usage, the worst-case scenario is not without its problems. First among these is that to introduce the "conservatism" of a worst-case scenario into an analysis is to introduce a deliberate error if the true intent of the investigation is to provide objective analysis. Second, given any worst-case scenario, an even worse case can, paradoxically, usually be defined. Thus, there may be no such thing as a true worst-case; we can always add "then an asteroid hits Earth" to any worst-case scenario. Third, the likelihood of a worst-case scenario may be so small as to lead to the waste of resources if there are efforts to reduce it. Fourth, there is an almost hypnotic appeal to thinking that if we have covered the worst-case we have covered everything. Failure in less than the worst-case world is still possible. Nonetheless, worst-case scenarios are likely to remain useful and popular failure scenarios to devise and investigate.

5.6.0 Types of Scenario Analysis

Here the language gets tricky. We have described scenarios as logics, plotlines, and descriptions of alternative futures that could be. This is all well and good. Analysis within scenarios is what we do when we move into a scenario and explore its details. This is the estimating, assessing, calculating, and all the analytical work that is done "inside" the scenario. Thus, we estimate habitat units for different species and scenarios, we calculate EAD, we forecast water quality parameters, and we estimate wildlife populations along with the user days for outdoor recreation enthusiasts. When there are uncertain quantities in a scenario, for example, there are tools available for addressing them. This practice is called scenario analysis in the literature. Monolithic, deterministic, and probabilistic scenario analyses are three different approaches to conducting analysis within a scenario.

The first of these involves the invention of language. I doubt you will find monolithic scenario analysis anywhere in the literature. Nonetheless, scenarios have always been used for decision making. It has long been common practice to develop and present a scenario as if it was a fait accompli, when in fact it is simply one of many possible stories about a conjectured condition. A single unchallenged scenario used as the basis for decision making that does not consider uncertainty in any explicit way is a monolithic scenario. Monolithic scenarios are essentially assertions of what an analyst believes to be true. Inventories without uncertainty are examples of monolithic scenarios. Politicians often assert a monolithic view of the future with their program or platform in place.

When we move from a single monolithic scenario to a few selected scenarios, we are engaging in deterministic scenario analysis (DSA). DSA defines and examines a limited number of specific scenarios. This can be a useful way to organize and simplify an avalanche of data into a limited number of possible future states of the system being modeled. The scenarios so identified are usually chosen for specific reasons. Planners tend to use DSA for devising without and with conditions.

As useful as DSA is, it is not always adequate for resolving uncertainties that remain within a scenario. Even when a single representative scenario can be used, smaller uncertainties remain and must often be addressed. This requires the use of probabilistic techniques. When probabilistic techniques are added to a scenario tool, the resulting technique is called probabilistic scenario analysis (PSA). PSA addresses significant knowledge uncertainty and natural variability that remain inside the scenario logic. Using a Monte Carlo process to characterize the effects of uncertain inputs in a project cost estimate is an example of a PSA.

5.6.1 Comparing Scenarios

The effectiveness of a planned solution is usually judged on the basis of changes in decision criteria observed through scenario comparisons. Consider the simplified example with two effects of interest in Table 5.3. If no action is taken, there will be 2,500 habitat units but there will be no costs associated with preserving or adding to them. If Plan A is implemented, habitat units will increase by 2,500 to 5,000 at a cost of $1 million. Plan B quadruples the change in habitat units to 10,000 but at a much greater cost.

Table 5.3 Without and With Condition Scenarios Compared

	Habitat Units	Cost
Without condition	2,500	$0
With Plan A	5,000	$1,000,000
Change due to Plan A	+2,500	+$1,000,000
With Plan B	10,000	$100,000,000
Change due to Plan B	+7,500	+$100,000,000

If scenario comparisons are to be useful for decision makers, they must highlight differences in scenarios that make a difference, that is, differences that are important and that matter to decision makers. In best practice, these metrics will reflect some or all of the planning objectives. Gathering evidence to construct scenarios and then to conduct analyses within them constitutes the most significant part of this ongoing process. We will return to the topic of scenario comparisons in Chapter 7. There the without and with condition comparison is discussed at length as part of the evaluation of solutions.

5.6 TAKE AWAYS

Lesson 1: Evidence gathering is a process that continues throughout the planning process; it consumes most of the time and money in a planning study.
Lesson 2: Reducing uncertainty is a primary purpose of evidence gathering; gather only the evidence you need to make better decisions.
Lesson 3. Planning should be based on the best available science; its analysis should be risk-based.
Lesson 4: Scenarios are important and the without condition scenario may be the most important of all the planning scenarios.
Lesson 5: The best available data are those data that reduce the uncertainty to a point that decision makers can feel reasonably comfortable with the decisions they are making.

REFERENCES

Ackoff, R. L. 1989. From data to wisdom. *Journal of Applies Systems Analysis* 16:3–9.
Albert, K. J. 1983. *The Strategic Management Handbook*. New York. McGraw-Hill Book Company.
Mangel, M. et al. 1996. Principles for the conservation of wild living resources. *Ecological Applications* 6(2):338–362.
Principia Cybernetica Web. 1992. What is systems theory? Accessed Febuary 6, 2012, http://pespmc1.vub.ac.be/SYSTHEOR.html.
Saritas, O. and J. E. Smith. 2010. The big picture—trends, drivers, wild cards, discontinuities and weak signals. *Futures* 3287(10):292–312.
Sullivan, P. J. et al. 2006. Defining and implementing best available science for fisheries and environmental science, policy, and management. *Fisheries* 31(9). Accessed February 6, 2012, http://www.fws.gov/wafwo/fisheries/publications/fisheries3109.pdf.
Swanson, T. A. 2007. Teaching students *about* information: Information literacy and cognitive authority. *Research Strategies* 20:4:322–333.
United States Environmental Protection Agency. 2012. "Role of Science at EPA." Accessed February 6, 2012, http://www.epa.gov/osp/science.htm.
Von Bertalanffy, L. 1968. *General System Theory: Foundations, Development, Applications*, revised edition 1976. New York: George Braziller.
Yoe, C. 2012. *Principles of Risk Analysis Decision Making under Uncertainty*. Boca Raton, FL: CRC Press.
Yoe, C. and K. Orth. 1996. *Planning Manual. IWR Report 96-R-21*. Alexandria, VA: Institute for Water Resources.

CHAPTER 6

Formulate Solutions

The primary purpose of this second planning step is to identify as many different potential ways to achieve the objectives and avoid the constraints, thereby solving problems and realizing opportunities, as possible.

6.1 INTRODUCTION

Here is the short story. You've identified problems you want to solve, and you've identified objectives that describe what success looks like. It's time to begin to think about as many different ways to meet these objectives as you possibly can. Ultimately, you are after the best solution possible. You cannot possibly trust that you have the best solution unless you have considered many solutions.

Think left and think right and think low and think high. Oh, the thinks you can think up if only you try.

Source: Dr. Seuss. 1975. *Oh the Thinks You Can Think.* New York: Random House.

Devising solutions to complex and wicked problems is the most creative part of the planning process. It's the most challenging, inventive, and rewarding part of a planner's job. Done well, it's the most fun you can have with your digital devices. Still, this is not an easy thing to do. Formulation strategies can help, and you'll find a couple dozen of them in this chapter. Choose one or invent one of your own, gather your team and others, and spend time formulating plans. Even though there may be specialists, understand that everyone on the team ought to be a plan formulator.

Here are a few more things to know before leaping into this chapter. You identified the problems and opportunities and then developed objectives and constraints in Step 1. You also verified the problems and developed a view of the future should you decide to take no new action as a result of your planning study. You've also begun to accumulate a good deal of data and information in many forms and formats. You need some ideas. It is time to begin to formally formulate alternative solutions, and it's time to put plans together.

Where do plans come from? They come from people. They come from moments of inspiration and from months of investigations, analysis, and meetings. They can also come from years of experience or from several iterations of the planning process, but they always come from people.

Here is what you get in this chapter. First, there is a pep talk encouraging you to be creative and to break rules. Then, there terminology needed to help us talk about formulation and solutions. This chapter concludes with many formulation strategies you can use.

6.2 PEP TALK: BREAK RULES AND BE CREATIVE

ROAD RUTS

Imagine an acid mine drainage project with many remotes sites in a watershed that could be treated to solve the water quality problems of a stream. Experience and agency practice dictate that an access road is needed for each site. The quality of the road will depend in part on whether there is one-time or periodic access required. The institutional practice, the "rule" in this case, is to build an access road.

Sometimes, the access road can be three-quarters or more of the site costs. The expert does not object. The novice asks about using pack animals, helicopters, off-road all-terrain vehicles, new and lighter materials. Sometimes these new ideas pan out, and sometimes they do not. But the ability to break the informal rules of plan formulation is an important quality of good formulation.

Here is why creativity is important for formulation. First, many of the easy problems have already been solved. Second, the world is complex and there is a lot of uncertainty. Third, lots of people with many different viewpoints care about planning issues. Fourth, resources are limited and we can't do everything. Therefore, to solve wicked problems in a complex and uncertain world with lots of stakeholders and limited resources, we need creative solutions. Adapting a famous quote by Paul Gauguin, I suggest good plan formulation is either plagiarism or revolution. Be a revolutionary. Break the rules. Unleash your creative self.

Organizations and individuals all have ingrained patterns of thinking. The older the organization or the more experienced we are, the more ingrained the patterns of thinking. We often mistake these patterns for the truth. They become our rules, and they usually exclude conflicting ideas. Thorpe (2002) says these rules are like railroad tracks: If you want to go where the tracks take you, they are very good. Other destinations cannot be reached with the rules in place, however. To take communities to new places, you may have to break the rules. Identify all the things that limit what you can do. These are your rules. Now break them.

Critical thinking is a wonderful skill, but not for formulation. It's what you want when you evaluate, compare, and select plans. Creative thinking is what you want for formulating plans. These matched pairs of attributes contrast critical (first) and creative (second) thinking.

- Analytic-generative
- Convergent-divergent
- Vertical-lateral
- Probability-possibility
- Judgment-suspended judgment
- Focused-diffuse
- Objective-subjective
- Answer-an answer
- Left brain-right brain
- Verbal-visual
- Linear-associative
- Reasoning-richness and novelty
- Yes but-yes and
- New ideas-choices
- Assesses-originates
- Evaluation-synthesis
- Reason-imagination
- Safe-daring
- Defends-challenges

Source: Adapted from Harris, R. 1998. Introduction to creative thinking. Accessed February 6, 2012 from http://www.virtualsalt.com/crebook1.htm.

Novices, unfamiliar with the rules, are quite proficient in breaking them. Over time, experience teaches us not to break the rules and we lose our novice's talent for rule breaking. Unfortunately, following rules enables these "inviolable laws" to defeat many promising ideas. Consequently, in many organizations, what we know is often a greater obstacle to problem solving than what we don't know. How many times has someone said, "We have no authority to do that," and that has killed an idea or a discussion? Stop that! Facts and expertise can be deadly to new thinking. Experts are great at conventional knowledge, but they do not have the best record of generating or recognizing great new ideas.

Ignore the answers that please authority and the solutions that suppress more questions. Embrace uncertainty and not knowing. Be at home with the unknown and friend to the paradox. Rely more on intense curiosity and an open mind than on following the rules. An open mind in the face of uncertainty can unleash the creative potential inside a planner. Seek a balance between science and art. Logic and imagination play equally important roles in plan formulation. Plan formulation is left-brained and right-brained. Plan formulators need both sides of their brains.

Planning is a rational, science-based process, but there is an important role for emotion and creativity in that process. Empathy, for example, is essential to good plan formulation. You have all the data but you must also be able to see through the eyes of others, to imagine what someone else is thinking and how they feel about the problems. Empathy for others when formulating plans is every bit as important as technical competence. In the early stages of plan formulation, it is more important. If you haven't invested properly in developing your decision context, it is likely that your plan formulation efforts will suffer. Narrowly defined problems and opportunities lead to narrowly defined solutions. Following the planning process is a good guide to successful planning. Nowhere is following the process more unwelcome than in plan formulation.

There may be laws, authorities, budgets, conflicts, politics, and even laws of the physical universe that may restrict what you are ultimately able to do, but nothing, nothing at all, should restrict what you can think. Identify the three most obvious solutions to your problems and opportunities. Declare them off-limits. Now solve your problems and realize your opportunities without them. Do not let anything, not even reality, limit your thinking. Free your mind instead. As you refine your plans and reformulate them, reality will rein you in soon enough. Let your formulated plans be revolutionary. You can always revert to the three most obvious solutions or do what you have always done in situations like this. If you've got a real solution, we'd all love to see the plan.

6.3 THE LANGUAGE OF FORMULATION

In this section, some basic language is introduced and then the various attributes of a solution are defined. This is followed by a consideration of multipurpose and multiobjective formulation.

6.3.1 The Basics

We need a vocabulary to make it a little easier to talk about plan formulation. A solution is a way to achieve all or part of one or more parts of planning objective. Solutions can be management measures, alternative plans, or programs. The words "solution" and "plan" are often used as synonyms. I favor the word solution because that is what a plan ultimately is, and it is broad enough to encompass the notion of a program, which can be a set of plans. An "alternative" is one of several solutions. An "array" is a set of alternative solutions. Formulation is the process of creating plans that solve problems. It has three basic phases:

1. Identify measures that meet planning objectives.
2. Combine these measures to build plans.
3. Change the plans as necessary.

A formulation strategy is a disciplined way to complete these phases to produce an array of solutions. Several are presented later in this chapter.

A measure is a means to an end. Measures are the elemental building blocks of plans and the starting points for plan formulation. They are sometimes called management measures. A management measure is defined as a feature or activity that can be implemented at a specific geographic location and at a point in time to address one or more planning objectives.

A feature is a physical entity or element that requires construction or assembly on site. It is a "structural" component of a plan. An activity is a management action; it's a policy, practice, or (a different) way of doing something or managing resources that does not require construction but has a measurable impact. It is a "nonstructural" element of a plan. An activity can be a one-time occurrence; it can be a continuing or periodic occurrence or an ongoing process. Examples of each are provided in Table 6.1.

A plan, according to *Webster's* (1971), is a "scheme for making or doing something." The working definition of an alternative plan is that it is a set of one or more management measures functioning together to address one or more planning

HOW DO YOU IDENTIFY MEASURES?

For common and recurring problems, planners can develop and use checklists. When checklists do not already exist, brainstorming measures is a common and effective technique. In fact, checklists can be constructed using brainstorming techniques. Alternatively, an individual can construct a checklist by looking at measures used in other reports, by reviewing the literature, and by talking to experts.

Previous iterations of a plan can be an excellent source of relevant measures. Some planning efforts have been preceded by prior studies and reports. A review of the measures considered in previous iterations of the planning process is often the best source of viable measures.

Table 6.1 Examples of Management Measures That Include Features and Activities

Management Measure Examples	
Features	**Activities**
Breakwaters, water pumps	Reallocating reservoir storage
Jetties, water control structures	Modifying water releases
Groins, fences	Seeding, cutting, and burning vegetation
Channel modifications, food plots	Applying pesticides
Dams, brush piles	Buying flood insurance
Detention basins, nest boxes, and baskets	Evacuating the floodplain during storms
Levees, roosting platforms	Building codes and zoning restrictions
Floodwalls, relocations	Grazing agreements

objectives. Most plans consist of more than one measure. Different plans have different measures or they combine the same measures in significantly different ways. For example, suppose we have a town with two creeks. Plan A channelizes Beatle Branch and builds a levee along Stones Creek. Plan B builds the levee along Beatle Branch and channelizes Stones Creek. Both plans consist of the same measures. The specific sites of these measures are sufficiently different to constitute two different plans.

Measures can be combined in any number of ways to make a plan. It could be an experienced engineer working with topographic maps, a straight edge, and a pencil. It could be a carefully structured strategy that leads a planning team through a formal formulation process. It could be a computer program using numerical techniques to examine all the possible combinations of measures. Or it could be anything between these approaches. Any process that produces an array of plans that meet the planning objectives and solve the problems is acceptable.

A project is an implemented plan. It's a set of features and activities that are constructed and otherwise carried out in reality. A plan is not a project. A plan is an idea documented in a report. You can read a plan; you can visit a project.

What's bigger than a plan? What do we call a large-scale solution with a series of plans all over a region or state? We're going to call it a program or system. Just as management measures can be combined to form plans, so, too, can plans be combined to form programs. In a planning context, a program is a set of one or more plans, usually located over a large geographic area, that function together to address one or more planning objectives.

6.3.2 Attributes of a Solution

Are a 32-foot deep channel and a 34-foot deep channel alike in all other respects different plans? Are a 10-foot floodwall and a 12-foot floodwall two different plans if they are alike in every other way? Or are these simply examples of different scales of the same plan? Language is important if we are to communicate effectively, so let's turn our attention to the different attributes of a plan.

A plan consists of measures, and these measures have many attributes, like scale, frequency, duration, dimensions, intensity, composition, dependence, mutual exclusivity, and alignment, to name a few. The array of plans that results from a good formulation process should include different plans as well as refinements of the better plans. Refinements result when you vary one attribute of the plan with all other things equal. Let's consider some more terms worth understanding.

Alignment. The manner in which features are positioned can vary from plan to plan. There are many ways to configure 1000 acres of habitat. Levee tie-outs can be positioned at various points between the waterway and high ground. The layouts of a breakwater, jetty, or groin are additional examples of the alignment attribute. The adjustment of parts in a physical feature of a plan can also be varied.

Composition. Features of a plan may vary in composition by using different construction materials. A flood barrier, for example, may be made of steel sheet pile, concrete, or earth. A trail can be macadam, gravel, woodchips, or earth. The activities of a plan could also vary by composition. Fish can be removed from a disposal area by commercial fisherman, by a paid contractor, or by private individuals in a fishing contest.

Dependence. Measures can vary in the extent to which they need other measures to function at all or to function optimally. At the other extreme, some measures function totally independently of other measures.

Dimensions. Separate from the geographic scale of a plan are the physical dimensions of some of its features. It pertains to how deep, how high, how heavy, what color, and related sorts of questions. This might include the size of a pumping station, the height of a fence, the thickness of a wall, and so on.

Duration. Another time dimension attribute is the duration of an activity. Planners may have some choice about how long an activity is kept up.

Footprint/design. Engineering design can vary for some features of a plan. For example, the side slopes on a levee or the nearshore and foreshore slope of a coastal berm can vary and produce different footprints. Some designed features of a plan have a wide range of options. Interior drainage in a flood risk management levee project can be handled by many different combinations of interior ponding area (storage) and pump station size.

Frequency. When a plan comprises one or more activities, the frequency with which those activities occur may vary. How often is a herd culled, water released, or a controlled burn used to achieve desired outcomes?

Intensity. Plan measures may differ in the degree to which a characteristic is present. The easiest way to conjure this notion is by the numbers of things. Plans may vary by the intensity or numbers of duck boxes or watering holes in an area. Wetland creation may plant grasses on 12-inch centers or more intensely at 9-inch centers.

Location. Where is the plan? What town is it in? What landmarks is it close to?

Measures. The measures themselves are the most obvious way to differentiate one plan from another. Plans that use different measures are different plans.

Mutual exclusivity. Some measures preclude the use of other measures either because they occupy the same space, perform the same function, or are otherwise incompatible. For example, land cannot be simultaneously managed as a fenced and open access area.

Name. Solutions need names so we can tell them apart and talk about our favorites. There is no universal convention for naming solutions. You can call them whatever you want, but it really helps if the names have some easily communicated meaning (see the "What's In a Name" text box). Numbering or alphabetizing plans is not generally helpful, but that is preferable to complex and opaque naming schemes like 290BC2.

Scale. Think of scale as the size attribute. In most cases, size is best described by geographic extent. A local plan is very different from a regional-or a landscape-scale plan. A watershed study is different from a localized study on a small tributary. In some cases, scale may be described by volume, density, weight, or other dimensions.

Sequencing. When it's time to implement a plan, what comes first, second, and so forth through completion? Will it be built from upstream to downstream, or vice versa, and why? Sequencing questions may or may not come into play at the plan level, but they are usually critical at the program or system scale.

Timing. Planners may have a choice about when to conduct activities that are part of the plan. There are times of day decisions, seasonal, cyclical, and decisions periodic in other ways for some activities.

Of these attributes, the ones that most commonly produce alternative plans, as opposed to refinements of a single plan, include alignment, location, scale, and measures. When these attributes of a solution are varied, you are more likely to have a different plan. When the other attributes are varied, you are more likely to have refinements of an existing plan, like variations on a theme.

One of the most common reasons for refinements is for economic optimization, that is, maximizing the net benefits of a plan. Thus, once the measures, alignment, locations, and scale of a levee have been determined, one might examine different heights (dimension), side slopes (design), or materials (composition) in search of a design that minimizes the costs of a given level of benefits, for example.

WHAT'S IN A NAME?

- Name alternatives after geographic features, such as neighborhoods, towns, villages, and land forms ("Downtown Plan," "Phelge Lake Plan").
- Consider using targeted species or resources ("Mottled Duck Plan," "Indian Burial Mounds Plan").
- It may be helpful to name a plan after its originator or sponsor ("City Council Plan," "Audubon Plan").
- The plan's dominant measure may suggest a name ("Channel Plan," "Levee Plan").
- Combinations of these methods might also work ("Downtown Channel Plan," "Garden Club Greenway Plan").

Source: Ken Orth, beer snob and planner. (Personal communication.)

OPTIMIZATION

Federal water resource agencies were required by the Principles and Guidelines (1983) to maximize net national economic development benefits in water resources planning. This often resulted in plan formulation becoming overreliant on variations of a single plan or two. Flood risk management studies often involved a levee plan that considered numerous heights or a channel plan that looked at several design flows. Navigation plans often comprised a single channel alignment with various depths. Devising solutions in this manner is more of an optimization process than a formulation process, which would result in distinctly different plans.

6.3.3 Formulation Concepts

There are at least two concepts that are important to distinguish before we talk about how to formulate plans. The first of these requires us to consider multiobjective and multipurpose planning. The second of these considers the three natural phases of formulation, that is, identify measures, formulate plans, and reformulate plans.

6.3.3.1 Multipurpose and Multiobjective

Good plan formulation is always multiobjective. Good projects are often multipurpose. Planning is multiobjective, and plans are multipurpose. A purpose is something that planners intend or hope to accomplish through a strategy or project. For example, until about the 1920s, most water resource projects were single-purpose projects. Gilbert White identified 11 purposes that were commonly considered in 1958 when the Southeast Basins study was started. They were navigation, flood risk management (called flood control at the time), hydroelectric power, irrigation, municipal water supply, municipal waste disposal, industrial water supply, recreation, wildlife conservation, low flow regulation, and soil conservation. You can think of a project purpose as an output.

A multipurpose plan includes two or more distinct natural resource purposes. A multipurpose plan may be a plan that has several individual measures (i.e., physical components or management actions) each serving different purposes combined in a single plan that serves multiple purposes, or it may have one or more single measures that simultaneously serve several purposes. The Hoover Dam on the Lower Colorado River was a prototypical multipurpose project.

Multiobjective planning refers to the consideration of more than one planning objective at a time during plan formulation. Major (1977) explained the difference quite well, saying

> The terms multiple objective, referring to the multiple economic, social, environmental and other objectives of water development, and multiple purpose, referring to the multiple functions, navigation, flood risk management, etc., of water projects, are not synonymous. Purposes can vary and still be aimed at the same objective, and one purpose can fulfill more than one objective. For example, projects for the purpose of navigation as well as those for the purpose of water supply can be designed for the objective of increasing the income of a region, while a navigation project can contribute both to the objective of regional income and to the objective of increasing national economic growth.

WHY DO MULTIOBJECTIVE PLANNING?

- Problems are complex and multifaceted.
- Stakeholders have varying perspectives.
- It's cost-effective.
- It is collaborative and involves diverse interest groups.
- It can preserve programs.
- It helps overcome budget cuts.
- Creative problem-solving synergies can arise.
- It's more suitable for watersheds, ecosystems, and larger areas.
- It provides a hedge against uncertainty.
- It's well-suited to adaptive management.

Plan formulation ought to be multiobjective, and it should produce multipurpose plans when it is appropriate to do so.

6.3.4 Formulation Phases

The three phases of plan formulation introduced earlier are (1) identify measures that meet your objectives; (2) combine measures to build plans; and (3) change (i.e., reformulate) plans as necessary. There are a variety of ways to conduct each phase. Each different way to conduct a phase of the formulation process comprises a formulation strategy. Many of them are described later in Section 6.4. But first let's look at how formulation progresses through these three sequential phases.

Each phase is conducted at least once during a study. Most of the time you'll repeat them many times before you're done. The phases usually overlap. You'll be identifying new measures at the same time you're changing some older plans. Sometimes the phases occur in such quick succession that it is difficult to distinguish among them.

Formulation begins with your objectives. The first formulation phase is done to, quite simply, identify ways to meet those objectives. Those ways are your management measures. These are the individual pieces or building blocks that meet your objectives that can be put together to form plans. Plans are usually built-up from measures, but occasionally you might identify the whole plan first and then break it down into its component parts. The output of this phase is a list of measures that achieve each of your objectives; the longer the list the better.

The second phase, often called plan formulation, requires you to match and mix these measures into a variety of plans. There are many reasons for combining any two measures, and several of these are listed in the following text box. It is during this process that the plan attributes of Section 6.3.2 become relevant details to consider. A plan rarely comprises only one measure. Your goal is to develop a variety of combinations of measures that meet different mixes of your objectives to varying degrees of satisfaction. The output of this phase is a set of potential solutions to your problems and opportunities.

WHY COMBINE MEASURES?

- A measure requires another measure to function (dependency). If this is true then they should be treated as a single measure since one would not be implemented without the other.
- A single measure may only get you part of the way to your objectives.
- To avoid constraints, including universal constraints (endangered species habitat, for example) and study-specific constraints, such as if the city council made it clear that no levee will ever be built on the south side of the river.
- Meet legal or policy requirements. A requirement for a nonstructural plan is the classic example. How do you get a nonstructural plan? Combine only nonstructural measures.

The third formulation phase is called reformulation. It might be subtitled, "Don't throw that plan away just yet." It is part refinement, part reinvention, and part salvage operation. Sometimes a first effort is not quite right and we know it. In these situations, reformulation is a special iteration of the formulation process aimed at refining a plan to make it better. Measures may be added, dropped, or any one or more of their attributes can be modified to produce a better plan. There are many reasons to reformulate a plan; some of the most common ones are listed in the following text box.

Once a plan has been formulated, it moves through the evaluation, comparison, and selection steps. A formulated plan may be found lacking at any one of

WHY REFORMULATE A PLAN?

Because it …

- Is all wrong
- Makes no sense
- Is too expensive
- Is not technically feasible
- Causes impacts that must be mitigated
- Is not complete
- Is not the least costly way of achieving the outputs produced
- Does not produce as much output for its cost as it could
- Is not acceptable to stakeholders
- Is not in compliance with federal or other policy or legislation
- Lacks economic feasibility
- Fails to address enough of the planning objectives
- Does not adequately solve the problem(s)
- Has been rendered obsolete by changed conditions
- Planning is iterative by nature and design

these points in the assessment process. The evaluation and comparison processes can shine a light on the shortcomings of a plan. This sometimes suggests the kinds of refinement needed to improve it. When that happens, planners have a simple choice to make. They can either delete the plan from further consideration, or they can try to fix what is broken about the plan. Not all plans are broken; some simply are not yet living up to their full potential. For example, maybe the bottomland in a water detention basin could also serve as a wetland and recreational trails could be added. Look for ways to reformulate that will make goods plans into better plans.

Planning is an iterative process, and plans are usually formulated in an iterative way. They begin as the seed of an idea that is grown into a more mature concept that is fleshed out with details. Before a plan is actually implemented, it must be brought to a design level with specific details. Reformulation is different from refining and iterating the details of a plan; it is fixing or improving a plan. Great ideas do not spring forth fully developed; they appear as infants and require tender loving care as well as tough love and every other child-rearing cliché you can imagine.

Here's an insider's tip for all new plan formulators: failed ideas are useful! They enable you to tweak, refine, and grow your ideas until they are robust enough to serve as good potential solutions to your problems. If you can identify the reasons why

WHO CAN HELP YOU FORMULATE?

The most immediate and perhaps best place to start a search for alternatives is right in your office. In-house personnel are frequently overlooked, but they can offer years of experience and familiarity with problems and what may or may not work to fix them. Talk with knowledgeable individuals. Hold a 1-hour brown bag brainstorming session for everyone in your office. Conduct an office-wide survey for solutions. When you do, don't leave anyone out. Call the hydraulic engineer who retired last year. Ask the new secretary. Now could be the time to call that planner from another office you met at the conference last year.

Professionals outside the office are also valuable formulation resources. Other public agencies at the federal, state, and local levels may be charged with similar problem-solving missions and can often provide formulation ideas. The academic community, consultants, and professional associations should also be considered, especially those located in the study area.

What types of solutions do hunters, homeowners, boaters, owners of businesses, anglers, and others with day-to-day familiarity and experience with the problems think will work? What alternatives would they like to see? Which ones do they oppose? A broader and more innovative array of alternatives can also be obtained by using published materials like professional journals, textbooks, and earlier planning reports. The Internet is a fantastic underexploited resource for planners. Ask a listserv, solicit ideas on your website, and then start a contest. Set up a Facebook page, Tweet about it. Find a window to the cyber universe, open it, and yell, "Help!"

a plan is not worthy of further consideration, then you know what needs to be fixed to make it work better and be acceptable. Sometimes it is not until a plan is subjected to the rigor of evaluation and comparison that its weakness and flaws become apparent. When a rejected plan can be fixed or made stronger through reformulation, stronger candidate plans result.

Some plans cannot be fixed. Even these failures teach us what not to do, and they often provide a clearer idea of where to explore next. That is progress and good work! Even if a plan was completely wrong, useless ideas are often fertile ground for new solutions. The original plan creates associations, connections, and ideas that can be used again on something that will work (Thorpe 2002).

6.3.5 Formulation Iterations

The first plan formulation iteration may do little more than identify some measures that are combined to make a plan. For example, the plan consists of a wider and more efficient conveyance channel for Watts Creek, a bike path along the creek, and a greenway through town with a walking trail.

A second iteration might more generally locate the solution's component measures on a map. This is when someone sketches in the bike path, defines the greenway, and so on. Subsequent plan formulation iterations would begin to consider solution attributes more carefully as the design detail is gradually increased.

Subsequent plan formulation iterations will also produce new ideas for additional solutions. Plan formulation is one of the most iterated tasks of the planning process.

6.4 FORMULATION STRATEGIES

Formulation, like the planning process, begins where it begins. Sometimes you'll begin with a solution. Following Hurricanes Katrina and Rita in 2005, Congress' basic response was to effectively say Louisiana needs category 5

THE PLAN FORMULATOR

One of the most disturbing developments in the author's planning world is the rise of the plan formulator. That would be one person on a planning team whose job it is to formulate plans. While I would applaud a move to place a person in charge of this important activity, that is not the trend. Instead, most other planners have ceded any responsibility for devising creative solutions to a community's concerns to the plan formulator. Ask these teams where plans come from and they will tell you, we have a plan formulator.

Plan formulation is a creative group activity. Everyone on the planning team needs to be a plan formulator. Everyone involved in the study at any level at all, from consultant through stakeholders to marginally interested members of the public, is a potential plan formulator.

hurricane protection and they funded a planning study to consider that solution. Many involved in the process immediately leapt to an idea that became known as "the great wall of Louisiana." A lot of planning begins this way, with a solution. "The problem is we need a bike path." Sometimes solutions appear before the problem is even known, much less agreed upon. Solutions can and do appear at many points in the planning process. All these ideas should be preserved and considered. But every planning process needs at least one and likely more periods of intensive activity focused on devising measures and building plans to meet the objectives. That is what is meant by formulation.

Only a fool would insist on trying to impose order on such a creative process. Only a bigger fool would attempt it without structuring the process in some way. Previously, we saw how formulation evolves through three general phases. This section presents several different strategies for formulating plans during these phases.

A formulation strategy is a disciplined way to produce one or more specific plans. Its discipline derives from the structure provided by a more or less orderly sequence of activities. A strategy usually consists of a set of tactics or conditional decisions that shape and guide the development of plans, and thus, strategies structure the "how to" of plan formulation. Strategies can be devised for any one or all the phases of the formulation process. Ultimately, they help the planner identify measures and they guide the planner in assembling individual management measures into plans.

In the pages that follow, you will find almost two dozen examples of methods that have been successfully used to formulate plans at one time or another. They are grouped by the three phases of formulation. I have put methods I think are most useful for identifying measures together and have done the same for methods to combine or reformulate measures. Those decisions were entirely arbitrary. Feel free to use any methodology in any formulation phase. Field-modify these methodologies in any way you like. Invent new ideas. The only real criterion for a formulation strategy is, does it work?

There are two golden rules for formulation. These, even for novices and revolutionaries, should probably not break. First, you must have a well-defined decision context to do good formulation. Second, you must spend time with others actually formulating plans.

The ultimate purpose of plan formulation is to devise good solutions. A good solution solves your problems and helps you realize your opportunities. Clearly, you must identify the problems and opportunities. Then you have to identify the objectives and constraints of your planning investigation. These are the cornerstones and keystones of the formulation process. Objectives state in broad but explicit terms what you intend to do about your problems and opportunities. Your constraints state what you will avoid doing. This is a critical step because you do not formulate directly to solve problems. You formulate ways to achieve your objectives while avoiding your constraints. Solutions that achieve your objectives will then solve your problems and realize your opportunities. Most, but not all, formulation strategies operate at the objectives level.

People build plans. Building plans is hard work. People have to spend time together doing that work. Specifically, the study team has to spend time together in the same place for the sole purpose of formulating solutions that will achieve your objectives and solve your problems.

6.4.1 Identifying Measures

I want to say that these strategies are in no particular order, but they kind of are. The first few strategies for identifying measures might be considered the low-hanging fruit of collecting the ideas of others before you begin to invent your own. I am biased toward working smart rather than working hard, so collecting the existing wisdom seems a reasonable way to begin.

6.4.1.1 Do Your Homework

Has anyone worked on this problem before? Did they produce a report or document their effort in some other way? If so, begin there. That is your reading assignment; do your homework. Previous iterations of a plan can be an excellent source of relevant measures. If no one has studied your problem before, has anyone considered problems similar to yours? Read their reports to see how they solved their problems. What measures were considered? Which ones found their way into the recommended plan? A review of the measures considered in previous iterations* of the planning process is one of the best sources of viable measures.

6.4.1.2 Ask People

Plans come from people, so ask people how to solve your study's problems and realize the opportunities. Start by asking those closest to the study—your study team. If these people are the insiders, everyone else is an outsider, and it can be helpful to get outsiders' viewpoints.

Involve stakeholders directly. Ask them how to solve the problems. Members of nongovernmental organizations (NGOs) will likely have an opinion, as will other interest groups. Professionals with knowledge of the problems can be helpful. Ask academics, consultants, and professional associations how to solve your problems.

Ask the people who are affected by the problems or interested in the opportunities. What do park rangers, shop owners, and bird watchers think will work? What measures would they like to see? Which ones do they oppose? Ask your children, your in-laws, a former professor, reporters, the person next to you on a flight, people in line at the drug store, the homeless. Ask anyone at all how they would solve your problem. Good ideas can come from anywhere.

* Public works projects often follow some version of a sequence of increasingly detailed planning studies such as reconnaissance, feasibility, and advanced engineering and design. The golden age of water resource development in the United States spawned a great many river basin studies (the 308 Reports), which provide initial iterations of the planning process.

A LIST OF NONSTRUCTURAL FLOOD RISK MEASURES

Moderating community susceptibility to floods:

Emergency preparedness
Flood forecast and warning
Flood insurance
Flood proofing
Information and education
Modifying equipment
Relief, recovery, and rehabilitation

Reducing hazardous uses of floodplains:

Building codes
Design and location of services and utilities
Evacuation
Housing codes
Public acquisition
Relocation
Sanitary and well codes
Subdivision regulations
Tax adjustments
Urban storm drainage
Zoning codes

Source: Institute for Water Resources. 2012. *Nonstructural flood damage reduction measures*. Accessed June 15, 2012 from http://www.corpsnedmanuals.us/FloodDamageReduction/FDRID094NonstrucFldDmgMeas.asp?ID=94.

Tap the local common knowledge and conventional wisdom. Did you see that editorial in the newspaper urging the corps to build a 50-foot navigation channel? What's the word at the local brewpub about the location for a new off-ramp from the interstate? Tap the wisdom of crowds. Find out what everyone "knows" and is talking about. This is a good way to begin formulation.

There are many ways to get information from people. Be creative not only with whom you ask but also in how you ask. Ask in private conversations, in the blogosphere, on a wiki, in a contest, in chat rooms, and on discussion boards. Provide people with as many opportunities for input to the formulation process as you can.

6.4.1.3 Checklists

Has anyone already thought systematically about your problem? Look for checklists of measures. They are sometimes hard to find but often worth the effort. For common and recurring problems, planners can develop and use their own checklists. You can research or brainstorm a checklist. The best checklists may be the ones you develop specifically for your own situation.

There are a number of generic checklists designed to help you identify solutions to problems. If there are no checklists of measures, examine your situation with a generic checklist in hand. The best checklists provide a comprehensive set of questions to ask, conditions to monitor, factors to consider, and the like that will help you identify solutions to your problems. Deep and Sussman (1997) provide a set of 140 lists to aid decision making. *That's a Great Idea* by Husch and Foust (1987) has several situation analysis checklists. VanGundy's (1985) *Techniques of Structured Problem Solving*, though out of print, is well worth the search, and it offers 105 different creative problem-solving techniques.

6.4.1.4 Brainstorming

Brainstorming by the study team is the most traditional and productive way to identify measures that could meet objectives. It is actually quite simple and logical. Take each objective one at a time and brainstorm measures that could accomplish it, making sure to identify at least one measure for every objective. There are a number of variations of this theme, but that is essentially all there is to it. The truly amazing thing is how infrequently planners do this!

The two most fundamental rules are (1) no evaluation of ideas (corollary: unusual ideas are especially welcomed) and (2) quantity counts (corollary: combine and improve ideas). Brainstorming is best practiced as a team exercise but can be a solo effort. It's probably your best hope for the 5 PM Friday afternoon formulation surprise when you're the only one left in the office and the boss needs a plan for the emergency county commissioners meeting at 9 AM Saturday morning. The Internet is full of examples of specific brainstorming methodologies. Try 3X Yeah, outlined in Chapter 4, Section 4.10.3, to identify measures, for example.

6.4.1.5 Measures That Solve the Problems

Four generic problem-solving strategies may help you identify measures. Use them simultaneously and congruently or individually, as fits the team's mood or needs. Focusing on problems rather than objectives gives you a slightly different way to identify measures. Here are the strategies:

- Obliterate the context in which the problem occurs.
- Modify or control the effect of the problem phenomenon.
- Prevent the problem from occurring in the first place.
- Remove objection to the phenomenon that causes it to be perceived as a problem.

SOLVING A FLOOD PROBLEM

- Obliterate—remove all developments from the floodplain.
- Modify effect—build a levee or wall to reduce flood damages.
- Prevent problem—restrict land use in floodplain.
- Remove objection—educate people to accept flooding as a natural process that provides ecologically significant value in the variability of flows.

6.4.1.6 Clues in Data and Models

The evidence you develop during the study process can provide useful insights into potential solutions to your problems and opportunities. An Ishikawa diagram (Chapter 4 Addendum) of your problem's causes and effects provides a lot of useful clues. The technical analyses you do to verify and understand the nature of your problems can reveal useful information about the specific nature of your problems and therefore potential solutions. For example, a flood problem characterized by a quick-spiking hydrograph suggests upstream detention basins or even dams would be useful measures. A long and low hydrograph, on the other hand, suggests that rising waters might be more effectively contained by walls or levees.

A habitat suitability index model analysis in an ecosystem restoration study may reveal that of the dozen or more habitat variables under consideration, dissolved oxygen in the stream is the limiting factor. This immediately points you toward measures that can increase dissolved oxygen levels in the stream, like aeration schemes and water temperature modifications.

FROM OBJECTIVES TO MEASURES

Let the planning objectives be

- Reduce acidity of stream.
- Increase diversity of instream flora and fauna.
- Reduce safety hazards due to mine subsidence.

Formulation begins by taking each objective in turn and identifying measures that can attain the objective. For example, one can reduce the acidity of a stream through one or more of the following measures that generate alkalinity:

- Compost or anaerobic wetland
- Aerobic wetland
- Open limestone channel
- Successive alkalinity producing systems
- Limestone ponds
- Limestone leach bed
- Slag leach beds
- Diversion
- Inundation/saturation
- Underground mine sealing
- Low-head dams to aerate streams
- Stream subsidence closures

The process continues by developing a similar list of potential measures for each of the other planning objectives. A good planning process will produce one or more viable measures for each of the planning objectives identified. Some measures may be listed for more than one objective.

6.4.2 Building Plans

The first three techniques that follow are pretty tried and true techniques. If you read them and find yourself saying, "Duh!" then we agree. The surprising or distressing truth, however, is that relatively few planning teams are doing formulation in these ways. Following these three strategies you will find several more. All but one of them are field tested. They work. I am looking for an opportunity to test that other one.

6.4.2.1 Objectives, Measures, and Plans, Oh My!

This is, in my experience, the most productive formulation strategy. It is simple and logical, and it produces plans. If you have not already identified measures that would achieve each of your objectives while avoiding your constraints (as described in Section 6.3.4), then this strategy begins there. Do that first. Objectives are the cornerstone of a good planning process, and they are essential to good formulation. Identifying candidate measures for each objective requires planners to think critically about how best to achieve them. This assures that proper attention is given to achieving objectives in the problem-solving process. For brevity, let's assume you have already rigorously identified sets of measures that address each objective, as shown by the letter X in Table 6.2.

Notice there are multiple measures for each objective and some measures contribute to multiple objectives. The team's job is to mix and match the measures, combining them into as many plans as possible, almost in a tinker toy approach. There will be some rules imposed by the physical universe. Some measures may be dependent upon each other. Maybe measure 2 requires measure 9 to be effective. Measure 9 may or may not need measure 2 to function effectively. Measures 2 and 3 may be mutually exclusive because they need to occupy the same physical space and

Table 6.2 Measures Identified to Achieve a Specific Objective

	Objective		
Measure	**A**	**B**	**C**
1	X	X	X
2	X		
3	X		X
4		X	X
5		X	
6		X	
7		X	
8			X
9	X		X
10			X

obviously cannot do so simultaneously, or they may be redundant, accomplishing exactly the same thing in different ways.

With this matrix and these kinds of logical rules in mind, the team can begin to build plans. Measure 1 looks attractive; it contributes to all three objectives. So we may construct a plan comprising measures 1, 2, 9, 4, 7, and 10. It is easy to imagine how a great many such plans can be constructed. This strategy does have a few rules: construct the matrix and identify dependencies, redundancies, and mutually exclusive measures. But it is also organic. How you put these measures together and how many plans you build is entirely up to the team's judgment.

Let's return briefly to the idea of a plan comprising measures 1, 2, 9, 4, 7, and 10. In future iterations, the team will begin to flesh out the attributes of these measures. Where will measure 4 be located, what will it be made of, how big will it be? This is all part of the "joy of planning" that awaits you in subsequent iterations of plan formulation.

At this point in the process you want plans that address your objectives adequately. You may even differentiate plans based on which objective you do and do not address and to what extent. If objective A is to reduce flood damages, objective B is to restore habitat adjacent to the river, and objective C is to increase urban recreation, you may formulate single purpose plans, for example, a primarily flood risk management plan, a primarily habitat increasing plan, and a primarily recreation plan. There may be a multipurpose plan to contribute to all the objectives. There are a limitless number of ways to build plans from measures.

It is not uncommon for objectives to contradict one another at times. Thus, it may be necessary to develop a wide variety of plans. It is not essential that each plan address every objective, but every objective needs to be addressed by at least one plan. Any strategy that has a significant organic element like this should take care to document the formulation process and rationale. You are not going to be considering every possible combination of measures in this strategy, so you want to capture the logic for the plans you do build as well as your logic for not building more plans. Write these reasons down and vet them in an appropriate manner.

The disadvantage of this approach is that if the planning team has not done a good job identifying problems and specifying objectives, it is not likely to be an effective way to formulate. This strategy requires a strong planning process and effective planners.

6.4.2.2 *Just Do It (Organic Formulation)*

Building plans tends to be a mix of mechanistic and organic activities. Mechanistic plan building follows an algorithm or a set of logic rules that automatically leads to the generation of plans. The all possible combinations strategy below is the best example of this. Organic formulation is old school, and you make it up as you go. Basically, planners enter a room and start combining measures into plans. There is no algorithm, no structure, and no software. It only takes knowledgeable people gathered in one place and a few ideas to formulate some plans.

Organic formulation draws on the experience, knowledge, wisdom, insight, inventiveness, and synergy of the planning team. Plans emerge through a collaborative work process that is difficult to typify. Judgment is exercised throughout an organic process, as opposed to codified in a mechanistic process. Organic formulation is a fluid, growing, and changing process. The only real requirement is a genuine effort to identify substantially different combinations of measures that contribute to the achievement of the planning objectives. If there is any such thing as a traditional formulation approach, this may be it. You show up and muddle through it.

6.4.2.3 *All Possible Combinations of Measures*

The most comprehensive way to develop plans is to make every possible combination of measures identified in the first formulation phase a separate plan. If you have been comprehensive in your efforts to identify measures, then you can take comfort in knowing the best plan is somewhere among the combinations generated. Assuming you apply the rules of dependence and mutual all the logically viable plans will be generated by this mechanistic approach.

When you have a small number of measures, this approach is straightforward and can be completed with pencil and paper. For example, if you only have two measures, 1 and 2, there are only three possible combinations or plans:

Plan 1 = measure 1
Plan 2 = measure 2
Plan 3 = measure 1 + 2

If we add a third measure, 3, the total number of plans increases from three to seven. That is a scary jump!

Plan 1 = measure 1
Plan 2 = measure 2
Plan 3 = measure 1 + 2
Plan 4 = measure 3
Plan 5 = measure 1 + 3
Plan 6 = measure 2 + 3
Plan 7 = measure 1 + 2 + 3

Perhaps the disadvantage of this strategy is readily apparent, and you are either going to need several pencils and a long piece of paper or some computer assistance. The number of possible combinations increases exponentially as the number of measures increases. The formula for the number of possible combinations (N) is $N = 2^M - 1$, where M is the number of measures you have. Ten measures can be combined into 1023 plans.* Twenty measures can be combined into over a million plans.

* The actual number of viable plans will be reduced depending on the nature of the dependencies and mutual exclusivities. It is common practice to consider each variation of a measure (based on attributes) as a separate measure when using this approach.

IWR-PLAN Decision Support Software was developed specifically to assist planners in formulating and evaluating all possible combinations of ecosystem restoration plans. Its input data include the measures identified, an estimate of the output of each measure, and an estimate of the cost of each measure. This software tool relies principally on the use of cost-effectiveness and incremental cost analysis rules to screen the results of a formulation effort from a set of potentially thousands down to a much smaller final array of plans that are called "best buy" plans.

Source: IWR Planning Suite. 2012. Accessed June 15, 2012, http://www.pmcl.com/iwrplan/.

Computer technology provides planners with the ability to use numerical processes to identify and even screen these large numbers of potential plans. The identification and screening of plans is usually based on a simple set of rules that specify how to combine measures. These include identifying those measures that must always be used together (dependence) and those that cannot be used together (mutually exclusive). Other rules constraining the combinations can be devised as well. For example, you might want to make sure duck boxes are included in every plan. This negates the need to consider plans without duck boxes. A computer algorithm is then used to generate all the possible combinations of measures that do not violate any of the formulation rules. This can still produce more plans than your team can cognitively absorb, so if the software may be augmented by specifying and applying some preliminary evaluation rules that are used to screen the large number of plans down to a reasonable number, the task becomes easier.

The advantage of this strategy is that the computer-assisted capability enables planners to consider a far more comprehensive array of plans than ever before imaginable. When the brute force of numerical techniques is needed, there is no better approach. You're guaranteed that the best plan that can be built from your measures, no matter what it may be, is somewhere in the mix of combinations. If you have used the right screening criteria in your computer algorithm, you will find that plan. When the data measurements are good, the algorithm rules are clear, and the interactions among measures are negligible, this strategy can be helpful.

Planning is an art based on science. There is no guarantee a technologically based approach is going to produce a final array of plans that reflects the artful nuance of an experienced planner who is balancing objectives, constraints, and stakeholder interests. If the objectives have not been adequately considered in identifying measures or incorporated into the screening algorithms, good plans may be lost in the numerical methods approach. New ideas are not as likely to evolve during a mechanized formulation process as they are in more organic strategies. Assembling measures into entirely new combinations "by hand" can often inspire insight into the ways ideas connect, which can lead to new ideas. An all possible combinations method is not capable of imagining new components.

6.4.2.4 *The Gift Plan*

It is not unusual for someone to hand you a plan at the start of your investigation as Congress did for Louisiana after the Gulf hurricanes. The reason you're doing a study is that there are conditions that have attracted people's attention and they want something done. Chances are someone has thought about what they would like to see done. Planners with the U.S. Army Corps of Engineers may have studies dating back to the 308 Reports of the 1920s and 1930s. There may be a local government interest, an environmental interest group, a retired planner, or a county engineer that hands you their plan soon after learning of your study. This plan is a gift, not a curse. You already have your first candidate plan and when you begin your own formulation, you begin with a solution you will either top or recommend. There may be pressure to implement a gift plan, depending on the influence of the gift giver. There is no obligation to implement a gift plan, however, especially if the planning process builds a better plan upon it and the giver is credited with the inspiration for the "even better" plan. While you are at it, take the opportunity to reverse engineer the gift plan. Deconstruct it into its component measures, add them to your lists, and credit the giver generously. If you do, chances are you now have a few measures to combine in new ways and a planning ally.

6.4.2.5 *Ask an Expert and Get Out of the Way*

If you organization has an expert, ask him or her to solve your problem. Give him or her a topographic map, a good description of your problems and opportunities, a pencil, a French curve, and a straight edge and then get out of the way. This expert may be the experienced wise sage who has quietly worked every square inch of your study area over a long distinguished career. Or it could be the whiz kid that graduated at the top of her class. Experts have the ability to say things like, "We can fence in this area, relocate this and that, create riffles here and there, and add a winding walking trail through the center." Voila! A plan is born! Such spontaneous formulation will probably be a once-in-your-study event, so savor the experience if you are lucky enough to be in the room when it occurs. The expert's plan, like the gift plan, becomes a threshold for the planning team to exceed. If you do so, you will have good reasons aligned for not recommending this early iteration of a solution. If you are unable to find a better solution, then you look like a genius for having asked the expert!

KEN HARTZELL

Ken was "the expert" in my early planning career. He was one of the few people in the office without a college degree. He simply spent years working the length and breadth of the Susquehanna River Basin. If he did not work on every planning study in the district, he advised those that did at some point. Ken knew the why of every solution ever formulated in the basin. If you know a Ken, or a Red Goodnoe, or a Jack Ross, be sure to ask him how he would solve this problem. There is a good chance he has already done so in his mind a few times while driving the area.

6.4.2.6 Jagger–Richards Approach

The Rolling Stones were a cover band until their manager confined Mick Jagger and Keith Richards to a room, telling them not to come out until they had written a song. They did. The rest is musical history. This is not a bad strategy to emulate. Pair each member of your study team with another person. That person could be another team member or a "guest formulator." Give them a space and as much of a part of the day as they need to come up with a plan. If you have eight team members, you'll have four pairs of plan writers and four plans by the end of the day. Change partners and repeat the process taking care not to violate the copyright of previous plan writers!

By the way, if it takes a day to come up with an early iteration of a plan, you may be getting bogged down in attribute specifics. Specifics come later as you continue to discuss and polish the plans you formulate in this step. Formulation often continues through implementation and beyond when monitoring, evaluation, and adjustment are part of the planning process. You should not expect a single formulation iteration to complete a plan.

6.4.2.7 Improvisation (Last Planner Standing)

Okay, this is one method I have never tried but I would like to. It was inspired, actually stolen is more like it, from Tina Fey's *Bossy Pants* (2011). (Note to reader, take a good idea wherever you find it and make it yours.) She offers four rules for improvisation. First, you must always agree. Say yes. Accept whatever your partner has created. If someone says the plan needs a paved road with a Starbucks along the old canal, say yes. Never let the first response to a measure be, "No, we can't do that." "No, it is too expensive/impractical/unpopular" is not an acceptable answer.

Rule two is not only to say yes, but say, "Yes, and..." If you just say yes, the formulation is slowing. But if you say, "Yes, and there is a secret trail that looks just like the canal looked in 1819," then you are making progress. Add something to the plan, and your efforts are worthwhile.

The third rule is to make statements. Don't ask questions all the time. Don't say how much will that cost or how are we going to do that? In other words, don't sit there pointing out obstacles. Find a way and make it work. No matter what the problem is, be part of the solution. Instead of saying, "Where are we?" you might say, "So here we are with a canal trail that has conveniences for those who want them and an authentic historical experience for those that want it." This might seem like a weak start to your formulation, but it leads to rule four, Tina's favorite.

Rule four is that there are no mistakes, only opportunities. Maybe you can formulate a bad plan, but if you do it won't get implemented. So don't be afraid to improvise. You may end up looking at a happy accident.

6.4.2.8 Cornerstone Strategy (aka First Added)

This strategy is useful in those cases where there is a single most important measure in a plan that everyone will agree is essential to a successful plan. That measure is the cornerstone. Alternative plans are then built, each upon this cornerstone. Additions to the cornerstone may come from the objectives as measures are added to meet the objectives not served by the cornerstone. Additions may come from the desire to provide something for every stakeholder, to craft specific sorts of plans (e.g., the sustainability plan, the recreation plan), or to identify the plan with maximum net economic benefits. In other words, this strategy is well-suited to being used in combination with some of the other strategies found in these guidelines.

The advantage of this strategy is that it begins with a firm foundation that contributes toward solving the problems and meeting the opportunities in a way that meets with the support of most, if not all, stakeholders. If sealing off the mines or building the outdoor amphitheater have to be part of whatever is done, then you have your cornerstone. If reducing flood damages is essential and offset levees do that best, then you have your cornerstone. It is a good place from which to begin to construct an array of plans. The disadvantage is that relying on a common cornerstone may limit the range of plans considered as potential solutions to the problems identified.

6.4.2.9 The Ideal Scenario

Scenarios, discussed in Chapters 5, 7, and 13, can be helpful in the formulation step. This strategy begins by envisioning an ideal or Utopian future condition for the study area. In other words, what does complete success look like? If a consensus vision of this future can be articulated in a simple scenario, then the planners' formulation task is clearly focused. What has to happen to make this future a reality? What are the different ways one could make this future a reality?

It is not easy to arrive at a consensus vision of an ideal future. Nor is it likely that there will be any one clear and unambiguous path to this future. Most scenarios are heavily influenced by factors that are beyond your control and your knowing, so uncertainty will be another hurdle in devising an idealized vision of the future. That vision is, of course, expected to reflect the planning objectives and constraints. Once the ideal scenario is narrated, however, it is a relatively straightforward process to brainstorm ideas for measures that would achieve the various aspects of this future.

This process uses convergent thinking at the outset. The team is asked to converge on a vision of an ideal future. Using the without project condition scenario can help focus the team's efforts. What is it about the without condition that you would like to change? Never mind how you would do that, just focus on what success looks like. Then the team exercises its divergent thinking skills by identifying as many ways to realize that future as possible. This is followed by selecting the measures capable of producing the ideal conditions.

Alternative plans are developed as planners approach this idealized vision from different directions. If planners are able to articulate quantitative measures of the ideal future, plan evaluation can be accomplished through gap analysis, described in the next chapter. When planners are fortunate enough to be able to articulate a vision of success that enjoys a consensus, this simple formulation strategy can be very effective with or without the quantitative measures of success.

6.4.2.10 Multiple Desirable Scenarios Strategy

This strategy builds on the ideal scenario strategy and relies on adding rich narrative descriptions of alternative future scenarios as the driving force for formulation. It may be more appropriate for planning investigations that focus more on opportunities than on problems. The basic idea is to identify different future scenarios for the planning area. For example, one might identify the recreation scenario, the industry scenario, the double population scenario, the downtown turns green scenario, the port doubles its cargo scenario, and so on. Planners develop as many interesting and appealing scenarios as possible. Then they devise a plan that will realize as much of that specific future scenario as possible.

Ordinarily, planners formulate a plan and then try to describe the future that plan will produce. With multiple desirable scenarios strategy, planners begin with the futures different stakeholders would like to see produced and then they develop plans to produce these different futures. An advantage of this approach is that it enables planners to develop rich narrative descriptions of desirable future states that can be true to the objectives. This outcome-driven plan formulation strategy can be very effective. A disadvantage is that planning for alternative futures may overlook some of the objectives if they are not key elements in the scenario narratives.

6.4.2.11 Something for Everybody Strategy

This is a pragmatic strategy that recognizes the importance of satisfying stakeholders. The team could set out to formulate plans that provide outcomes that will satisfy all the known stakeholders. In a well-integrated planning process, formulation strategies that focus on objectives, futures, or stakeholder interests should all lead to about the same place.

There are at least two ways to address stakeholder interests. First, a plan can be developed for each stakeholder group. There could be a recreation plan for outdoor recreation enthusiasts, an ecosystem restoration plan for environmentalists, and a hunting and fishing plan for these interests. Alternatively, plans could be developed to ensure that each stakeholder group finds some element of interest to them in each of the formulated plans. This will be more likely to happen if the values and interest of stakeholders have been captured in the objectives. Thus, each plan would have some recreation, some ecosystem restoration, and some hunting and fishing enhancements.

The advantage of this strategy is that it produces plans that people can support. It is stakeholder-driven planning. The disadvantage of the strategy is that it is a stakeholder-driven strategy. It gives people what they like and want. This is not always going to lead to the same result an objective-driven planning process will lead to when stakeholder interests and objectives diverge. In theory, a good planning process will produce objectives that reflect a balance of stakeholder values and sound scientific principles. In practice, stakeholder values and science may not always be compatible.

6.4.2.12 Directed Plan Formulation

Sometimes a higher authority tells you what to do. It may be a specific one-time directive, like provide category 5 hurricane protection for Louisiana, or it may be a standing policy like the national economic development plan for federal water resource planning agencies. Federal water resources planning policy currently requires a national economic development plan.* In the past, federal water resource planners have been directed to develop environmental quality, regional economic development, nonstructural, locally preferred, and no action plans. When policy or practice dictates the formulation of specific kinds of plans, finding these plans becomes its own formulation strategy. This formulation strategy identifies the specific types of plans desired in the final array and mounts an effort to formulate those plans. Measures are combined for the specific purpose of producing a specific kind of plan.

This strategy differs significantly from others in that it is not driven by the desire to identify as many alternatives as possible. Meeting objectives takes a backseat to a different strategic focus. The obvious advantage of this strategy is that it produces an array of plans that is meaningful from a policy perspective and that is important to higher authorities. The disadvantage of this strategy is that it could overlook achieving those objectives that are not reflected in the specific plans that are formulated.

6.4.2.13 Divergent Thinking Strategy

This strategy is based on the presumption that if we identify enough plans, we will identify some good plans. The central idea of this strategy is to continually expand the number of possible solutions. Imagine a checklist of measures with dams on it. Instead of eliminating alternatives with dams, force yourself to identify the best dam plan possible. When you have one, try to modify it and expand it to produce other alternatives. Consider other locations and alternative materials and sizes. Do this sort of thing for every measure. Divergent thinking is intended to expand the universe of options. Assessment is not initiated until a rather large universe of ideas has been identified.

An advantage of this strategy is that it produces a large number of plans. A disadvantage is that it is not easy to work with obviously inferior ideas to develop additional ideas. It is difficult not to make screening decisions as one proceeds.

* This plan basically maximizes positive net benefits to the national economy subject to the environmental laws of the nation.

6.4.2.14 Convergent Thinking Strategy

This strategy tries to focus on the best plan as soon as possible. The central idea here is to limit the number of alternatives and to zero in on the best one as quickly as possible. It is the opposite of divergent thinking. Limiting the number of alternatives is generally regarded as anathema to what has traditionally been considered a good planning process.

This strategy requires planners to say what they mean by best. The secret to best is presumably found in the decision criteria identified and refined earlier. With this strategy, planners look for good reasons to eliminate measures. Aspects, threshold setting, and other rationales can be used to eliminate measures and plans. The best plan is the last one remaining. Alternatives are those eliminated late in the process.

The advantage of this process is that it does not spend time considering ideas that will not be implemented. The disadvantage of this strategy is that it may not give ideas sufficient opportunity to grow into viable solutions.

6.4.3 Reformulation (Making Plans Better)

The strategies that follow can be used for building plans or for reformulating them. I have often thought of reformulation as like a dry dock operation. Dry docks haul a ship out of the water and fix the things they were asked to fix. But they also go over it with a fine-tooth comb and fix whatever needs fixing to make the vessel seaworthy again. Some plans that are rejected in the evaluation and comparison planning steps go right to the junk heap. Others are sent to dry dock, where they tend to arrive with identified shortcomings and other possible issues. In my experience, most plans are fixed one at a time. For example, the county loved the recreation plan but it gave the wildlife management people major heartburn. See if you can make the recreation plan more wildlife friendly. There is no reason we cannot have more structured strategies for reformulation, so I have requisitioned a couple of plan-building strategies and call them reformulation strategies. Keep in mind that a plan is being reformulated because it has been evaluated and found lacking in some way or it has the potential to be a better plan. Thus, you should always have a pretty decent idea what it will take to revive a rejected plan.

6.4.3.1 Rapid Iteration Prototyping

Microsoft Word 2010 is a powerful word processing package with an incredible array of features—one might say an almost unimaginable array of features. It was not formulated in its current form. Someone developed a program that produced letters and spaces in an endless line of text in cyberspace when keys were struck. Then someone said make it fit on a page so I can see it. And they did that. Then someone said make it change fonts. And they added that. Make it copy and paste, and so on. And gradually the software evolved into what we see today.

Rapid iteration prototyping is a technique used to develop software. You develop a program that does what you want it to do and then gradually expand the quantity and quality of the things you want it to do. You can approach reformulation in a similar way. On day 1 of a planning investigation, formulate a plan that solves the problem as you understand it. Do it as quickly as possible. Then each time your understanding of the problem evolves or the planning objectives change, quickly reformulate your plan or formulate another plan. Solutions, as simple and inadequate as they may be, are improved frequently and incrementally. It is the constant iterating of these ideas that grows these prototype ideas into viable solutions.

This strategy tends to go right to plans, skipping over the development of measures. This makes it an especially useful reformulation strategy. If this plan is lacking in wildlife management features, fix it quickly! When a plan is evaluated and found lacking, you know why it is not working. Fix it. Formulate a solution. Does it work now? If so, you're done. If not, what is still missing? Add that to the plan.

This strategy exploits the iterative nature of planning. The idea is to develop and refine plans quickly. Do not spend weeks on them. Develop as complete a plan as you can. Critique it and revise it, as quickly as possible. This can be a helpful way to begin to consider the attributes of a plan. If you're not sure whether to make the path asphalt, concrete, wood, crushed stone, or dirt, pick one. A good evaluation process will point out the serious shortcomings, and if this is one of them you can fix it readily enough. At some point, the process will slow down of its own weight as it grows closer to a finished plan.

An advantage of this strategy is that you always have a plan or plans. Plans are developed quickly with a high tolerance for ambiguity, and plan formulation is a constantly ongoing process. A disadvantage of this approach is that plan formulation is a constantly ongoing process.

6.4.3.2 Mutation Strategy

This strategy requires planners to identify at least one decent plan by some means. It can be derived from any reasonable method, or it can be a plan that has failed some aspect of the evaluation process. The basic reformulation idea is to mutate this one plan in as many ways as possible to find the best fix.

Plans can be mutated in any way you can imagine. Trigger words have been used to guide the mutation. Examples include subtract, add, transfer, empathize, animate, superimpose, change, scale, substitute, fragment, isolate, distort, disguise, contradict, parody, prevaricate, analogize, hybridize, metamorphosis, symbolize, mythologize, fantasize, repeat, combine, and so on. The advantage of this technique is that it combines some creative techniques—for example, using trigger words to generate new ideas—with a very structured approach—to reformulate the failed plan.

6.4.3.3 The Fast Plan

Fast planning is the topic of Chapter 15. It suggests that the entire planning process be iterated in a 5-day period. The resulting recommended plan is considered the best plan possible. The sequence of events necessary for this to be the best plan possible is identified. This is usually closely related to the objectives to be met and the constraints to be avoided. This sequence of events is turned into planning hypotheses. Each event is translated into a planning hypothesis and is subsequently tested. When a hypothesis is rejected, the plan is modified to make it better than the tested plan.

For example, suppose a plan has to maximize net national economic development benefits to be the best plan in a decision context. A suitable planning objective would be that the best plan does not maximize net benefits. An economic analysis might be planned to test the hypothesis. If the analysis rejects the hypothesis, the plan is modified to a plan that does maximize net benefits and the next hypothesis is tested. If the analysis fails to reject the hypothesis, the best plan stands until a necessary condition hypothesis can be rejected.

6.5 FORMULATING FOR UNCERTAINTY

How should uncertainty affect formulation? Phased implementation of a solution and adaptive management strategies are two options that ought to be considered when formulating with significant uncertainty. When there are measures that could become rendered unnecessary, ineffective, or counterproductive depending on the manner in which the future unfolds, your formulation strategy should reflect these possibilities. Do you build a nuclear power plant in a coastal lowland that some experts project will be underwater within a century due to sea level rise? Do you double the width of a navigation channel for traffic increases that may or may not develop? Will changes in water quality and quantity bring back enough birds to require construction of bird islands? Do you purchase land now to maintain the option to use it as a future wildlife refuge? How will cattle grazing affect critical habitat for a sensitive species? How will off-road vehicle use and expanding a campground affect sensitive species?

There will be times when an uncertain future makes you unsure what the best thing to do right now is. Some of these times there will be some plan measures you are absolutely sure are needed now and with the passage of time you'll have a better idea whether the other measures are needed or not. That is the ideal situation for phasing the implementation of a plan over time. Implement measures that are most certain to be effective and phase in other measures as conditions warrant. Phased implementation is especially useful when the major uncertainty is concerned more with the actual timing of events rather than with their occurrence or not. When the uncertainty is not amenable to further reduction through gathering more data, research, experiment, testing, or other measurement, formulating plans with phased implementation of its measures is a reasonable hedge against uncertainty.

USE ADAPTIVE MANAGEMENT WHEN...

- There is a management decision to be made.
- Management objectives can be identified.
- Stakeholders are engaged and committed.
- Information value is high, and there is an opportunity to learn more about the problem and measures.
- Uncertainty can be expressed as testable hypotheses.
- A monitoring system can be established to reduce uncertainty.

Source: U.S. Department of the Interior. 2010. Adaptive Management Working Group. *Adaptive Management: The U.S. Department of the Interior Technical Guide. Accessed June 15, 2012 from http://www.doi.gov/initiatives/Adaptive* Management/whenused.html.

In instances where the fundamental direction of the future is in doubt, adaptive learning and adaptive management strategies may be most useful. The U.S. Department of Interior (2009) has been especially helpful in articulating these ideas for planning purposes. In part they have said: "Adaptive management forces stakeholders to confront unresolved uncertainties that can significantly influence management performance. An adaptive approach provides a framework for making good decisions in the face of critical uncertainties, and a formal process for reducing uncertainties so that management performance can be improved over time" (p. 3).

The Bureau of Land Management (Huntsinger and Sorenson undated) has described adaptive management as a scientific approach to flexible natural resource management. It is a tool based on the scientific research process. It requires a measureable objective, monitoring to determine the effectiveness of the management practices in achieving the objective, evaluation to determine if the objective is being reached, and adaptation based on the results. The purposes of adaptive management include learning something about the uncertainty that remains and the effectiveness of management measures. Adaptive learning comprises systematic and intentional efforts to learn about and reduce uncertainty in the planning process.

As a decision process, adaptive management promotes flexible decision making that is adjustable in the face of uncertainties. As carefully monitored outcomes from management actions and other events become better understood, they advance scientific understanding and enable managers to adjust policies or operations as part of an iterative learning process. Adaptive management is not haphazard "trial and error," it is a science-based process that emphasizes learning while doing. Neither is it a way to short-circuit the planning process. "Let's do this much and push the rest of the planning process off into an adaptive management process" is not an option.

It is important to understand that adaptive management is not an end in itself, that is, it is not a true measure or part of a plan. Instead, it is a means to make more effective decisions under conditions of significant uncertainty. It helps decision makers maintain flexibility in their decisions while they reduce the remaining uncertainty that exists. In so doing it enables decision makers to change direction as new information (i.e., reduced uncertainty) dictates. Adaptive learning and adaptive management are strategies that help assure progress.

6.6 TAKE AWAYS

Lesson 1. Formulation is the most challenging, inventive, creative, rewarding, and fun part of the planning process.

Lesson 2: Formulation is not always easy, and there are a lot of iterations.

Lesson 3: There are three phases in the formulations process. These are

1. Identifying measures that meet your objectives and avoid your constraints
2. Combining measures into plans
3. Changing plans that do not work

Lesson 4: Everyone on the team is a plan formulator; many people not on the team are as well.

Lesson 5: Although formulation is a creative process, it is best served by a strategic approach; develop and use one.

REFERENCES

Deep, S. and L. Sussman. 1997. *Smart Moves: 140 Checklists to Bring Out the Best in You and Your Team*, revised ed. Cambridge, MA: Perseus Publishing.

Fey, T. 2011. *Bossy Pants*. New York: Little, Brown, and Company.

Harris, R. 1998. Introduction to creative thinking. Accessed February 6, 2012 from http://www.virtualsalt.com/crebook1.htm.

Huntsinger, R. and P. Sorenson. Undated. Adaptive management: The scientific approach to flexible natural resource management. Accessed February 21, 2012 from http://www.blm.gov/pgdata/etc/medialib/blm/wo/Planning_and_Renewable_Resources/presentations.Par.83536.File.pdf/Adaptive_Management_2.pdf.

Husch, T. and L. Foust. 1987. *That's a Great Idea*. Berkeley, CA: 10 Speed Press.

Institute for Water Resources. 2012. *Nonstructural flood damage reduction measures*. Accessed June 15, 2012 from http://www.corpsnedmanuals.us/FloodDamageReduction/FDRID094NonstrucFldDmgMeas.asp?ID=94.

IWR Planning Suite. 2012. Accessed June 15, 2012 from http://www.pmcl.com/iwrplan/.

Major, D. C. 1977. *Multiobjective Water Resources Planning*. Washington, DC: American Geophysical Union.

Seuss, D. 1975. *Oh the Thinks You Can Think*. New York: Random House.

Thorpe, S. 2002. *How to Think Like Einstein*. Naperville, IL: Sourcebooks.

U.S. Department of Interior, Adaptive Management Working Group. 2009. *Adaptive Management: The U.S. Department of the Interior Technical Guide*. Washington, DC: DOI.

U.S. Department of the Interior. 2010. Adaptive Management Working Group. *Adaptive Management: The U.S. Department of the Interior Technical Guide*. Accessed June 15, 2012 from http://www.doi.gov/initiatives/Adaptive Management/whenused.html.

VanGundy, A. B. 1985. *The Product Improvement Checklist (PICL)*. Point Pleasant, NJ: Point Publishing.

Webster's 7th New Collegiate Dictionary. 1971. Springfield: G. & C. Merriam Company.

CHAPTER 7

Evaluate Solutions

The primary purpose of this third planning step is to screen solutions for comparison in the next step by identifying the viable solutions from among the many potential solutions formulated. The most important effects of the viable plans are identified in this process.

7.1 INTRODUCTION

Here is the short story. You've formulated a number of solutions, and now you need to know if any of them are good enough to be implemented. Figuring that out is the evaluation step, and it has three major pieces. The first piece is assessing the plan's effects, the second piece is judging those effects as good enough or not, and the third piece is screening out solutions that do not pass the qualifying criteria.

Evaluating a plan can be a lot of work. Don't begin it until you are ready. The first iteration of evaluation will tell you whether your solution is ripe enough or not. If your set of assembled measures is complete, effective, efficient, and acceptable enough, then call it a plan. It is ready for the next iteration of evaluation.

The second evaluation iteration begins by identifying the evaluation criteria, which ideally were identified among the decision criteria in Step 1, and developing a with condition scenario in which you can analyze the plan's effects on the evaluation criteria. Once this analysis is complete, you compare the effects of your plan to the effects of the without-condition scenario. All these evidence gathering effort comprises the assessment task of evaluation.

Once the effects have been assessed, you must appraise them and decide if the plan you are evaluating produces enough desirable effects to be considered qualified for further consideration as an implemented solution. In other words, will this plan become a finalist that is considered in Step 4 (comparison)?

Here are a few more things to know before leaping into this chapter. Evaluation is both analysis and judgment. It's time to begin to understand how your solutions are likely to perform if implemented. That means it's also time to make some important planning decisions. Specifically, which of your many formulated solutions are worthy of serious consideration as a solution to the problems and opportunities identified in Step 1. In this evaluation step, the planning process turns from the creative, divergent thinking of formulation to the critical, convergent thinking needed to figure out how best to meet your objectives and avoid your constraints so you can solve your problems and attain your opportunities.

Evaluation is a qualifying step. This is the first time you begin to scrutinize your formulated solutions to see which ones belong in the final array of plans, from which a recommended solution will be chosen. There are seven evaluation tasks in this step:

1. Determine whether each solution is ready for evaluation (iteration one).
2. Prepare one or more with condition scenarios for each solution (begin second iteration).
3. Conduct the necessary analysis within the with condition.
4. Compare with and without condition values for evaluation criteria.
5. Assess, that is, describe quantitatively where possible, all important differences between the two conditions.
6. Appraise, that is, judge the solution's effects.
7. Qualify the plan for further consideration, reformulate it, or drop it from further consideration (end second iteration).

This chapter describes how to accomplish these tasks. It begins by discussing what evaluation is and why it is done. Most of the chapter describes how it is done. The primary output of this planning step is a set of viable solutions to your problems that you can begin to compare.

7.2 WHAT IS EVALUATION, AND WHY DO WE DO IT?

Evaluation is a screening task. Each solution or plan is evaluated individually, without respect to any other solution. First, planners assess the effectiveness of a solution, that is, they estimate and quantify its ability to meet the objectives and avoid the constraints. Second, they appraise that assessment, that is, they make a subjective judgment about whether the solution does enough to be considered as a finalist for selection and implementation.

PLANNERS WHO LOSE THEIR WAY

Engineers like to design things. Biologists will make lists of critters in the study area. It's easy for economists to think their job is to do economic analysis. Analysis is not the purpose of this step. The purpose of the analysis is to produce the information needed to qualify some plans and eliminate others from further consideration as solutions. Be a good analyst, but do not lose your way. There is a reason for your analysis, you use it to make good decisions.

The assessment of a solution's potential effects should be quantitative whenever possible. Qualitative assessment may be acceptable in earlier iterations of the planning process or when quantitative assessment is impractical or impossible. The assessed effects of a plan are appraised to determine if it is making sufficient progress toward achieving your objectives and avoiding your constraints. If it can solve your problems and realize your opportunities, it qualifies for further consideration as an implementable solution. Solutions that don't make significant progress toward the objectives and constraints fail to qualify for further consideration. They should either be reformulated to better contribute to objectives and constraints or be dropped from further consideration.

You evaluate each plan because you want good solutions. It's important to make sure your candidate solutions are good ones. Advancing a bad plan could waste vital resources, it could solve the wrong problem, or worse, it could lead to undesirable outcomes if it is ultimately chosen. Using a formal and structured evaluation process helps minimize the risk of a poor plan being implemented. The purpose of the evaluation step is to produce an array of distinct plans, any one of them is a viable candidate for solving the problems and achieving the opportunities in an acceptable manner. If planning was a beauty contest, evaluation is where the names of finalists would be read. This set of candidate plans will constitute the final array of plans that will be carried forward to the next step in the planning process, where they will be compared to one another.

There are usually three general types of iterations of the evaluation process, which is itself a screening task. The first screening task is to make sure the solution is not lacking any essential element (think preevaluation). Then there is the actual

KEY TERMS

Evaluation consists of assessment and appraisal. Assessment is the quantitative measurement or qualitative description of a plan's effects. Appraisal is the subjective weighing of the significance of the assessed effects. Effects can be assessed without appraisal when the intent is to simply describe plan effects without judgment.

screening of plans (think evaluation). Finally, there is the evaluation of a reformulated plan (think reevaluation).

No attempt is made to compare one plan to another; that will occur in the next step. A plan's effects are, generally, not compared to absolute targets as often as they are compared to threshold values. Evaluation results in a simple judgment that a specific plan is or is not a viable solution to the identified problems and opportunities. Those that are will be carried forward to the comparison step (Step 4). Although any number of iterations of the evaluation step is possible, the three iterations described next are quite common.

7.2.1 First Evaluation Iteration

How do you know when your ideas have become a plan? When your assembly of measures is complete, effective, efficient, and acceptable, you have a plan. These criteria usually include the "must do" hurdles each plan has to leap to become a complete plan. Once it is established that a solution meets these criteria, they tend not to be explicitly considered again in subsequent planning decisions. The first evaluation iteration makes that determination and is described in Section 7.3.

7.2.2 Second Evaluation Iteration

Unlike the first evaluation iteration, this task does not have a prescribed set of criteria. The criteria will vary from context to context. To objectively evaluate the individual plans, planners must discern which criteria make a real difference, and which criteria are simply "nice to have." Once it is established that a solution meets the qualifying criteria that make a real difference it will then be compared to other qualifying plans.

The challenge at this point is to identify criteria that will enable you to accurately determine the plan's chance of success if you have not already done so in the first planning step. Then you must identify metrics for these criteria that enable you to make that judgment. Decision making is fundamentally and ultimately a subjective task, but the evaluation process should be as objective as possible. This means basing it on good evidence, good models, and good analysis and tying the decision to the objective assessment of this information. Ideally, the evaluation criteria will be included among the decision criteria identified in Step 1. Constraints tend to be more prominent as evaluation criteria than in comparison and selection, where objectives and other decision criteria are more prominent.

7.2.3 Third Evaluation Iteration

Plans that fail to qualify for consideration as a possible solution to the study's problems and opportunities may be discarded or fixed. An unqualified plan would be one that fails to meet the subjective approval of planners for some very specific reasons. A plan may be reformulated, that is, tweaked or polished in ways that address the shortcomings that led to the plan's rejection. Once a plan has been reformulated, it would be reevaluated, the third iteration. Reevaluation may focus exclusively on the plan's originally perceived shortcomings rather than on a complete evaluation. The three iterations of the evaluation step may, themselves, be repeated as many times as necessary.

7.2.4 Criteria Revisited

Step 1 is where the decision criteria are first identified. That task, like most others in planning, is iterative, and it is not likely you'll get it right on the first try. But now you begin the planning steps, where these criteria become increasingly important, so it is time to revisit them. Figure 7.1 provides a graphic depiction of the potential relationships among the various criteria and effects identified in the analytical tasks of the planning process. Ideally, planners identify as many of these criteria as possible in Step 1.

The largest ring represents the evaluated effects of plans. These include benefits and costs, physical dimensions, construction noise, and virtually everything and anything that can be described or quantified in the analyses along the way. Notice the darkest ring inside the large ring. Those are the selection criteria, where this discussion is ultimately headed. It's important to understand how all these concepts hang together, and you'll be hearing about them over the next few chapters.

Some of the evaluated effects are used as evaluation criteria (the size of the rings are not meaningful in any ratio way). Another subset of the evaluated effects is shown as the comparison criteria. This is drawn to suggest these criteria may or may not overlap with the evaluation criteria. The decision criteria are a subset of the evaluated effects, and they may include some of the evaluation and comparison criteria. They may also include effects that were not included in either of those prior criteria sets.

The selection criteria are those that lead to the identification of the selected solution. Ideally, these are a subset of the decision criteria identified in Step 1. The circle is located in such a way as to suggest some criteria in this set may or may not have

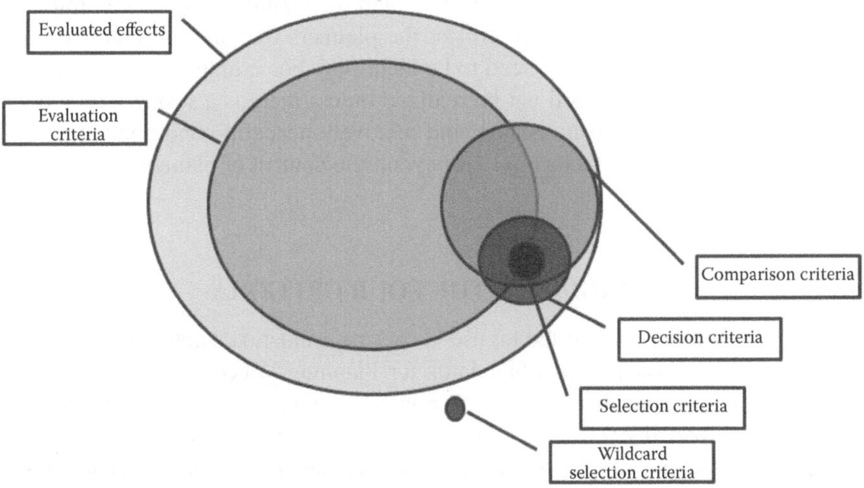

Figure 7.1 The potential relationships among effects and criteria in the planning process.

been considered in previous decision steps (evaluation and comparison criteria or evaluated plan effects). Note the possibility that a criterion might emerge that was not anticipated by the study team, these are called wildcard criteria and they are generally identified by decision makers outside the planning process. Here is the take-away point: the criteria you use for evaluation may be used only for evaluation and never considered again, or some (or all) of them may be reused for subsequent decision making in the comparison or selection steps.

7.3 HOW TO EVALUATE PLANS

The seven evaluation tasks presented in the introduction to this chapter are discussed at length in the pages that follow.

7.3.1 First Evaluation Iteration (Is It a Plan Yet?)

How do you know when a solution is complete? When is it a finished plan? This very first evaluation iteration evaluates a formulated candidate solution against some preliminary criteria to determine if it is finished and ready for a full quantitative evaluation. The first iteration evaluation criteria are completeness, effectiveness, efficiency, and acceptability (see the following text box). This first evaluation is likely to be qualitative.

Completeness means the solution will work. A complete plan is one that is well thought out. All the necessary measures and implementation actions needed to make the plan successful have been accounted for in the formulation process. Ask yourself, is everything present that is needed to make this plan work as desired? Will this plan produce the desired outcomes? If so, it is complete. If not, it needs to be returned for more formulation. A complete plan is implementable.

Completeness can also mean considering things beyond the planners' control as well as things beyond the scope of the study and the planning entity's commitment. If there are factors beyond the control of the planners that are necessary to make the plan's effects a reality, they need to be identified. For example, if the benefits of a channel-deepening plan will not be realized unless there is a strong international economy, dredging of private berths, and relatively peaceful conditions in the oil-producing nations, these factors are all beyond the control of planners and must be

ORIGINS OF THE FOUR CRITERIA

The four criteria suggested for use in the first evaluation iteration were introduced in the "Principles and Standards for Planning Water and Related Land Resources" of September 10, 1973, as revised September 29, 1980 (U.S. Water Resources Council 1973). This policy guidance for federal agencies involved in water resources planning required that alternative plans be formulated in consideration of the four tests of completeness, effectiveness, efficiency, and acceptability.

identified as significant sources of uncertainty. Decision makers must understand the plan is not complete enough to produce the desired results unless these conditions are met.

Effectiveness. Effectiveness means the solution will make progress. An effective plan is responsive to the wants and needs of people. It solves problems and achieves opportunities. An effective plan contributes significantly to planning objectives. The most effective solutions make significant contributions to all the planning objectives. A viable plan has to be "effective enough," an imprecise matter of degree that is most often a subjective judgment. A plan that fails to address the more important objectives is not effective. Neither is a plan that violates important constraints.

Efficiency. Is there a less-costly way to accomplish the same planning objective outcomes? Is there a way to accomplish more progress toward the planning objectives for the same cost? If the answer to either of these questions is yes, the plan is not efficient. An efficient plan is one that does not waste resources. It does not cost more than the value of the outputs it produces and it is cost-effective. There is no obvious way to get the same results with fewer resources and no way to get greater results for the same resources. A plan that meets a given level of objectives in the least-costly fashion or that maximizes contributions to objectives for a given level of resources is an efficient plan. This is also likely to be a subjective evaluation intended to avoid egregious and obvious violations of this efficiency criterion.

Acceptability. Is everybody (reasonably) happy? Is the plan reasonably acceptable to the planning partners? Is it acceptable to those stakeholders whose support is needed for implementation? Is the plan compatible with existing laws, policies, and authorities? Acceptability means there is no clear reason the solution will be stopped—it raises no red flags. A plan has to be doable. It does not mean everyone must like it. There are many legitimate factors that can render a plan infeasible. These factors can be technical (engineering or natural-world limitations), economic, financial, environmental, social, political, legal, and institutional. Opposition to a plan does not make it unacceptable, that simply makes it unpopular. Nonetheless, there is no sense to give serious consideration to a plan that is unacceptable to someone whose support is needed for implementation.

There is no reason to further evaluate a solution if it fails to meet any one of these criteria. A solution that is incomplete won't work. There is no point to an ineffective plan. Resources are too scarce to consider an inefficient plan. An unacceptable plan is a nonstarter. The real challenge is in determining thresholds for these criteria. They are usually qualitative, if not totally subjective, criteria that are going to be judged on a continuum rather than quantified in any meaningful way. When planners judge a plan to meet these criteria, they should carefully and concisely document their rationale and the evidence it is based on in a simple narrative. Any solution that fails to meet these criteria clearly does not qualify for serious consideration as a solution to your planning problems and opportunities. Plans that fail to meet these criteria should be fixed by reformulation or dropped from further consideration.

7.3.2 Prepare With Condition Scenario(s)

The with condition scenario describes what the future will look like if a specific solution is implemented to correct the problems and realize the opportunities. There is a separate and distinct with condition for each solution. Thus, Plans A, B, C, and D will lead to the forecast of separate and distinct future scenarios. Scenario A will describe how the study area will look if Plan A is implemented; Scenario B will describe the future with Plan B in place, and so on.

Although each solution has a different with condition, the same without condition is used to evaluate each plan considered in evaluation. That is, each plan's performance is compared to the very same without condition scenario. Developing a with condition scenario for each plan is a significant evidence gathering and analytical task in the evaluation step. One of the more important practical judgments planners must make is how many with condition scenarios are required to adequately bound a single plan's uncertain impacts on the future. When it is not possible to adequately represent the uncertainty about the future with a single scenario, multiple with condition scenarios may be used for each plan evaluated. Most of the time a plan is well represented by a single most likely future scenario.

The with condition should be noticeably different for each plan if you've formulated them well. When the results of the with condition analyses look very similar, that's often a hint that your planning process might benefit from some more divergent thinking. There is always going to be uncertainty in your scenarios and the analyses done within them. Significant uncertainty about your plan's effects should be described and accounted for whether you use a single most likely scenario for a plan or a scenario planning framework (see Chapter 13) with multiple scenarios.

When major uncertainties could significantly affect the nature and direction of the without condition future, they may also affect future with condition scenarios. For example, consider a planning effort with sufficient uncertainty about climate change and energy that it identifies the four future without condition scenarios shown in Figure 7.2 rather than a single one.

It is likely these same drivers could influence the performance of your plans. If you've already decided to use a scenario planning approach to handle future uncertainties, it is important to consider whether there will be multiple with condition

A SIMPLE WITH CONDITION EXAMPLE

Consider a flood risk management study. The without condition calls for a future with repeated flooding, a shaky economy in the floodplain, and deteriorating social conditions in the neighborhoods most frequently flooded. Now imagine a with condition that calls for a levee to prevent flooding. The levee is also expected to lead to regentrification of two of the three neighborhoods. The economy will stabilize but not grow. Quantifying these differences or describing them qualitatively is an essential part of the evaluation process.

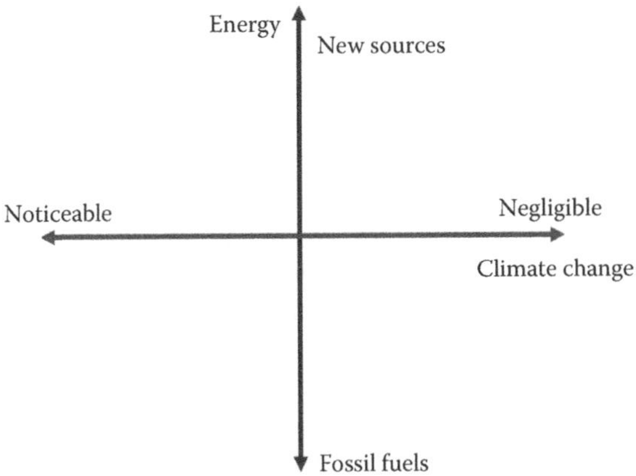

Figure 7.2 Four potential without condition scenarios defined by two drivers of future uncertainty.

scenarios or not. This raises special challenges in qualifying a plan that are taken up in Section 7.3.4. The outputs of this task include a decision to use a single most likely scenario or multiple with condition scenarios as well as the development of each scenario needed.

7.3.3 Conduct Necessary Analysis

Once you develop the basic future scenario, it's time to "occupy" that scenario and conduct the necessary analyses within that scenario framework. As was the case for the without condition, planners must conduct relevant analyses of problem, opportunity, resource, and other conditions in this altered future. A common focus for analysis includes the benefits and costs of an alternative plan. This will require significant evidence gathering and analysis. In most instances, the with condition analysis will parallel the without condition analysis. So for example, if future land use was projected for the without condition, it would be projected for the with condition if it would change with your plan in place. Hydraulic and hydrologic analyses, economic and environmental investigations, and the like conducted in the without condition would all be reanalyzed if affected in any way by the plan. Presuming that the analysis done for the without condition scenario concentrated on important resources, impacts, and decision criteria, that same analysis would be repeated for each with condition scenario.

A SIMPLE EXAMPLE OF ANALYSIS

Imagine the with condition scenario associated with a levee to reduce flood damages. Hydraulic analyses will have to be redone to reflect the presence of the levees. Expected annual damages (EADs), the economic measure of flood damages, must be re-estimated to reflect the presence of the levee and resurgence of the two neighborhoods. Housing values and tax revenues will change. The job base will differ by the numbers and types of jobs preserved by the plan. All these things must be estimated as plan effects.

To help envision the analyses conducted within the with condition scenario, there are some helpful questions to consider. What aspects of the study area's future will your plan change? What will be different in the future because of your solution? Which of these differences are important to your stakeholders, publics, and decision makers? How will measures of your objectives and your decision criteria be affected by this solution? Answering these questions will point you toward the effects you need to quantify in your analysis. Ideally, these effects will be anticipated and identified in Step 1.

Don't forget your evaluation criteria. Be sure to analyze the state of your evaluation criteria and other important metrics with your solution in place. These often include things a plan must do that can then be set aside once a plan is qualified as a finalist. For example, a plan should not adversely affect endangered species. Once you have established that it does not, you need not consider this criterion further in the later planning steps.

Evaluation criteria are essentially qualifying criteria. What must a plan do to warrant serious consideration as a solution? You need to make those criteria explicit. Evaluation is an important part of decision making in planning. It should not be a loose, undocumented process. Tell us why you qualified some plans but not others.

7.3.4 Compare Without and With Conditions

This is the scenario comparison task you use to assess the differences a solution makes. A plan is qualified for consideration as a potential solution based on changes in evaluation criteria observed when comparing between the without condition and with condition scenario values of a criterion for a specific solution.

Imagine a hypothetical Plan A for which we have identified three evaluation criteria: habitat units (HUs), expected annual flood damages, and cost. A qualitative assessment of this plan might say it reduces habitat and flood damages at a substantial cost to the community. HU losses are expected to decrease, flood damages are expected to decrease, and costs are expected to increase. The true purpose of this step is to identify the important points of difference and similarity between the without condition and with condition scenarios.

For scenario comparisons to be useful for decision making, they must identify differences in scenarios that make a difference; that is, they must show things that are important and that matter to decision makers. That means planners, with input and feedback from stakeholders and decision makers, need to carefully identify evaluation criteria. These are the measurable criteria upon which the plan qualifying decisions of this step are based. As noted previously, these metrics may reflect some of the planning objectives. As the evaluation step plays an important screening role in decision making, it is possible, if not likely, that the evaluation criteria will differ from the final decision criteria. Scenario comparisons must begin by comparing things that matter, and different things may matter at different decision points in the planning process. That is, evaluation criteria may differ from comparison criteria, which may differ from the selection criteria. Or, they may all be the same.

As noted earlier, this assessment task does not have a prescribed set of criteria in the way the first evaluation iteration does. The evaluation criteria used should enable planners to objectively evaluate the differences an individual plan will make in the future as compared to a future without a plan in place. As we have noted, the effects should be quantified whenever possible. The challenge at this point is to identify criteria that will enable the planning team to accurately determine the plan's chance of success.

In best practice planning, success is directly tied to the achievement of planning objectives and the avoidance of planning constraints. When these objectives are met and the constraints are avoided, the identified problems will be solved and the opportunities realized. Thus, it makes sense that a plan be qualified for further consideration based to some extent on its objective attainment and constraint avoidance. It is indeed common for evaluation criteria to include some subset of metrics based upon objectives and constraints.

The evaluation criteria may be a subset of the decision criteria identified in the decision context framing of Step 1, or it may be a different set of criteria. If the formulation of solutions is based on objectives, the planning team will, pragmatically, have a good feel for the extent to which a candidate solution will contribute to objectives. Consequently, it is not unusual for the planning team to use a set of criteria for the individual assessment of a plan's contributions to realizing a more desirable future that do not heavily reflect the objectives. The criteria assessed in this screening task need not be the same as the criteria that will ultimately be used to select a plan.

There are three basic comparison methods. To simplify the discussion, let's focus on a single decision metric. In actual practice, a scenario comparison involves multiple evaluation criteria, each of which would be compared in the manner described next.

The simplest scenario comparison is the before and after comparison shown in Figure 7.3. This takes an estimate of an evaluation criterion before any plan-related

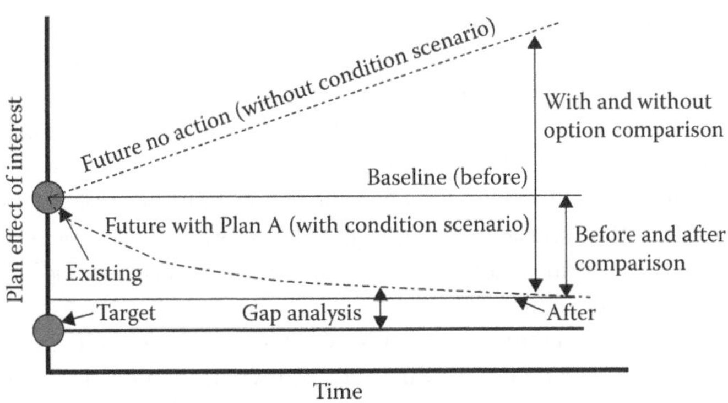

Figure 7.3 Methods of scenario comparison.

changes are implemented (the baseline scenario) and compares that value to the criterion value that would occur after the plan is in place and functioning (the with condition scenario). The difference between these two values is calculated in a before and after comparison.

The use of this comparison method is widespread. Its greatest weakness is that a baseline (before) value does not take into account changes in the effect that would naturally occur over time. In dynamic systems, effects may increase or decrease with environmental or other factors. A solution's effectiveness may also vary depending on the phase-in period for the plan and, for example, compliance rates related to the plan. In the case of mechanical or other physical solution measures, their effectiveness may change over time with wear and tear.

To account for these kinds of changes, the without and with comparison, seen in Figure 7.3, is preferred. The figure shows that if no additional action is taken the effect will increase and, in this example, grow worse over time. This is called the "without condition." This without condition value is obtained by analyzing the effect of interest under the conditions described in the without condition scenario.

The "with condition" shows a future in which it takes some time to realize maximum reductions of the undesirable effect. This could be because measures that comprise the solution are phased in over time or it could reflect the fact that those required to implement the plan comply at different times. It is obtained by analyzing the effect of interest under the conditions described by the with condition scenario.

Under the without and with condition scenarios comparison, a proper analysis would have to estimate the changes in the effect over time. The original baseline estimate provides a reference point. If, for the convenience of this argument, we consider the baseline the before condition and the lowest position of the with condition as the after condition, it is easy to see that the previously presented static before and after view of the solution's performance provides a different picture than the without and with condition comparison. In general, the without and with condition scenarios comparison is more accurate and preferred.

A third kind of comparison is predicated on some higher authority or the planning team establishing a target for the effect of interest. For example, a legislative body might establish a water quality standard. Once a target is established, planners try to hit the target by creating a plan that yields a with condition scenario that is closer to the target than the baseline or without conditions. When the target is ambitious, some solutions may fall short of the target, establishing a gap between the desired level of performance of the plan and its actual performance. Gap analysis is a comparison technique that focuses on the distance between the desired target and the actual performance. Frequently, when an option falls short of the target, additional measures will be considered to mitigate this short fall. The effect paths shown in Figure 7.3 were arbitrarily selected and many other patterns, including exceeding the target effect, could occur.

Without and with condition scenarios comparisons may occur in a number of different ways. The most common situation is shown in Figure 7.4. There we have a single most likely future without a plan in place and a single most likely set of future

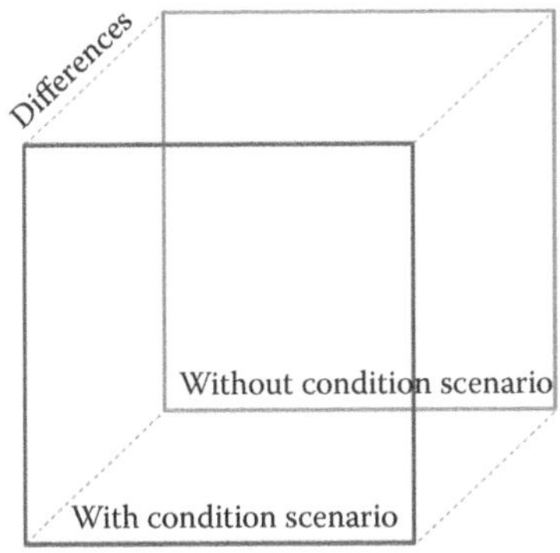

Figure 7.4 Mapping a single with condition scenario to a single without condition scenario.

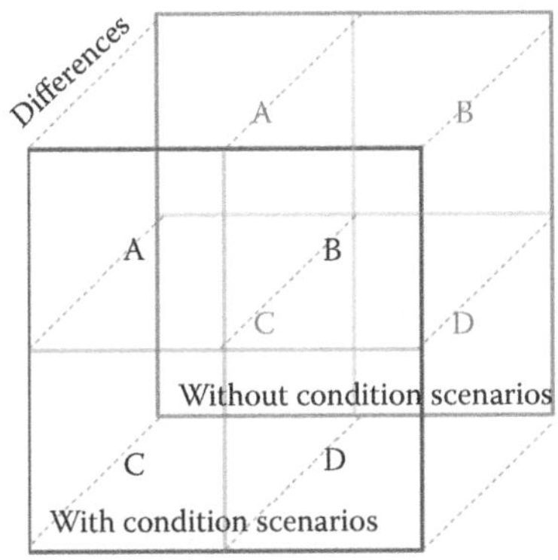

Figure 7.5 Mapping multiple without and with condition scenarios in a scenario planning context.

conditions with a plan in place. There is one scenario for each condition, so comparing them simply maps differences in effects from one scenario to the other.

The second most likely comparison of without and with condition scenarios is shown in Figure 7.5, where we see multiple scenarios (A, B, C, D) for both conditions in a scenario planning situation. In this instance, planners will align the like future scenarios with one another and complete a four-part assessment and appraisal. With condition A gets mapped into without condition A, and so on. It is conceivable that a situation could arise when any one of the four with condition scenarios could map into any one of the four without condition scenarios, giving rise to 16 potential comparisons. Pray this never happens to you.

Another conceptual possibility would be the asymmetric alignment of these conditions, as seen in Figures 7.6 and 7.7. Figure 7.6 shows a single with condition scenario and multiple without condition scenarios. This outcome will be realized when a plan has the same basic effects no matter what the future scenario is.

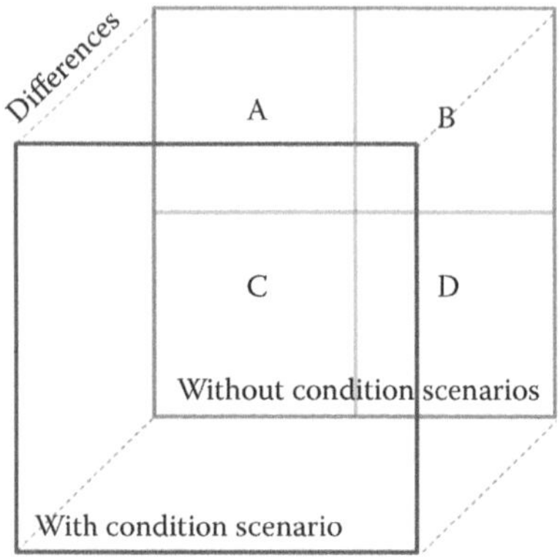

Figure 7.6 Mapping a single with condition scenario into multiple without condition scenarios.

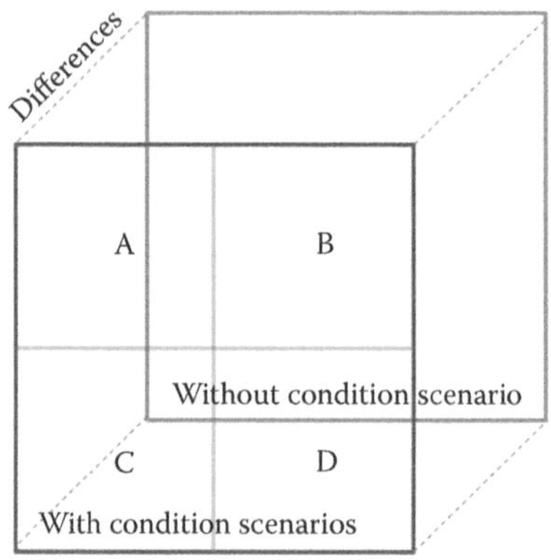

Figure 7.7 Mapping multiple with condition scenarios into a single without condition scenario.

This is not easy to imagine but the possibility is allowed, as is the possibility shown in Figure 7.7, where there is one without condition but multiple with condition scenarios. This hypothetical situation is a little easier to imagine. New technology with significant uncertainty concerning its performance, especially the potential for unanticipated consequences, could create multiple future with condition scenarios.

7.3.5 Assess the Differences

The assessment task requires planners to measure changes identified by one of the comparison methods described above. Assessing is a measurement activity. It is an evidence-based task where planners say how much things change quantitatively, if possible, and qualitatively, if not. This analytical task calculates the estimated effects of interest. The without and with condition scenarios usually provide the comparison framework. Assessing the differences in evaluation criteria is a relatively trivial task once the values of interest have been estimated for the without and with condition scenarios.

Table 7.1 provides an example of a simple assessment of the differences between two scenarios for a single plan using three evaluation criteria. Without a plan in place, the most likely future for the study area includes 2500 HUs, expected annual flood damages of $10 million, and with no plan there is no additional cost. Plan A reduces flood damages at a cost of $50 million and it destroys habitat to do so.

The HUs and EAD are analyzed first under the conditions that define the without condition future of the study area. Plan A would alter conditions that define the future, producing a new scenario, the with (Plan A) condition scenario. Using those scenario conditions as the starting point, HUs and EAD would be analyzed anew. If Plan A is implemented, HUs will decrease by 2000 to a total of 500. Plan A reduces flood damages significantly from $10 million in EAD to $1 million. The plan costs $50 million.

Comparing values for the two scenarios, the changes noted in Table 7.1 are observed. HUs decrease by 2000, flood damages decrease by $9 million in EAD, and costs increase by $50 million. Differences between the without and with condition scenarios like these need to be estimated across the entire range of evaluation effects so the planning team can judge the most likely effects of the plan. The differences are the effects evaluation relies upon to qualify and eliminate solutions.

This assessment is an analytical step that precedes and supports a significant planning decision. As such it should be as objective as possible. That means it should be based on good evidence, good models, and good analysis. It should also reveal the significant effects of uncertainty on the evaluation of plans.

It is common for the evaluation criteria to include threshold effects at times. National, state, and local legislation may require a planning activity to meet certain standards and to adhere to specific laws. These thresholds might include such things as not violating the provisions of the Endangered Species Act, the Wild and Scenic Rivers Act, the Clean Water Act, and the like. In general, the evaluation criteria are a larger set than the decision criteria, in part because plans may be qualified on the basis of contributions to objectives and meeting critical threshold requirements. Selection decisions tend to be made on a narrower set of criteria.

Estimates of plan effects should be presented in a way that enables decision makers to identify the evaluation criteria and the changes in them attributable to the

Table 7.1 Simple Scenario Comparison

	HUs	Expected Annual Flood Damages	Cost
Most likely without condition	2500	$10,000,000	$0
Most likely with condition for Plan A	500	$1,000,000	$50,000,000
Differences attributable to Plan A	–2000	–$9,000,000	+$50,000,000

SAMPLE UNCERTAINTY RATING DEFINITIONS

High—There is little to no concrete evidence available, or there is a very broad range of possible outcomes that include extremes.

Medium—There are some good evidence and some significant data gaps, or there is a broad range of outcomes that does not include extreme values.

Low—Good evidence is available, data gaps are not significant, or there is a limited range of possible outcomes.

None—All relevant facts are known.

plan being evaluated. Displays of effects should help decision makers understand the significance of the uncertainty that remains in the assessment of plan effects. This uncertainty can be conveyed qualitatively or quantitatively. Qualitatively, it is usually done in a narrative or by using an uncertainty/confidence reference. A narrative typically identifies the critical uncertainties that could affect the solution's performance. This would include a description of the key assumptions that have been made to get around these uncertainties as well as any efforts underway to further reduce the uncertainty. An uncertainty/confidence reference is a notation that accompanies each assessed plan effect. A simple example of a confidence reference would be: HU = high uncertainty, MU = medium uncertainty, LU = low uncertainty, and NU = no uncertainty. Such scales are generally ad hoc measures that are of dubious value unless accompanied by an evidence-based explanation for the rating.

Uncertain plan effects can also be presented quantitatively. Figure 7.8 summarizes the extent of the uncertainty attending the estimate of HUs lost as a result

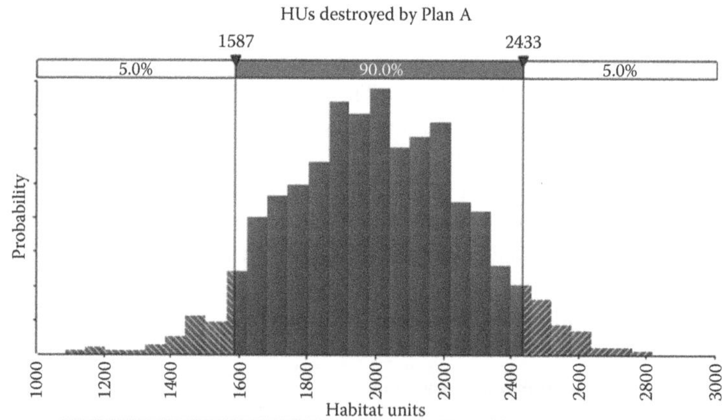

Minimum	25th percentile	Median	Mean	75th percentile	Maximum
1082	1816	1996	2000	2183	2820

Figure 7.8 Variation in the estimated number of habitat units destroyed by Plan A.

of Plan A. The point estimate reported in Table 7.1 was the mean. The distribution of Figure 7.8 provides a much more complete picture of the uncertainty about that value. The summary shows that HUs destroyed could range from 1082 to 2820. The assessor's confidence bands, shown by the vertical delimiters reveal 90% confidence the actual number is between 1587 and 2433 HUs lost. Effects assessed with uncertainty accounted for provide much more honest estimates of the effect of a plan. They also present decision makers with a more challenging appraisal decision.

7.3.6 Appraise the Differences

Appraising the assessed differences between a plan's with condition scenario effects and the without condition scenario effects is a judgment task. It amounts to a subjective weighing of the objective assessment of differences. Subjective judgments require some sort of reference point to support the judgment. The reference points may be formal targets established by a process external to the planning study. Water and air quality standards provide easy examples, as does a benefit-cost ratio of at least 1. Existing or without condition scenario values also provide convenient reference points. Beneficial effects of plans provide more of the good things and less of the bad things than the existing or without conditions do. Adverse effects of plans provide more of the bad things or less of the good things.

Let's begin with the point estimates of Table 7.1. If Plan A is implemented, HUs will decrease by 2000 to a total of 500. This is an undesirable outcome and is one argument against the plan. On the plus side, Plan A reduces flood damages significantly from $10 million in EAD to $1 million. This is an argument in favor of the plan. The plan comes at a hefty cost of $50 million. Evaluation requires planners to weigh these effects and make a subjective determination whether it is a viable solution or not. In this hypothetical case, planners might decide the adverse environmental impacts disqualify the plan from further consideration in its current state. It could be reformulated by adding some habitat enhancement features or eliminating conditions that lead to the adverse consequences.

Introducing uncertainty to the evaluation can complicate matters. Consider the uncertain information of Figure 7.8 for the moment. Is a plan that destroys between 1582 to 2433 HUs with 90% confidence good enough? Does the range of impacts matter? Appraisal means looking at the assessed effects and judging them, either individually or in balance, to be acceptable or unacceptable. Even the best-case loss of 1082 may be too substantial for a plan to proceed to the finalist list, depending on the deciders' reference point. If the reference point (an informal target) is no loss, this is unacceptable. If the benefits of a plan that causes these losses are substantial, even the maximum loss may be acceptable. It is not difficult to imagine a situation where a low-end estimate may be tolerable while a high-end estimate is not. Imagine a different example where the plan could either cause a loss or a gain in habitat—then decision making gets more difficult.

Now, let's change the subject. When explicit thresholds (or targets) for plan effects exist, state them. Federal water resources planning, for example, for years required a plan to produce net positive benefits to National Economic Development. If this threshold was not exceeded, a plan generally* would not qualify for consideration as a viable solution.

Planners have been known to establish their own more or less arbitrary thresholds for some effects. All evaluation thresholds should be identified and explained. For example, the team might decide a plan should destroy a maximum of 500 HUs or, when uncertainty is accounted for, they might say a plan must have at least, say, an 80% chance of destroying 500 or fewer HUs. The hypothetical plan shows a 100% chance of destroying 1082 or more HUs.

If there are no explicit thresholds or rules for appraising the assessed effects of a plan, then planners should describe how they judged the assessed effects to qualify a plan for further consideration as a solution. Describing whether the effects were considered individually or if the solution's overall effects were judged in a subjective balance is an important distinction to make when explaining the appraisal method used. Documenting the evaluation step should include identification of the criteria, thresholds used, if any, and an explanation of how a solution was appraised and the final appraisal, addressed in the next section.

7.3.7 Qualify or Not

The evaluation of an individual plan can end in a number of ways:

1. It fails to pass the first evaluation iteration. These plans are incomplete, ineffective, inefficient, or unacceptable. They are returned to the formulation step for finishing or fixing, or they are dropped from further consideration. Plans that pass the first iteration are screened as described in this chapter.
2. Plans appraised as having desirable effects on the future and acceptable levels of uncertainty are qualified for consideration in planning Step 4, the comparison of solutions. These plans are the finalists.
3. Plans appraised as having desirable effects and unacceptable levels of uncertainty are returned to the evidence gathering step to either reduce the uncertainty or to develop more effective means of addressing it.
4. Plans appraised as lacking in desirable effects can be returned for reformulation with instructions on where and how desirable effects ought to be increased.
5. Plans appraised as lacking in desirable effects can be disqualified from further consideration.

Disqualified plans are those that cannot be modified to produce enough desirable effects to warrant consideration as a recommended solution. All plans eventually should be either qualified or disqualified.

* There were policy exceptions to this rule, but they add more confusion than enlightenment.

7.4 TAKE AWAYS

Lesson 1: Evaluation is a screening step.
Lesson 2: Evaluation is an iterative process.
Lesson 3: Evaluation consists of assessment and appraisal of plan effects and elimination of unsatisfactory plans.
Lesson 4: Evaluation criteria probably will reflect objectives, but they need not.
Lesson 5: Evaluation criteria may include all, some, or none of the eventual selection criteria.

REFERENCE

U.S. Water Resources Council. September 10, 1973. *Principles and standards for planning water and related land resources.* 38 *Federal Register* 24778–24862.

CHAPTER 8

Compare Solutions

The primary purpose of this fourth planning step is to compare viable solutions to each other for the purpose of highlighting the differences among them that will make a difference to decision makers.

8.1 INTRODUCTION

Here is the short story. Evaluation has given you a set of plans, any one of which could be a viable solution for your problems and opportunities. You want to choose and implement the best plan. You can't get from a bunch of individually viable plans to the best plan without comparing them to one another. Comparison requires you to systematically look at what is alike and what is different about your plans. To be most useful, comparison looks for differences that make a difference to people and it displays them and the trade-offs among plans in a way that enables decision makers and others to see and understand them. There comparison stops. Comparison does not say what is better or worse. That judgment is reserved for the last planning step, decision making.

There are a few more things to know before leaping into this chapter. The comparison of plans enables planners to reveal the relative strengths and weaknesses of the candidate solutions. It is only by comparison that a once-viable plan is no longer good enough, or that a good plan becomes the best plan.

In this step, planners update the comparison criteria identified in Step 1 that will be used to compare the plans that have survived the evaluation process. When all the criteria are measured in the same units, like dollars, the comparison can be simple. Financial decisions are often based on choosing the alternative with the largest net benefits or smallest total cost. More realistically, multiobjective plan effects will be measured in a combination of dollars, habitat units (HUs), species abundance, water quality changes, noise levels, mitigated effects, navigation safety, changed erosion rates, or a host of other units, tangible and intangible. When that happens, the comparison task becomes more complex and inevitably involves trade-offs. A good comparison process is explicit and objective. It should also be transparent to informed planners and ideally to the planning effort's stakeholders. It should also be replicable and easy for the public to follow and understand.

This chapter begins by considering what it means to compare things in planning. It then moves directly into a discussion of a method for comparing plans.

8.2 COMPARING IN PLANNING

Webster's (1971) says comparison is "an examination of two or more items to establish similarities and dissimilarities." A good planning process produces an array of plans. That takes care of the two or more items. So what is it one might compare among plans? Let's begin with similarities. Find any characteristics the plans have in common. These comparisons can be qualitative or quantitative. Here are some logical things to compare:

- Physical features, location, and timing.
- Plan effects.
- Plan outputs.
- Plan's effects on decision criteria.
- Plan's contributions to planning objectives and constraints.
- Plan outcomes—will they solve problems and achieve opportunities?

Having passed the evaluation screening tests, all your plans are complete, effective, efficient, and acceptable. None of them violate the inviolable environmental statutes. That is all very important, but not very interesting. Knowing that the important hurdles have been leapt during formulation and evaluation, we need not consider them again. Although they identify what is alike about plans, similarities are often not very interesting.

Comparison is usually most enlightening when it highlights the differences among plans. Differences can be tangible or intangible. They can include things present in one plan but absent in others or effects that vary in their magnitude, frequency, or duration. Variations in the physical attributes of plans or temporal differences may also be interesting differences.

There is a virtually infinite supply of things to compare in a planning study, and quite frankly, no one cares about a lot of the differences. Will the side slopes, moisture content, or organic material content of a filter berm be important to people? Let's hope so. But let's also hope these things are only important to the people designing the filter berm. Few plans will be chosen because one has a filter berm with 30% moisture content by weight, while another has 60%.

You are going to find the differences among your plans is a very large set indeed, so it is useful to see subsets of differences. Figure 8.1 suggests some of the possibilities. The figure is suggestive, so don't read any significance into an overlap or lack of overlap. Note there are a great many things people do not care

DIFFERENCES

What physical features does one plan have that are missing in others? Does one plan produce effects or outputs other plans do not? Do some plans have no effect on a decision criterion, objective, or constraint that most other plans do affect? Will one plan have an outcome no other plan will have?

The same questions can be asked as a matter of degree. How do the attributes of one plan differ from another? Does one plan have a bigger, longer, more frequent effect on water quality than another plan? Will one plan produce more or different opportunities for recreation than another? Do the plans make varying contributions to your social justice objective? How much will species abundance change with this plan as opposed to that one? Do the species vary?

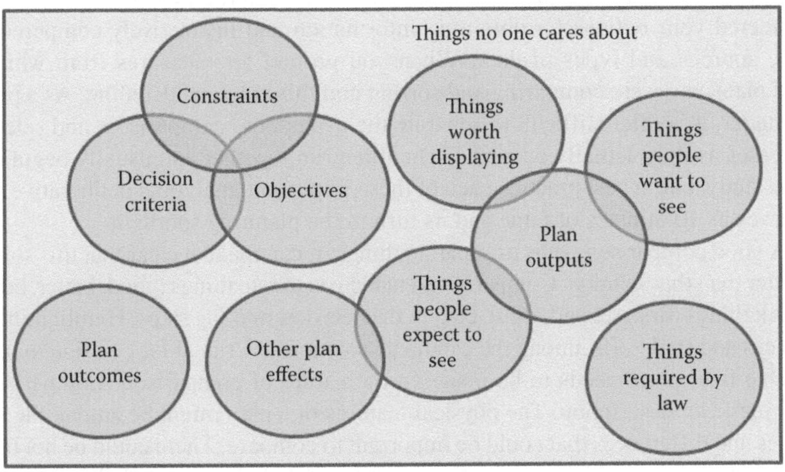

Figure 8.1 The set of differences among plans and some subsets they might form.

about, and this subset is indicated by the space outside the circles. Ideally, a good amount of the focus of your comparison step will be on the decision criteria. Even if you made an honest effort to identify them at the start of your study, there is a good chance some new ones will have made an appearance along the way. That is a normal part of an iterative process. One of the more difficult parts of the comparison step is making sure you compare the right things. Make that set too big, and you cause yourself a lot of work and may confuse the decision-making process. Make that set too small, and you will make enemies or maybe even run afoul of the law.

So, drum roll, here comes the punch line. The key to good comparisons in planning is to find the differences among plans that make a difference to people. In other words, the planner's job is to find and display important differences. Important differences are differences people care about. People care about things because experts have established that they are in fact important to care about. Other times people care about things just because they believe they are important, facts be damned. Then there are things that organizations have established as important through legislation, policy, administrative rule, and advocacy.

If you have done a good job establishing the decision context, you'll arrive at this step with a pretty good idea of what is and is not important to people in your planning investigation. Your decision criteria have to be in the comparison. Done well, they will reflect your objectives, and if your constraints were not taken care of during evaluation they may be reflected here as well. There will also be other effects and things people want to see, like costs, for instance. These need to be included as well.

When do you compare? It is natural for humans to compare things, it is one of the most common ways we think, and you have likely been making comparisons throughout the planning process. Comparison takes place consciously and subconsciously all the time. Back in Step 1 as you articulated your problem statement, you compared different ways to say things. When you gathered evidence, you no doubt considered your options for obtaining information and instinctively compared and chose sources and types of data. When you groped for measures from which to build plans you were comparing and sorting and thinking and deciding. As a practical matter, it is often difficult to separate the evaluation, comparison, and selection processes as they actually occurred. The three run together and usually begin early in the study, but in best practice each of these important analytical deliberative steps deserves its 10 minutes of fame and its turn in the planning spotlight.

A good comparison helps us separate things. It can make it clear that this solution is better than that solution. Comparisons enable you to rate things (good, better, best) or to rank things (first, second, third, etc.) in the decision-making steps. Highlighting differences and trade-offs among the candidate solutions is critical for decision making.

One final point needs to be made on the nature of comparison. Comparison is different from description. The physical features of a plan might be among the similarities and differences that could be important to compare. There could be hot button features for some interest groups that must be included (or excluded) and perhaps they need to be compared. However, there are many plan features, both tangible and intangible, that need to be described so people can understand the differences among the plans. That is not comparison, it is description. Summary sheets describing the various aspects and attributes of a plan may incidentally include similarities and differences. That is not comparison either. Comparison is a purposeful and intentional act to say, hey look over here, here are some things that are different (or the same) that you really need to know about. Do not confuse comparison with description.

8.3 HOW TO COMPARE

Comparing things is so instinctive to us it seems a tad arrogant to prescribe a process for doing it. When you compare things for yourself, you're free to use any method you like. When you make comparisons for others or in some stewardship role, as planners usually do, it is important to have a process so that comparison is evidence-based and transparent. Here are the tasks that comprise a reasonable comparison process:

1. Identify the things to be compared.
2. Find the differences among them.
3. Compare the differences.
4. Contrast the differences.
5. Identify trade-offs.
6. Explain the differences to people.
7. Display the differences with their attendant uncertainty.

8.3.1 Identify the Things to Be Compared

Ultimately, you are comparing alternative plans, but you need to identify the plan attributes, effects, and the like you will consider to do this. Let's call those things your comparison criteria. There are a lot of sets of criteria floating around a planning process. In Step 1, we talked about decision criteria. In the last chapter, we talked about evaluation criteria. Now we add comparison criteria. These may be different from the evaluation criteria, and they may be larger than the set of selection criteria* that will be used in decision making.

Decision making in planning is both iterative and evolutionary. You decide what's in and what's out of your study. There are many incidental decisions to make along the way, like what data to collect and what to ignore, what measures to use, how to combine them, how to involve the public, and the like. There are also more formal decision-making parts of the planning process where you decide what plans are worthy to consider as solutions and which solution is best. Many sequences of decisions are made but they all point toward the final choice, making a decision about which plan to recommend for implementation. And when a change or a surprise comes along, you may go back to the start and begin it all again.

* Selection criteria are, ideally, some proper subset of the decision criteria. If you were very fortunate, the decision criteria you identified in Step 1 include the evaluation, comparison, and selection criteria.

Some criteria may persist throughout the many decision iterations; cost may be one of them. Some may wax and wane like a full moon throughout the process. It does not matter how it happens so long as all the important criteria are considered at some point. Some criteria stand as hurdles; a plan cannot proceed unless the hurdle is cleared, but once it is cleared there is no longer a need to consider it. Planning constraints often function that way.

The bottom line is that once you get to this formal step, you need to drive a stake into the ground and identify those criteria that will be used to compare the plans. Some of the revealed differences will be important for decision making—these are the must-have criteria. They are the selection criteria, the set of criteria that will actually be used to select a plan for implementation. Your original decision criteria may include the selection criteria, or the selection criteria may evolve as the study progresses. In any event, these are most important to include in the comparison of plans.

You need criteria that measure progress toward each of the planning objectives that are not explicitly included among the selection criteria. Remember, your planning objectives define a successful planning process. You need to be able to objectively describe how different plans contribute to each objective. It is only through the attainment of your objectives that problems are solved and opportunities are realized.

If there are mandates, legislative or otherwise, that require certain effects to be shown, these must be considered. To these you can add as many "nice to have" criteria as you can afford to prepare. Remember to include things that are important to people. Some stakeholders will want to see the different effects of the final plans on the values of interest to them. For example, miles of walking trails may be very important to some stakeholder groups. Even if this is not one of the criteria decision makers will use to select a plan, it ought to be presented in the comparison because it is important information for stakeholders.

When identifying your comparison criteria, keep in mind that comparison is important for at least two reasons:

1. To support decision making, specifically the selection of a plan.
2. To inform and satisfy the public's varied interests in the plans.

At the completion of this task, let people know what you are going to compare and why. Explain the rationale behind your choice and the grounds for comparison. Let stakeholders know the comparison is deliberate and meaningful, rather than a collection of random effects that can be compared and contrasted. If you find yourself looking for things to include in your comparison, you have missed the point. The desired outcome of this task is a list of specific tangible and intangible criteria that will be compared and not simply described.

8.3.2 Find the Differences among the Comparison Criteria

To compare things they must be measured in the same way. When you evaluated each plan, you found the difference that plan would make between the without-condition future and the with condition future. This is an evaluated effect. These evaluated effects of plans are your primary source of information for comparison and decision making. To include a criterion for any kind of comparison, its effects must have been previously assessed* for each plan under consideration. These are the differences you must cull to find the differences that make a difference to decision makers.

It may sound obvious, but someone must make an effort to recognize the differences among plans for the comparison criteria. This can be as simple a task as juxtaposing evaluated plan effects, as is done in the example of Table 8.1. Simply putting facts together enables us to see they are different. Other comparison criteria may not lend themselves to such easy identification of differences. If one comparison criterion is the Irvington Community Association's preferences for each plan, there may be more evidence gathering to be done and it may be less revealing once obtained. For example, an association vote might only reveal they like Plan B best, which is sparse information about differences.

A desired outcome of this task is a table or other display that shows the differences among plans for each of the comparison criteria. Imagine that the team identified 15 variables to compare in the preceding comparison task. Members of the team would analyze and assess the criteria for which they are responsible to identify the differences. Differences must be searched for and found.

8.3.3 Compare the Differences

Once you have identified the significant differences, the first step is to compare them. In what ways are the differences alike? In Table 8.1, all the plans reduce expected annual flood damages and all the plans increase costs. The most obvious similarity in differences is in the direction of the differences. Order of magnitude is another attribute of the differences that is important to consider. Some differences tend to be the same order of magnitude, like the costs in Table 8.1. Suppose HU changes for three plans were –2000, –1997, and –2011. We might fairly say the differences are not important.

Similarities among the criteria can be summarized by statistical values like their mean or median values. The mean cost of the plans is about $38 million. Is that helpful? Two plans have above-mean costs, and one is below the mean. Sometimes, the differences can be bounded by the range. There is a $25 million range in plan costs. This indicates the plan costs are clustered within $25 million of each other. It can be very instructive to consider how the plan effects tend to be alike before focusing on contrasts.

Table 8.1 Finding Differences in Comparison Criteria by Juxtaposing Evaluated Plan Effects

	HU Changes	EAD Changes	Cost
Plan A	–2000	–$9,000,000	+$50,000,000
Plan B	+1000	–$1,000,000	+$25,000,000
Plan C	–300	–$6,000,000	+$40,000,000

* Recall that not every assessed effect is appraised.

Looking first at how the differences tend to be alike provides a more balanced perspective. Sometimes a careful look reveals that what appears to be a significant difference on paper may not be so significant in actual fact. Context can help as well. If the hypothetical study area of Table 8.1 started out with 10 million HUs, the changes shown in the table may suddenly be rendered trivial. Context is important and comparing differences before contrasting them can help provide a more complete context.

Comparisons are often more appealing to less quantitatively oriented people. Saying Plan A costs twice as much as Plan B can be more revealing than saying Plan A costs $50 million and Plan B costs $25 million. Using ratios is one way to compare differences.

8.3.4 Contrast the Differences

Another sense in which differences can make a difference, already alluded to above, is when they are significant differences. Differences in sign and orders of magnitude of effects can be important. There may be times when small differences in comparison criteria are significant. There may be some water quality criteria for which the differences in evaluated effects look trivial but are actually quite significant in terms of their impacts on marine life. Context can help people better appreciate the significance of the differences they are seeing.

Contrasting differences essentially relies on being able to see how and why the differences matter. One can say that Plan A reduces EAD from flooding by $8 million more than Plan B does, but is that a significant difference? Is it dramatic? What does it mean for people and their lives?

Numbers are numbers, and they do not tell the same story to everyone. To contrast the significance of the differences evaluated, a simple narrative may help. An example follows:

> Plan A is expected to prevent damage from a flood like the devastating flood of 2010 as well as any lesser flood. Plan B will prevent heavy thunderstorms from flooding low-lying roads in town but it will do nothing to prevent damages from any of the four floods the town has had since 1970.

Telling a story about what the numbers mean (and not worrying so much about what the numbers are) is one of the most effective ways to help people understand where the differences they see make a difference to things they care about. The output of the compare and contrast tasks is a clear indication of the differences among the plans that make a difference in terms of their relative magnitudes. Knowing what differences are (contrast) and are not (compare) important is an essential outcome of the comparison step. This output requires a combination of numbers and narrative.

8.3.5 Identify Trade-Offs

No one plan is likely to be better at everything. The usual case is for one solution to make more-desirable contributions toward some comparison criteria and less-desirable contributions toward others. When a solution gives you more of some things you want and less of other things you want than another solution does you face a trade-off. A trade-off is defined as a situation in which the only way to get more of one thing you want is to give up some of another thing you want.

In Table 8.1, Plans A, B, and C offer trade-offs. We'd like a plan that reduces flood damages and increases HUs. Plan A offers more flood damage reduction than B or C, but it also destroys more habitat. To get fewer adverse habitat impacts we can move from A to C, sacrificing flood damage reductions for habitat damage reductions. We can move further from C to B, again offering flood damage reductions in trade for habitat improvements. Plan B offers more habitat and less flooding than the without condition. The contributions to each are modest and there is no way to gain more of both of these values from the available plans. A trade-off has to be made to gain more flood damage reduction.

The example of Table 8.1 is unrealistically simplistic. That helps us see the trade-offs. In the messier real world, there are likely to be more criteria and far more subtle trade-offs. To examine trade-offs, the best way to begin is to look for dominant or dominated solutions. A dominant solution is one that offers more desirable effects for each of the selection criteria. In our hypothetical example, that is a solution that provides the most flood reduction at the lowest cost with maximum habitat creation. It would be unambiguously better on each criterion. This solution, when it exists, is going to be the selected solution. In fact, if you have a dominant solution, it is likely you need another iteration of the formulation process because you offer no choice in your array of final solutions.

Dominant solutions are rare. Slightly less rare are dominated solutions. A dominated solution is one that fails to make a more desirable contribution to at least one of the selection criteria when compared to another solution. Imagine a Plan D that destroyed 2500 HUs, cost $51 million, and reduced EAD by $0.5 million. There is not one single plan it scores higher on for any criteria. There is no reason to consider a dominated solution further. These should be deleted from consideration or reformulated to make them better on at least one criterion than another solution is.

When trade-offs exist, they should be identified. Pointing out plans that maximize desirable contributions to the selection criteria is useful. Identifying plans that minimize desirable contributions can also help. Plan A is the best at reducing flood damages, and Plan B is least costly and best for the environment. Plan C is less damaging to the environment and less costly than Plan A, but it provides less flood damage reduction. Plan C is more damaging to the environment and more costly than Plan B, but it provides more flood damage reduction. Someone is going to have decide what is more important, but that happens in Step 5.

Quantifying the trade-off can be helpful when it makes sense to consider trade-offs two criteria at a time. Plan B provides 3000 more HUs than Plan A at a loss of $8 million in EAD reduction. That means every additional HU costs $2667 ($8 million/3000 HU) in EAD. Now, the terms of trade are established. If we look at the cost of the trade-off between A and C, we see an increase of 1700 HUs at a cost of $3 million in EAD. This yields an HU at a cost of $1765 EAD. The cost of additional HUs in terms of flood damages is now explicitly known. The eventual choice may be made easier with such information. This kind of quantitative pairwise trade-off information can be useful when the trade-offs of greatest interest are known. Generating all the possible trade-off information for many plans and many criteria could overwhelm rather than help decision makers, however. Documentation formats that enable interested parties to manipulate the data and find the trade-offs of interest to them may be a practical value-added compromise between providing all this information and none of it.

8.3.6 Explain the Differences to People

If this seems redundant it is intentionally so. It is well and good for the planning team to prepare a comparison of the solutions under consideration, but they also have to help people understand it. A comparison should explain to people the importance of what they are seeing. They need to understand that if you are want to create more habitat you are going to have to live with more continuing flood damages than you would if you were willing to trade some habitat for less flooding. Tell the story of your comparison succinctly.

8.3.7 Display the Differences with Uncertainty

A large table that describes each plan is not a comparison table. Planning studies are rife with these tables. If the reader has the time, patience, and sufficient expertise, one can indeed derive a comparison from most of these descriptive tables. A comparison display should be different. The display or the text around it ought to figuratively say, "Hey, over here, look over here. This is important for you to know."

The table itself does not have to be complex. It could be as simple as an expanded version of Table 8.1. The purpose of a comparison table is to extract the important details from all the evidence gathered and distill it for the interested reader and the decision maker. This is a simple idea that will take a great deal of work and effort to accomplish.

What makes this task truly challenging is to represent the uncertainty associated with the assessed effects in the display of the information. This is an art in its infancy in planning. Most plans I have worked on or have seen still tend to treat evaluated effects with more awe than they deserve. In almost every instance, you will not know the true values of the numbers you attempt to display in your comparison, so admit it and help the reader while you are at it by letting her know how unsure you are.

Let's consider the HU effects of the hypothetical plans as an example. Table 8.1 indicated that Plan A lost 2000 HUs, Plan B gained 1000 HUs, and Plan C lost 300 HUs. These were point estimates that did not convey any uncertainty in the estimates. Figure 8.2 presents the HU changes per plan as uncertain numbers produced using a Monte Carlo process that captured the knowledge uncertainty and natural variability in these values. Notice the means are as indicated in Table 8.1.

What was not clear before was the extent of the uncertainty that remained in these estimates. Notice that Plan A could have losses substantially more or less than 2000 HUs. Figure 8.2 suggests it is even possible that Plan A could have losses smaller than the losses from Plan C, although this is unlikely. Plan C also shows some chance (26%) of actually producing gains in HUs. This is certainly a development that was not clear from Table 8.1. Likewise, Plan B could surprise people with HU losses. There is a 3.8% chance Plan B will result in negative changes. Graphics like Figure 8.2 could be used to replace the numbers in Table 8.1.

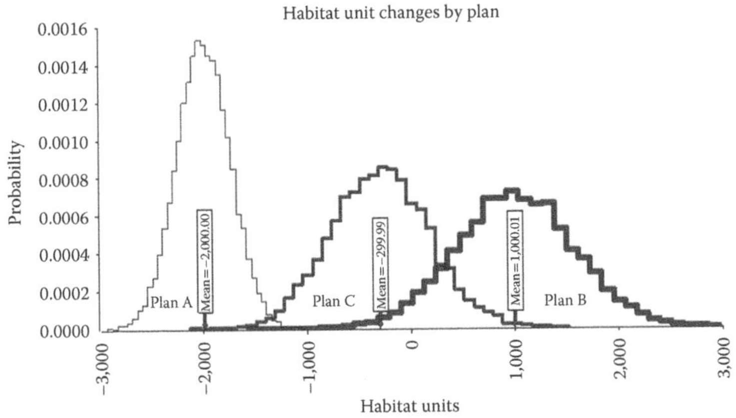

Figure 8.2 Comparison of evaluated habitat unit effects of three alternative plans.

Comparisons become more difficult when the uncertainty attending the effects is represented. The comparison step is an important opportunity for exploring and understanding the potential changes that are possible from the final array of plans. It is a time to make sure that you fully understand the effects of the plans that you're considering. It is especially important to consider unintended and "unexpected" consequences, as well as the expected effects. This is best done by considering the uncertainty that attends the evidence gathering efforts.

A well-constructed comparisons display can help decision makers and interested parties understand often complex information. When you compare things, you show how they are alike and how they are different. These similarities and differences can be displayed most effectively in a probabilistic assessment of the uncertainty attending a plan's evaluated effects.

8.4 TAKE AWAYS

Lesson 1: Comparison is the bridge between evaluation and deciding.

Lesson 2: Comparison criteria must include the selection criteria and ought to include contributions toward all the objectives.

Lesson 3: Comparison provides decision makers with the information they need to choose a solution.

Lesson 4: Comparison assures people see differences among plans that are important to them.

Lesson 5: Differences that make a difference are important. Consider what is alike about the differences (compare) and what is different about them (contrast).References

REFERENCE

Webster's 7th New Collegiate Dictionary. 1971. Springfield: G. & C. Merriam Company.

CHAPTER 9

Make a Decision

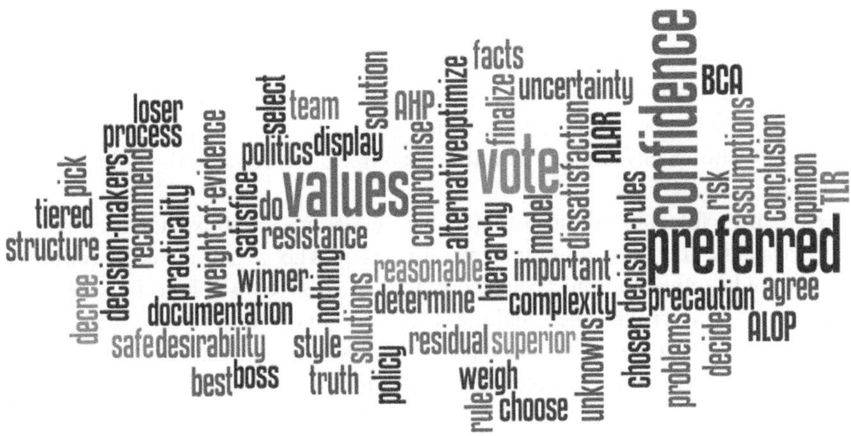

The primary purpose of this fifth planning step is to weigh the differences among solutions and select the best one for achieving objectives and avoiding constraints, thereby solving problems and realizing opportunities.

9.1 INTRODUCTION

Here is the short story. Making a decision is a political act. It is subjective. There is usually no one right answer for everyone. Today's decision may not be the same as next year's decision. It is political art, not science. It is sophisticated and informed guessing, but it is still guessing. This is usually the least analytical and most value-based step in the planning process.

A good comparison will present decision makers with a well-focused view of the important differences among the plans. Now someone must say which differences matter the most, how much they matter, and which plan offers the best solution. The tasks in this step are simple and deceptively complex. Decision makers must be found—hopefully, you've already done that—the selection criteria

FEDERAL RESOURCE PLANNING

National Environmental Policy Act (NEPA) regulations (40 CFR 1502.14(d)) require that no action always be considered a viable alternative. The no-action alternative is the default choice. The planning process is, in a sense, built on the default assumption that the federal agency should do nothing. The federal agency should become involved in a project of some type only if doing something is better for society than doing nothing. Hence, the planning process must convincingly establish that federal involvement in some project is preferred over no action. Do not overlook the importance of the first decision to be made at this step: should something be done?

must be confirmed, subjective weights may have to be assigned to the criteria, the importance of the remaining uncertainty needs to be considered, and a decision is made. What could be simpler? Sometimes, when it comes to values and subjective judgments, almost anything in planning is simpler. Welcome to the political realm of planning.

Here are a few more things to know before leaping into this chapter. This is, in many respects, the most important planning step. The wrong decision could lead to some undesirable and unanticipated outcomes. It's impossible to be sure you've made the right decision when making a decision under uncertainty. That is what makes being honest about what is and is not known throughout the study so important. Explicit consideration of the uncertainty does not erase it, but it does make us aware of its significance, and that can significantly increase the likelihood of making a good decision or, at least, avoiding a bad decision.

This chapter proceeds by considering the decision to be made in this step and then turns to the question of just who the decision makers are and how they make decisions. The bulk of the chapter is devoted to the development of a nine-step method that can guide the decision-making process.

9.2 THE DECISION

This decision comes in two parts. Is the status quo acceptable? If it is not, what should be done? If the status quo is not acceptable, is your best plan better than the status quo? Will it take the community to a future that is better than the without condition scenario? Is your plan better than the no-action choice? The final planning decision begins with this go/no-go decision. Take no action unless the analysis convinces you that doing so will bring the community to a better future. The first choice is always to do nothing. The second choice is to do something else. If you recommend a solution, which solution will it be? It's that simple. The decision is selecting a solution.

One way to consider what to do is through Beckhard and Harris' (1987) change equation, developed to help decision makers think about the strength of the case for change. They suggest that for change, that is, the do something option, to be successful the following equation must hold:

$$\text{Dissatisfaction} \times \text{Desirability} \times \text{Practicality} > \text{Resistance to Change}$$

If any one of the three left-hand-side variables is zero or low it will be difficult to implement a successful plan for change. Affected interests must feel dissatisfied with the current situation before a successful change can take place. The motivation to change will be weak without dissatisfaction. The proposed solution must be attractive to people. They need to understand what it is. If the planning team cannot project a clear vision of what things will be like with the plan in place, stakeholders are not going to want it. Furthermore, if the plan is not executable or if its effects are not realistic, the plan is not practical.

The resistance to change includes people's beliefs in the limits of the plan ("The situation is too far gone, this plan is not going to help"), stubbornness toward any change ("We don't intend to stop hunting"), and general inertia or lack of interest at the beginning ("Who cares?"). To make the go/no-go decision, determine whether there's enough dissatisfaction with doing nothing to justify making changes and whether change is desirable to people and practical to implement. If any of these three things is missing, or if resistance to change is very high, that means a difficult road ahead for any plan that is recommended.

If the first decision is to go ahead with a plan, then the next question is, "Is there is a law, a prescriptive policy, or a distinct process to follow?" If so, decision making is easy. Obey the law, follow the policy, and conform to the process. If the council vote decides the matter, that is how the decision gets made. If not, try the process that follows, unless you have a better method.

9.3 WHO AND HOW?

It's virtually impossible to work on a planning study and not have an opinion about the best course of action to take. Planning teams are often looked to for advice and maybe even a recommendation, but they do not often have final authority to make a decision. That actually belongs to someone else. The first part of this discussion considers some of the who's that make decisions. How they decide is sometimes a mystery; this mystery of how is the second part of the discussion.

9.3.1 Who Makes Planning Decisions

Let's begin at the most basic level. The planning decision is made by planners or someone else. As a planner, you probably won't make the decision. Worse, the decision maker may ignore your recommendations. That's okay, you're a planner not a decision maker. Never take it personally. Your responsibilities to the decision maker are as follows:

- Know what's important to the decision maker and provide the desired information in a useful form.
- Show what the plans do in a way that is understandable, honest about uncertainty, and useful.
- Be prepared to make a recommendation and to back it up.

If planners are not making planning decisions, then who is? The usual suspects include organization leaders. If, for example, a planning authority is conducting the investigation, the planning team may at some point hand the decision making over to the director of the authority. The director might be a planner, but he could be a civil servant or a political appointee.

Decision makers could come from the world of elected politicians. Governors, mayors, county executives, city councils, assemblies, senates, Congress, and even presidents could be making the final decision. Occasionally, the court system may be involved in the decision-making process. This kind of public-sector decision making has its own rich history. Here is a heads up for all you idealists out there: politicians tend to make political decisions. That does not mean they can't make good decisions or bad decisions, but, it may mean they are not going to be as impressed by the great job you did in evaluation and comparison as you are. They may be underwhelmed by the probabilistic analysis you used to quantify and describe the range of uncertainty you struggled honestly with. That's okay too. The big "P" of Politics beats the small "p" of planning every day of the week.

Some decisions may be made by elected, appointed, or otherwise assembled boards or panels that reflect various mixes of interests and expertise. There are a great many possibilities for subject matter experts and bureaucrats to become involved in the decision-making process. In some instances, the public may be empowered to make planning decisions. Those decisions may arise organically in a public involvement process or they might emerge in formal referenda.

Here is another perspective on these decision makers. The decision maker is really just another team member who just happens to have a special responsibility to decide. Let's call that role the decider. In best practice, the decider is not limited to selecting the best plan from among your final array of options. In fact, this book argues it is a mistake to limit their role to this final step.

Deciders have a role throughout the planning process. Table 9.1 suggests several points where decision-maker involvement would be useful. A large checkmark indicates a larger role for the deciders. Decision makers have an important role in establishing the decision context. They should be providing input and feedback throughout the problem identification tasks. Ideally, decision makers will have a large role in establishing the study objectives. One of their most critical roles will be to identify the most likely decision criteria as early in the planning process as possible. They need to be aware of the extent of uncertainty in the planning process because they might play an instrumental role in allocating resources to reduce uncertainty.

Table 9.1 Potential Roles for Decision Makers Throughout the Planning Process

Step	Decider's Role	Input	Feedback	Awareness
1	Identify problems and opportunities	√	√	√
	Identify objectives and constraints	√		
	Decision criteria	√		
	Uncertainty		√	√
2	Measures	√		√
	Plans		√	√
3	Plan effects			√
	Evaluation of plans	√		√
4	Decision criteria	√	√	√
	Differences that matter	√	√	√
5	Selection criteria	√	√	
	Formal weighting process	√		√
	Select a solution	√		

Decision makers need to be aware of the measures and plans under consideration to make sure that potential solutions important to them are being considered. Feedback on the initial set of plans formulated could be useful. During evaluation, decision makers should be made aware of any potentially significant effects of the plans. In some studies, they may have a role to play in deciding what constitutes a viable solution and hence in qualifying or disqualifying plans from further consideration.

During the comparison step, deciders can be of most help if they help confirm the criteria that will be used for decision making. Over the last two chapters I have begun to use the term "selection criteria" to make the point that decision criteria originally identified back in Step 1 may or may not be the actual criteria used in Step 5. The decision criteria are those criteria the team believes to be critical for making this final planning decision. The selection criteria are the criteria that are actually used to make the final decision. They are likely to be a subset of the decision criteria you identified in Step 1. They may or may not be equal sets depending on how much the decision process has been mandated in advance of the decision, how involved deciders are in the study, and whether the things that are important in your study are fixed or fluid. As the planning team zeroes in on the differences that matter to people, decision-maker input, feedback, and awareness become critically important. Decision-maker involvement is important throughout this last planning step.

No matter who the decision makers are, be prepared for problems. You will not always, or perhaps even often, get the kind of decision-maker involvement you'd like. Top-level officials and elected leaders are often too busy to remain involved in your study efforts and your options for forcing your way onto their agendas may be limited. But, even when deciders are engaged with you, there can be problems, because these deciders can change throughout the life of a study. New deciders can render old agreements meaningless. They bring new values. They like their ideas. They ask questions you did not anticipate and want information you do not have. A changing decision context has frustrated many planners. Keep in mind this is a valid part of the iterative nature of planning.

Even when the decision makers do not change, they can change their minds. They change their selection criteria, and they don't always tell you. They don't always know what those criteria are themselves. More than one leader has liked to lead from a gut feeling. Worse, there may be no such thing as the decision maker. It may be a body or an amorphous group with shifting identity, leadership, and interests.

Decision making may be tiered. In my career with the U.S. Army Corps of Engineers, the Chief of Planning had to be satisfied before he went to his boss, the district's Colonel, who had to be satisfied before he went to the division's General. The General went to the Chief of Engineers, who went to the Assistant Secretary of the Army, the Office of Management and Budget, and eventually Congress and the President. Each tier may operate by its own criteria and you will not have access to many tiers in most cases.

Decision makers may ignore what is important to you and to others. They may make unholy compromises that you do not understand. That is all part of the job. Planners plan. Deciders decide.

9.3.2 How Decisions Are Made

How is the selection made? Sometimes, it is a single event by one person: "The Governor said she wants to...." Other times it is a group event. The board of directors, county commissioners, city council, or some similar group reaches a decision. Other decisions are more tiered; consider them like a sequence of reviews and confirmations, like the Corps process described earlier. There is no one way a decision is made. The point is you should invest some time in understanding how decisions will be made for your study. This is an important part of the decision context.

The selection is usually based on some version of rule, debate, or vote. Mandated rules, such as to maximize contributions to National Economic Development (NED) subject to environmental constraints, may be the basis for selecting a plan. That rule guided water resources planning at the federal level from 1983 to the time this chapter was written. Rational debate is a common method for decision making. This allows for arguments like "Plan A is better than Plan B because..." These techniques can be more or less formal. In democratic processes, putting things up for a vote is a common method. Committee votes and public referenda are two of the most common voting methods encountered. There are countless ways to make a decision. Section 9.4 describes one of them.

9.4 MAKING THE DECISION: SELECTING A SOLUTION

Most decision processes can be deconstructed into a sequence of tasks or steps. The following nine-step decision process is a generic method that is reasonably well suited to selecting a solution. I think it is fair to suggest that some version of these steps, with the possible exception of number seven, are at least implicitly considered in most plan selection decisions. The more transparent, if not the better, processes consider these steps explicitly:

1. Confirm or identify the selection criteria.
2. Create a decision matrix.
3. Identify the decision-making process or style.
4. Weigh the criteria.
5. Apply decision model if one is used.
6. Order the alternatives.
7. Consider the uncertainty.
8. Choose a course of action.
9. Document basis for decision.

9.4.1 Confirm or Identify Selection Criteria

At this point in the planning process, the team will have generated estimates and descriptions of a wide variety of effects expected to result from your various solutions. Some of these will have been used to evaluate and qualify plans; some will have been used for comparing the plans; others may have been used simply to provide information to interested parties. Some subset of these criteria will actually be used for decision making; let's call them the selection criteria. Ideally, one focus of the comparison step will be on these selection criteria. In best practice planning, the team will have identified decision criteria that included the selection criteria back in Step 1. That is done to focus the data collection and analysis on the things that are most important to decision makers from the outset of the planning study. More than likely that list of decision criteria will be iterated and amended several times before you arrive at this decision point. The final iteration of the decision criteria yields the selection criteria.

Selection criteria are the criteria most relevant to decision makers. They meet the critical information needs of the decision makers. Although a planning study will generate many variable measurements and estimates of effects, only those that directly influence the decision makers' thinking are considered selection criteria. These criteria may vary from step to step within a study. Evaluation, comparison, and selection criteria can comprise distinct sets of criteria that may or may not overlap. Occasionally, perhaps rarely, the same set of criteria may be used for all three planning tasks.

During the deliberative steps (Steps 3–5) of the planning process it is essential that the criteria reflecting the objectives and constraints be considered at some point. These need not be the basis for the selection criteria, although the most important objectives will often be reflected in one or more selection criteria. If the earlier planning steps establish that each plan makes a significant contribution to the objectives while avoiding the constraints, the final decision need not always be based directly on the objectives.

Some selection criteria may be set by policy. A good example, mentioned previously, is found in the *Principles and Guidelines for Water and Related Land Resources Planning* (P&G) (1983), which prescribes that federal water resources planning agencies will make decisions based on net contributions to national economic development (NED). This assured that net NED benefits were always among the selection criteria for U.S. Army Corps of Engineers planning investigations. Thus, higher authorities may circumscribe some of the selection criteria. In a good robust planning process, however, selection criteria are likely to include study-specific criteria.

If these criteria were identified in Step 1 of the planning process, they should be confirmed as still relevant. If the criteria have changed or if decision makers will decide based on different information, the criteria ought to be amended. If the criteria have not yet been identified, the planning team needs to coordinate with decision makers and identify the subset of analytical measurements that will be used as the basis for decision making. The sudden appearance of new criteria may initiate new iterations of earlier steps. Selection criteria should always reflect the primary purpose(s) of the study. Some measure of cost is almost always among the selection criteria.

One of the most useful and, oddly enough, most rare things found in a planning study is a simple paragraph that says, "The selection criteria used to select the plan recommended for implementation are...." The reason this rarely appears is that even after all the work that was done the team either does not know how the plan was selected or it has simply failed to convey that information. The simplest way to improve the transparency of your decision process is to identify the selection criteria that were actually used. That does not mean presenting an elaborate table or system of accounts with dozens of different effects described or compared. It means saying something to the effect that once the plans cleared the preliminary hurdles of evaluation, the actual choice of the best plan was based on these "X" criteria.

9.4.2 Create a Decision Matrix

The decision matrix, different from the comparison table, is a handy device to summarize the performance of each alternative for each selection criterion. A decision matrix is defined by rows that reflect the decision criteria and columns that reflect the alternative solutions* under consideration. The cells created by such a matrix are then populated with the most objective measurements or descriptions of the plan's contributions to each selection criterion. The decision matrix entries should include a characterization of the relevant remaining uncertainty that attends any decision criterion's estimate.

Ideally, the decision matrix will be presented in a single unified table. Some table values may be accompanied by narrative descriptions. The goal of the decision matrix is to provide a succinct and transparent summary of the information decision makers consider necessary for decision making. The decision matrix is often called by other names, including effects table, performance table, multicriteria tableaux, effects matrix, and so on. Table 9.2 provides a sample decision matrix developed for

* In decision theory, the entries in the rows and columns are usually reversed. This is far from the most important point to worry about, however.

Table 9.2 Sample Decision Matrix

	Plan A	Plan B	Plan C	Plan D	Plan E	Plan F	Plan G	Plan H
Flood risk management ($1000s EAD)	$10	$800	$60	$80	$25	$50	$100	$900
Annual HU's	46	0	50	90	85	75	116	116
Annual user days of urban recreation	700	0	850	2000	1000	800	3000	3000
Property value effects	2	3	2	1	3	1	2	3
Cost ($Millions)	$15	$15	$23	$33	$20	$27	$53	$68

an urban stream in the DC metropolitan area. Note that it provides a visual comparison of critical differences among plans.

9.4.3 Identify the Decision-Making Process or Style

Let's revisit a point from earlier in this chapter. What will be the planning team's role in the decision-making step? Will they make the decision, provide a recommendation, prepare input for the decision, or will they defer to higher authority and step aside? Will the decision process be explicit and spelled out or will it remain an implicit and opaque mystery? Are decision makers autocratic, consultative, or collaborative? Will they consider all criteria or are they hierarchical? What planners do for this task very much depends on the answers to these questions.

Let's begin by considering the available information and how it will be used by decision makers. Planning frequently relies on a "maximizing" process. This occurs when the planning team collects and analyzes enough information to identify the best solutions in any given decision context. Maximizing is contrasted with "satisficing," which requires collecting only enough information to make a satisfactory decision. Satisficing may not lead to the best plan or the best outcomes. These two data collection extents represent a continuum rather than a discrete choice in data gathering. A maximizing process identifies a variety of acceptable solutions. A satisficing process tends to identify a single solution that, if not considered best, is considered good enough.

Of equal importance to the amount of data is where decision makers focus when making a decision. Even when a variety of decision criteria are identified and analyzed during a planning process (e.g., a comparison table), it is not unusual for decision makers to focus in a hierarchical manner on a single criterion or a limited number of decision criteria. Multicriteria decision making may be the stated goal, but it is not often the reality. Decision criteria may be identified in Step 1, but some decision makers may make decisions on a much narrower basis, that is, the selection criteria may be a very small set.

Throughout the last two decades of the twentieth century and through the first decade of the twenty-first century, water resources planning in the United States at the federal level has been characterized by hierarchical decision making, often with a solitary focus on net NED benefits. Despite the varied interests of many publics and the vast array of project impacts evaluated, the U.S. Army Corps of Engineers was directed to recommend the plan that maximized net NED benefits.* This narrow focus on one aspect of the data stands in stark contrast to the broad focus of multicriteria decision-making methods described elsewhere in this chapter.

Decisive decision makers consider a limited amount of information. They are usually interested in a good (not necessarily the best) solution and they act quickly. This is an autocratic style of decision making that relies on an implicit decision calculus that may never be made explicit. Decisive deciders often rely on their personal experience and gut feel for a situation. They tend to be unwavering and stick to a decision once it is made.

Flexible decision makers tend to use limited information as well, but they are more flexible in their choice of solution than decisive deciders. This is a more collaborative decision-making style. They are also more willing to change their decision based on new information.

Hierarchical decision makers rely on a maximization process, but they tend to focus narrowly on the decision criteria. For them some criteria are more important than others. The Corps of Engineers decision making might best be described by this style of decision making. It is an explicit decision-making style that relies on "rules" from a higher authority to direct the decision making. They collect large amounts of information and then select one best solution, often based on one or a few criteria. This is another autocratic style of decision making.

ADVANTAGES AND DISADVANTAGES OF CONSENSUAL DECISION MAKING

- Can produce a high-quality decision.
- Can produce strong commitment to implementation.
- Makes best use of group resources.
- Gains full benefits of group interaction.
- Future problem-solving ability of group is enhanced.
- Useful for serious, important, and complex decisions that affect a lot of people.
- Takes a great deal of time and energy.
- Time pressure must be minimal.
    - Places major demands on group members' skills.
- Requires rich exchange of ideas and information; the group needs to be informed prior to reaching the decision.
    - Hard to use in large groups.

Source: Johnson, D. W., and F. P. Johnson. 2000. *Joining together: Group theory and group skills*, 7th ed. Boston: Allyn and Bacon.

* As noted in an earlier footnote. This is an oversimplification of the truth. There are circumstances under which other plans could be selected. These are omitted, I believe without doing any great harm, to keep the discussion simple.

Integrative decision makers rely on the data from a maximization process and they consider that data broadly. All the decision criteria are used to select the best solution from among an array of viable solutions. This style is well aligned with the planning process described in this chapter. It is a collaborative decision-making style. The actual decision process may remain implicit with decision makers or it can be made explicit through more structured and transparent decision processes.

Figure 9.1 introduces a somewhat different taxonomy for five related decision styles. The styles are based on the answers to a series of seven questions shown in Figure 9.1. Beginning at the top, answer these questions for your planning study and the model suggests the appropriate range of decision styles.

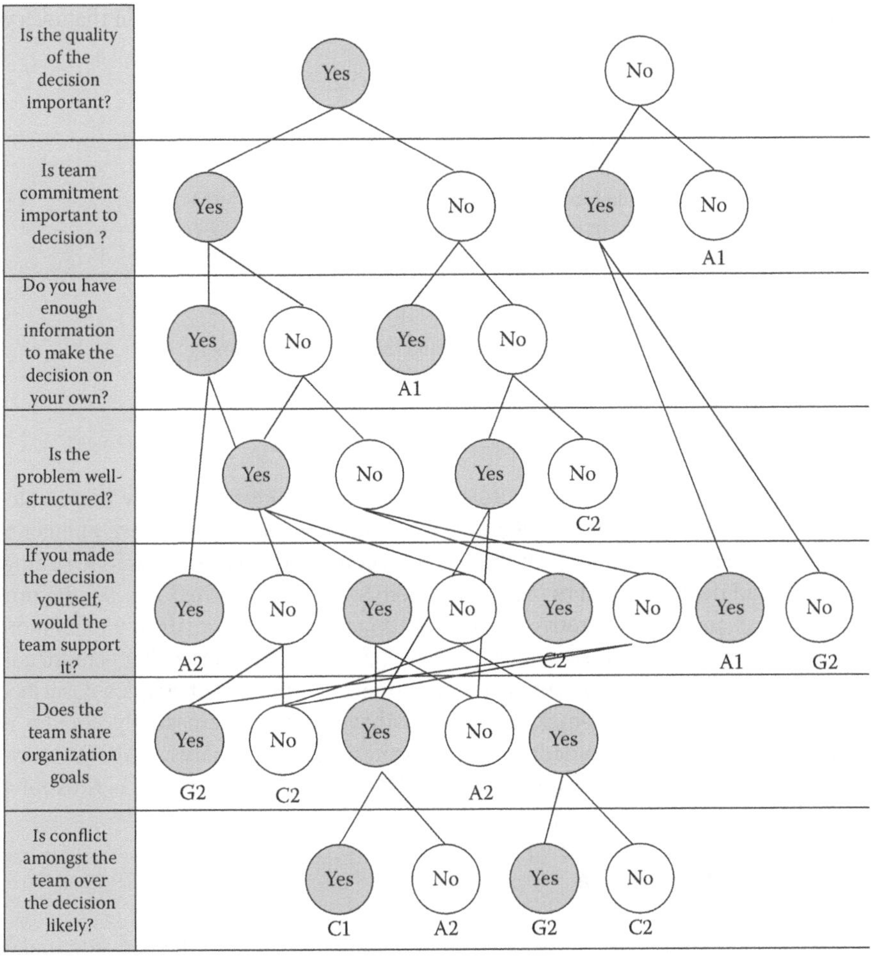

Figure 9.1 Vroom-Yetton-Jago decision model. (Mindtools. 1996–2012. *The Vroom-Yetton-Jago Decision Model, Deciding how to decide.* http://www.mindtools.com/pages/article/newTED_91.htm accessed June 16, 2012. Mind Tools Ltd, 1996–2012. All rights reserved. Reproduced with permission.)

Mind Tools (2008–2010) describes these decision styles as autocratic (A1 you make the decision), consultative (C1 you gather information from the team and then make a decision), and collaborative (G1 you work with the team/group to make the decision by consensus). A1 decision makers rely on their own knowledge, experience, and information to make the decision. They then inform the team. A2 decision makers request specific information from the team and use it to make their own decisions. C1 decision makers gather information from individuals, ask for opinions, and tell team members why they need the information. Once they have enough information, they make the decision on their own. C2 decision makers gather information in a team or group setting. They ask for suggestions and perspectives and use what they learn to make the decision. G2 decision makers facilitate the decision-making process and help the team find a solution that everyone can agree on.

A consultative or collaborative style is best when

- The problem definition is complex.
- Information from others is needed to solve a problem.
- There is enough time to manage a group decision.
- Team members' buy-in to the decision is important.

An autocratic style is most efficient when

- The decision maker has more expertise on the subject than others.
- The decision maker is confident about acting alone.
- The team will accept the decision.
- There is little time available.

When the planning team constitutes the decision makers, they may make autocratic decisions. Frequently, the planning team functions in a consultative manner to decision makers or they may treat the public as consulters. In a collaborative setting, the team and decision makers seek to harmonize the decision making. As the process becomes more collaborative, decision making becomes more difficult and more time-consuming. The next several steps describe a decision process that is more collaborative, explicit, and conceptually transparent than most of the styles described previously. Group decision-making processes have become increasingly popular in natural resource planning practice and multicriteria decision analysis (MCDA) techniques, in particular, are much in vogue. A simplified decision process is illustrated in the addenda to this chapter.

9.4.4 Weigh the Criteria

When a decision is made using more than a single criterion, the decision maker always weighs the criteria. Most of the time this weighing process is implicit, meaning the decision makers never consciously assign explicit weights to the criteria. When you choose a college at which you study planning, you likely consider multiple criteria, such as reputation and lengths of the program, faculty interests, cost, assistantships, location, and other things. You may not have even been aware of the relative importance of some or all of these criteria to you. It is not likely that you assigned explicit weights to the criteria. That is an implicit weighting process. Decision makers use implicit weights when they recognize differences in importance and they act on them to arrive at a decision that rarely has to be explained or defended.

When decisions that will affect the public or other large groups of people are made, we are usually lucky if we get a vote count and a few sound bites from the deciders explaining their reasons for voting as they did. Explicit weighing of decision criteria is a relatively rare occurrence. Planning studies that attempt to genuinely consider the concerns of varied stakeholder groups are more likely to seek explicit weights for decision making.

When the comparison step of the planning process shows the presence of tradeoffs among the decision criteria, there is no dominating alternative, that is, an alternative that is clearly better than every other plan for every decision criterion, then decisions makers are going to have to weigh the criteria. In the rare instance where these weights have been established as a matter of policy or by some higher authority, the problem is trivially simple: follow the policy. In all other instances, weighing the criteria can be a sticky problem. Whose weights will you use?

There is no use to search for the true weights; these are value parameters. They have no true value. There is no truth out there that can be discovered no matter how clever the methodologies employed. Criteria weights are subjective values. Some groups or individuals will prefer one set of weights, while other individuals or groups will prefer other weights.

Weights can be assigned either to make the decision process more transparent or as part of a more sophisticated decision process, for example, MCDA. When the planning team has decided to assign weights to the criteria, someone must by some means drive a figurative stake into the ground and say, "These are the weights." Criterion B is twice as important as criterion A, which is three times as important as criterion C. Or given 100 total points, criterion Z equals 40, Y equals 10, X equals 30, W equals 5, and V equals 15. Weights can be determined in any number of ways, from simply asking a decision maker to assigning them through a wide variety of more- or less-sophisticated derivations of the weights. Addendum 1 to this chapter provides an example of how weights can be derived and used with the analytical hierarchy process (AHP).

9.4.5 Apply a Decision Model if One Is Used

This is where the magic begins. Decision models are used to rank the candidate solutions, thus identifying the highest-ranking solution. Decision makers who use an implicit decision-making methodology reach within their opaque preference centers and emerge with a relative ranking of the selection criteria. When an explicit methodology is used, there is considerable work to do. The second part of Addendum 1 shows how the subjective weights of the decision maker can be used to get to a decision. Addendum 2 illustrates a somewhat simpler alternative to the AHP.

9.4.6 Ordering Alternatives

To get from many plans to a best plan, someone must put the plans in some sort of hierarchical order. Putting plans in order requires subjective judgments about which selection criteria are most important. Those judgments are part of decision making. MCDA techniques do this quite handily, but not every planning process will use MCDA. Putting things in order is an instinctive process used to aid decision making. We use it all the time. We choose a meal from an extensive menu. We find a show from the 500 cable stations we subscribe to. It is very instinctive, but if pressed to say why we chose as we did, we would be able to give reasons. Whether those reasons would satisfy others or not is entirely another question. When choosing the best plan jointly or as stewards of some public trust, it is important to make the reasons for the choice explicit. These reasons comprise the evidence, values, and rationale upon which the decision will be based.

Ordering techniques at this stage of the planning process include screening, rating, and ranking. Screening or sorting is the most basic ordering technique. It is the process of separating solutions into explicit categories through a systematic evidence-based process. It is used to identify the best solution and the rest of the solutions, or the implementable solutions and the nonimplementable solutions. In general, screening is not the tool to be used to find the best solution. It is the tool to be used to create the piles of qualified and unqualified solutions during the evaluation step.

Rating is an activity that individually scores or rates each solution of interest to the decision maker. It goes further than simple screening by systematically separating elements into multiple categories of varying degrees of interest. Items with like ratings are gathered into like groups where the groups usually, but not always, have an ordinal logic to them. Think of the Motion Picture Association of America rating system for movies: G, PG, PG-13, R, and NC-17, for example. Solutions may be rated good, better, best, or implementable now, implementable in the future, not implementable, or any other way that suits the decider's needs.

Ranking is an activity that distinguishes differences among individual solutions and assigns a position to one solution relative to others. It is a systematic process used to put solutions in an ordinal sequence when used in a qualitative setting. Ranking can also be a cardinal or scalar ranking when appropriate data (e.g., net NED benefits) are available.

No one is likely to know more about a planning issue than the planning team. In most instances, it would seem good common sense to elicit the planning team's opinions if not its recommendations on the relative desirability of the plans or at least an identification of the best plan. For a ranking example, see Addendum 1.

9.4.7 Consider the Uncertainty

One way to look at decision making in planning is as a process of reducing uncertainty sufficiently to allow a reasonable choice to be made about the best way to move toward the most desirable future. Planning, done well, stresses the evidence gathering and uncertainty reducing (or uncertainty illuminating) function of decision making. Few if any planning decisions can be made with absolute certainty because complete knowledge is rarely available. Thus, uncertainty is generally reduced rather than eliminated by planning studies. Because there is bound to be residual uncertainty, every planning decision involves some risk of loss or risk that an opportunity will not be fully realized. Things may not turn out as planned. If there is no uncertainty, decision making is a whole lot easier. Absent uncertainty, the plan's effects are all known and you simply implement the plan to bring about the known result.

When choosing the best solution, there are a lot of facts that may not be known (uncertainty) and deciders must often consider a variety of interrelated factors (complexity). The consequences of making a wrong choice under these uncertain and complex circumstances may be significant (high risk). Each plan will have its own risk profile. The varying perspectives of the many stakeholders assure that interpersonal issues will further complicate decision making. Considering all these difficulties, the best way to make such a complex decision is to explicitly consider the risks and the uncertainties that give birth to them.

What are the risks associated with the choice of a course of action? If they have not previously been identified in the evidence gathering process, it is time for planners to look at each alternative solution and to ask and then answer the following questions about each plan:

- What can go wrong?
- How can it happen?
- What are the consequences?
- How likely the consequences will occur?

If the planning team has not explicitly identified the risks associated with each solution, take time to consider them. Seek the team's help in this, if necessary.

ASSUMPTIONS

Explicit assumptions are the assumptions we know we are making. It is possible to make a list of these and to explicitly test the sensitivity of analytical results to these assumptions.

Implicit assumptions are those we do not even know we are making. These are often the most dangerous assumptions for precisely that reason. Our education teaches us the assumptions of our disciplines, which we rarely ever question. Our training often teaches us the assumptions of our trainers. The organizations we work for fill us with implicit assumptions about how things ought to be done. Our experience teaches us to make subconscious assumptions all the time. An independent peer review by multidisciplinary experts is often the best way to identify these assumptions we do not know we have made.

In a good planning process, planners will have analyzed and identified the most significant remaining uncertainties. It is not enough for planners to have conducted a scenario analysis or to have used probabilistic or other methods to address uncertainty in the plan. This information must also be effectively communicated to decision makers so that they are aware of this information and its implications for decision making in order to consider it appropriately.

If a critical assumption, for example, might affect a decision maker's choice, they need to be made aware of that assumption and its implications. If we assume settlement of stone rubble breakwaters at a harbor has ceased, this may lead to a very different decision than if we assume further settlement is possible. Decision makers need to know what critical assumptions could affect their decisions.

Decision makers need to understand there is no such thing as "the number" when it comes to answering questions that are uncertain. Most questions are uncertain. Uncertain results should always be presented as uncertain and never as a point estimate. It is the planners' job to identify and address uncertainty in the inputs to the planning process. It is the decision maker's job to account for and consider this uncertainty in the decision-making process.

The values presented in the decision matrix should reflect uncertainty in an appropriate manner. It can be helpful for planners to express the level of confidence they have in the decision criteria (see the following text box).

9.4.8 Make a Decision: Choose a Course of Action

It is time to make a decision. The goal of the "decision" step is to decide the best action to be taken to meet the objectives and avoid the constraints, thereby solving problems and realizing opportunities. That goal is accomplished in this choice task. There can be a single decision, multiple decisions, or a phased-in decision. Ultimately, decisions are based on judgment after weighing all the facts, evidence, information,

UNCERTAINTY RATING TERMS

Decision making might be aided by appending uncertainty or confidence ratings to decision criteria and other important study results. Two sets of examples follow.

High—there is little to no concrete evidence available.
Medium—there are some good evidence and some significant data gaps.
Low—good evidence is available; data gaps are not significant.
None—all relevant facts are known.

An alternative set of definitions is found below.

High—there is a very broad range of possible outcomes that include extremes.
Medium—extreme outcomes are not possible.
Low—there is a limited range of possible outcomes.
None—all relevant facts are known.

uncertainty, policies, and historical precedents. Decisions can be mandated or steered by a higher authority, such as a legislative or judicial body, or an executive. A decision can also be driven by emergency circumstances, expediency, costs, time, and knowledge. It may be necessary to make a decision quickly even though uncertainty is great.

The "decision maker" identified at the beginning of the planning process is responsible for making the decision. That includes choosing and recommending a course of action. It may also include vetting or clearing the recommended action through any and all higher authorities that need to approve the course of action before it is implemented. In some situations, decision making may include developing an implementation strategy.

The decision-making style (discussed earlier) may influence the decision process. In a collaborative decision-making process, the work of the planning team will be the decision maker's most valuable asset. In more autocratic decision processes, the decision maker may opt to follow his or her own decision process.

An adaptation of the ladder of inference (Senge et al. 1994) leads to a useful set of questions (Mind Tools 1996–2012) to help decision makers ensure they are thinking clearly about a decision and have considered enough perspectives. They are the following:

- Why have I chosen this course of action?
- Were there other actions I should have considered more thoroughly?
- What belief led to this action?
- Was the belief valid and supported by evidence?
- What conclusions did I draw?
- Were these conclusions sound?
- What assumptions have I made?
- How valid are these assumptions?
- What data have I collected and used?
- Were the data collected properly?
- What are the facts I used in my decision?
- Are there other facts I should be considering?

Senge et al. (1994) caution that we all live in a world of self-generating beliefs that remain largely untested. Decision makers should remain aware of their beliefs and always test them to assure their decisions are grounded in factual reality to the greatest extent possible.

9.4.9 Document Basis for Decision

Why did the deciders decide as they did? Capture this logic. Make sure the document at some points says, "The criteria upon which the decision was based are…." It needs to be part of the planning document and it is important for a number of reasons. First, it informs your stakeholders. It documents which decision was made by whom, when, and why. This provides a record of accountability. Sometimes, a formal decision document is required by law. When it is, your documentation should follow policy. Absent a more formally prescribed process, Modern Servant Leader (2011) offers seven primary components for a decision document.

It should begin with a description of the decision to be made, stated as clearly and succinctly as possible. Second, the current status of the decision needs to be clear. A very brief history of how you came to the decision point is useful. This should include the who, what, when, and where of the situation requiring a decision. Third, include a brief description of each of the alternatives that could have been chosen. This should include bullet list levels of detail on the strengths and weaknesses of each solution, the decision matrix, or comparison analysis. The fourth step is to identify the recommendation of the planning team. This is followed by the actual decision and its status. In best planning practice, this would include a rationale for the decision that is accessible to your stakeholders. If it is a final decision, say so. If the decision must be cleared by a higher authority, indicate that. Sixth, the documentation should describe what actions happen next when those actions are germane to the process. The final component is the formal sign off on the decision. This can take a variety of forms, from a formal signing to a press conference to assent at a meeting. Documenting the decision, especially the rationale, is an important and often overlooked part of the record of your planning study.

9.5 RISK-BASED DECISION-MAKING STRATEGIES

Planning becomes an increasingly risk-informed decision-making process as planners continue to become more intentional about addressing uncertainty in the planning process. To the extent that planning takes risks more into consideration in the decision-making process, this opens the range of decision-making strategies. Risk-informed decision making has been used outside the planning realm for quite some time. A number of decision strategies have been used to make risk-based decisions, some of which may be useful to planners. The more common principles for making risk-based decision include the following:

- Policy
- Zero risk
- Weight-of-evidence
- Precautionary principle
- ALARA principle
- ALOP principle
- Reasonable relationship
- Safety standards

Each is considered briefly in turn in the next sections.

9.5.1 Policy

Some decisions have already been made for the decision maker by people higher in the decision-making hierarchy. This might include agency management, executives, legislative bodies, Congress, the president, or other elected officials. In the United States, for example, authorizing legislation may prescribe what an agency can and must do. When this happens the decision maker's job is to figure out the best way to execute their mission.

Some planning issues may be resolved by a court decision. Decisions made in the public sector, especially by agencies and organizations acting as stewards of a natural resource or other public trust, will often be constrained by policy. Working with a government agency means dealing with their policy restrictions and requirements. International treaties and agreements may identify solutions or limit options as well.

9.5.2 Zero Risk

Making risk taboo and banning risky activities has been popular in the past. Years ago, the limits of our knowledge and of scientific detection methods made it possible to find comfort in laws that appeared to legislate safety as a matter of zero risk. That is not so easy anymore. By the middle of the 1980s, decision makers began to abandon the notion of zero risk in favor of more realistic versions of negligible risk. The one-in-a-million standard that captured our imagination as a reasonable threshold for increasing the chance of cancer over a lifetime evolved from a notion of "de minimis" risk. Think of de minimis, or negligible, risk as "practically zero."

Policy makers have, by and large, abandoned the idea that zero risk is a realistic measure of acceptable risk. Choosing solutions that establish a level of de minimis risk is a viable option for choosing a solution in many settings.

9.5.3 Weight-of-Evidence

The truth is not always easy to see in an uncertain world. Data gaps and conflicting evidence can complicate planning decisions. In a weight-of-evidence approach to evaluating solutions, decision makers assess the credibility of conflicting evidence about issues, especially risks, in a systematic and objective manner. A formal weight-of-evidence process might require a diverse group of scientists to examine the evidence in order to reach a consensus view.

Weight-of-evidence is an ongoing activity that attempts to balance the positive and negative evidence of important effects based on relevant data. Thus, decision making is conditional on the available evidence and subject to change as new evidence becomes available. When there is uncertainty about the nature of a risk, a weight-of-evidence approach may be useful in helping decision makers decide if a risk is acceptable or tolerable.

9.5.4 Precautionary Principle

Precaution, as used here, means refraining from action if the consequences of the action are not well understood. It is prudent avoidance exercised through the no-action option. The precautionary principle is broadly based on the notion that human and ecological health are irreplaceable human goods. Their protection should be treated as the paramount concern for organizations and government. All other concerns are secondary.

The precautionary principle is controversial and heavily influenced by culture. In a very loose and informal sense, the precautionary principle suggests that when there is significant uncertainty about a risk, we should err on the side of precaution, if we are to err at all. That means those activities that could give rise to catastrophic outcomes should be prohibited (e.g., building refineries on estuaries). It also means that if inaction could give rise to catastrophic outcomes (e.g., waiting for proof of the magnitude of sea-level rise) we should act, not wait. The precautionary principle is generally considered to be most appropriate in the early stages of an unfolding risk problem, when the potential for serious or irreversible consequences is great, or when the likelihood of occurrence or magnitude of the consequence is highly uncertain. The desire for precaution is usually positively related to the amount of uncertainty in a decision problem. The precautionary principle can be invoked for decision making whenever uncertainties are large or intractable.

9.5.5 ALARA Principle

ALARA is an acronym for As Low As Reasonably Achievable. This is one criterion for establishing a tolerable level of risk (TLR). Technology and cost present two realistic constraints on what it is possible to achieve in terms of solutions that target risk reduction. Once a risk is as low as we are reasonably capable of making it, what choice do we have but to tolerate the risk that remains?

Best available technology (BAT) is a related concept. It differs in a potentially significant way, however, as BAT says use the best available technology with no further qualification. ALARA introduces the idea of reasonableness and this opens the decision-making door to the consideration of other factors such as cost and social acceptability. BAT does not consider these other factors.

9.5.6 Appropriate Level of Protection

An appropriate level of protection (ALOP) defines or is defined by the risk society is willing to tolerate. Despite the promising sound of this principle, it is rather circular reasoning because it presumes one has found a way to identify the holy grail of what is "appropriate" for society. The significant contribution of this concept is that it flips the focus from risk to protection, where we might think of protection as akin to different degrees of safety. The factors used to determine the ALOP for a risk typically include the following:

- Technical feasibility of prevention and control options
- Risks that may arise from risk management interventions
- Magnitude of benefits of a risky activity and the availability of substitute activities
- Cost of prevention and control versus effectiveness of risk reduction
- Public risk reduction preferences, that is, public values
- Distribution of risks and benefits

9.5.7 Reasonable Relationship

This principle suggests that costs of managing a risk ought to bear "a reasonable relationship" to the corresponding reductions in risks. It is not a benefit–cost analysis (BCA), but it is an attempt to balance nonmonetary benefits (i.e., risk management outputs and outcomes) and the monetary costs of achieving them. Cost-effectiveness and incremental cost analysis are often used as the bases for determining the reasonableness of a solution.

9.5.8 Safety and Balancing Standards

Safety standards encompass a bundle of standard-setting methods that rely ultimately on some degree of subjective judgment. For example, the zero-risk standard is one possible safety standard. Zero just happens to be one of many potential thresholds that can be established to define safety. Any nonzero level of risk can be stipulated as safe, acceptable, or tolerable. In fact, the term "tolerable level of risk" has been dangled as one such tantalizing threshold standard in some of the literature. If we could develop a TLR for dam safety or for local extinction of a species, decision making would be much easier; just plan to meet the standard.

Many determinations of a TLR require a subjective balancing decision. Risks of uncertain potential gain or benefits may be best served by using some type of balancing standard. For example, risk–benefit trade-off analysis generally implies that greater benefits mean we are willing to accept a greater level of risk in exchange for those benefits. The risk–benefit trade-off explains why we are all willing to assume the risk of driving in a modern society.

Comparative risk analysis (CRA) ranks risks for the seriousness of the threat they pose. It began as an environmental decision-making tool (U.S. Agency for International Development 1990, 1993a,b, 1994; Environmental Protection Agency 1985, 1992a,b, 1994) used to systematically measure, compare, and rank environmental problems or issues. It typically results in a list of issues or activities ranked in terms of relative risks. The most common purpose of CRA is to establish priorities for a government agency. The concept is perfectly adaptable to planning, where solutions could be ranked by comparative risk.

Benefit-cost analysis (BCA) is another kind of balancing standard used to determine what is acceptable or tolerable. BCA attempts to identify and express the advantages and disadvantages of a risk or planning solution in terms of dollars. It is considered a useful measure of economic efficiency.

In addition to threshold and balancing standards, procedural standards are sometimes used to define what is acceptable or tolerable. Procedural standards typically identify an agreed-upon process, which is often the result of negotiation or a referendum of some sort. If the agreed-upon process is followed, then the results of that process are considered acceptable or at least tolerable.

9.6 TAKE AWAYS

Lesson 1: Choosing the best solution is a political act.
Lesson 2: Decision making is fundamentally a subjective exercise.
Lesson 3: Selection criteria are the criteria decision makers actually use to make their choice.
Lesson 4: At some point, decision makers must decide the relative importance of their different selection criteria.
Lesson 5: The rationale for choosing a solution should be documented in a transparent manner.

9.7 ADDENDUM 1: ANALYTICAL HIERARCHY PROCESS

This addendum provides an example that shows how subjective weights can be derived (Section 9.7.1) and applied to decision criteria (Section 9.7.2).

9.7.1 Deriving Subjective Weights

Thomas Saaty created the analytic hierarchy process (AHP) in the 1970s. It combines the order and structure of mathematics and the subjective intuitiveness of psychology to evaluate information and make decisions that are easy to explain and defend. Consider a simple example where you want to identify the plan that best reconnects an urban community to its local stream ecosystem.

Assume you'll use water quality (WQ), habitat units (HU), and cost (C) to make this decision from among three candidate plans (A, B, and C). The AHP can use tangible and intangible criteria as well as quantitative and qualitative information. It combines these different types of decision criteria measurements and converts them to a standardized numerical scale. This approach can be used to develop subjective weights along the way so that a choice can be made more explicitly and objectively, while including all the selection criteria and their trade-offs.

The AHP requires you to begin with a decision matrix and a question you are trying to answer. Using pairwise comparisons, AHP can help determine your criteria weights. Figure 9.2 shows the relevant hierarchical model for this example. Assuming there is more than one decision maker, each decision maker rates the relative importance or her preference for each criterion using a pairwise comparison approach. To do this, prepare a matrix to compare each criterion to the others. In this case, a 3×3 matrix, as shown in Table 9.3, is needed.

The AHP enables the decision maker to rank the importance of each criterion relative to the others, using the scale shown in Table 9.4. The even numbers 2, 4, 6, and 8 provide the halfway positions for these values. Thus, a 4 is halfway between somewhat more important and definitely more important. For the purposes of Table 9.5,

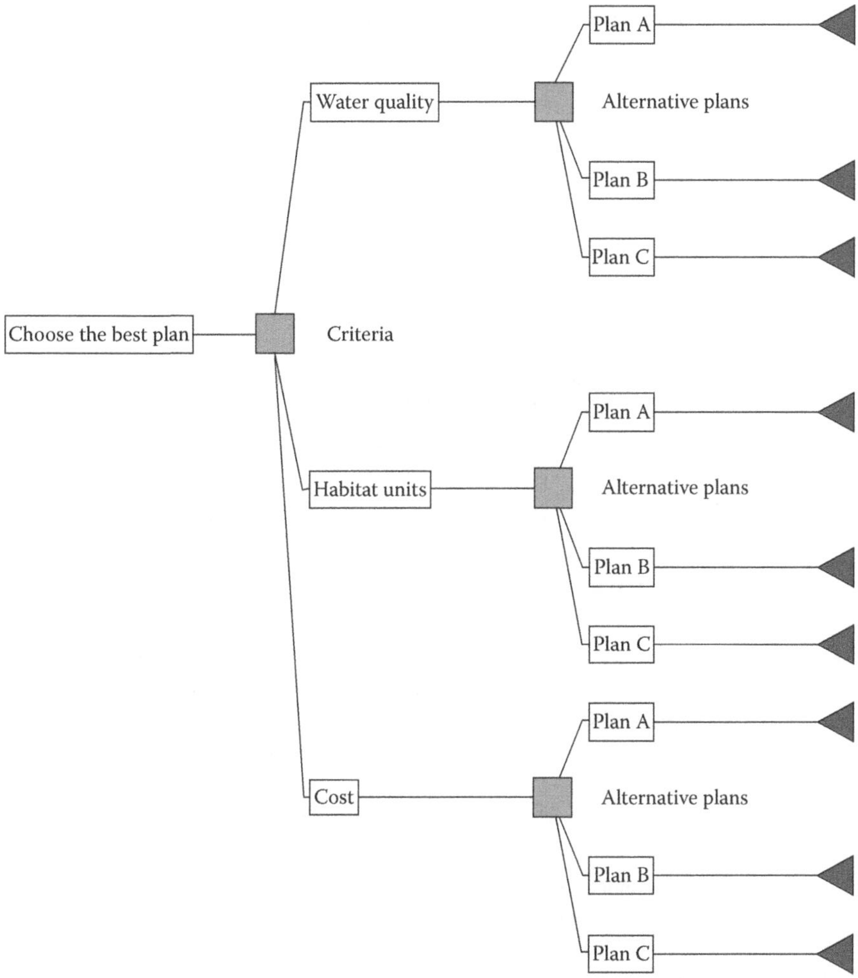

Figure 9.2 Simple hierarchical model for three criteria and three alternatives.

Table 9.3 Blank 3 × 3 Criteria Ranking Matrix for Determining the Decision Maker's Criteria Weights

	Water Quality	Habitat Units	Cost
Water Quality			
Habitat Units			
Cost			

Table 9.4 Saaty's AHP Scale for Determining the Relative Importance of Criteria in a Pairwise Ranking

Relative Importance	Value
Equal importance/quality	1
Somewhat more important/better	3
Definitely more important/better	5
Much more important/better	7
Very much more important/better	9

Table 9.5 Completed Criteria Ranking Matrix for a Hypothetical Decision Maker

	Water Quality	Habitat Units	Cost
Water Quality	1	7	3
Habitat Units	1/7 = 0.14	1	1/9 = 0.11
Cost	1/3 = 0.33	9	1
Total	1.47	17	4.11

the number chosen is based on the row as compared to the column value. A 7 means the row criterion is much more important than the corresponding column value. When the row value is less important than the column value, use the reciprocal. A rating of 1/7 means the column value is much more important than the corresponding row value.

Suppose the comparison for the decision maker in this case is as follows:

- Water quality is [much MORE important] than habitat units.
- Water quality is [somewhat MORE important] than cost.
- Cost is [extremely MORE important] than habitat units.
- Habitat units is [very much LESS important] than cost.
- Habitat units is [much LESS important] than water quality.
- Cost is [somewhat LESS important] than water quality.

The preferences/judgments are shown in Table 9.5.

The last row shows the sum of the weights and provides an estimate of the overall weighting for each criterion for a given decision maker. To normalize the weights, divide each entry by its column total. Thus, WQ is 1/1.47 = 0.68. Normalized weights are shown in Table 9.6.

When people enter subjective ratings, inconsistencies can arise. To smooth out these inconsistencies, the average of each row is calculated as seen in the priority vector column of Table 9.6. This is the final weight for each criterion. These weights suggest water quality represents about 61% of the final decision, habitat units account for 6% of the decision, and cost represents about 33% of the decision. Note that subcriteria can be considered, enabling planners to break WQ, HU, and C down into

Table 9.6 Normalized Weights

	Water Quality	Habitat Units	Cost	Priority Vector
Water Quality	0.68	0.41	0.73	0.61
Habitat Units	0.10	0.06	0.03	0.06
Cost	0.22	0.53	0.24	0.33
Sum	1.0	1.0	1.0	

finer elements, if desired. Subcriteria are not used here in order to avoid complicating the example unnecessarily.

The pairwise comparison described here is an essential part of the AHP process, one of many MCDA techniques. Thus far we have only looked at how the subjective weights have been determined. The second part of the AHP process that moves to a ranking of alternative plans is described in the next section. In the meantime, know there are many different methods for indirectly developing a decision maker's subjective weights.

9.7.2 Getting to a Decision

In Section 9.7.1 subjective weights are derived from a pairwise comparison. Table 9.6 showed the weights were WQ = 0.61, HU = 0.06, and C = 0.33. With that information, we now rank the alternative plans. If there are subcriteria in your model, the process begins with the subcriteria, that is, the lowest level in the hierarchy of decision criteria. For simplicity we begin at the criteria level. Step 1 is to prepare a plan comparison matrix for each criterion. Step 2 is to use a pairwise

SUBCRITERIA

Suppose our example has subcriteria. Specifically, imagine that water quality is divided into dissolved oxygen and pH. Weights would be determined for the subcriteria in a similar fashion. Suppose the ranking matrix looked like this:

	DO	pH
DO	1	3
pH	0.33	1
Total	1.33	4

The normalized weights become DO = 0.75 and pH = 0.25. WQ was 0.61. So DO = 0.75 × 0.61 = 0.46 and pH is 0.25 × 0.61 = 0.15.

When comparison of alternatives begins, it would begin at the DO and pH levels and aggregate upward, consistent with the methodology shown in the second part of the AHP example that follows.

comparison method.* Using the same 1–9 rating scale, as described in the preceding section, planners determine how each plan compares to the others, based on each criterion. Let us suppose the following is true for WQ:

- Plan A is somewhat better than Plan B.
- Plan A is very much better than Plan C.
- Plan B is much better than Plan C.

These judgments and the mathematical process previously described for the weights yield the results in Table 9.7. Repeating the process for HUs, suppose we obtain the information in Table 9.8. The corresponding information for cost is provided in Table 9.9.

The next step is to combine the various weights from Tables 9.7–9.9 and calculate the plan final scores. Figure 9.3 provides the structure for organizing these weights. The calculated criteria weights are entered above the criteria branches.

Table 9.7 Water Quality Comparison Matrix and Calculation of Normalized Weights for WQ Contribution of Each Plan

Water Quality	A	B	C	Water Quality	A	B	C	Priority Vector
A	1	3	9	A	0.69	0.72	0.53	0.65
B	0.33	1	7	B	0.23	0.24	0.41	0.29
C	0.11	0.14	1	C	0.08	0.03	0.06	0.06
Total	1.44	4.14	17					

Table 9.8 Habitat Unit Comparison Matrix and Calculation of Normalized Weights for HU Contribution of Each Plan

Habitat Units	A	B	C	Habitat Units	A	B	C	Priority Vector
A	1	0.14	0.14	A	0.07	0.02	0.11	0.07
B	7.00	1	0.2	B	0.47	0.16	0.15	0.26
C	7.00	5	1	C	0.47	0.81	0.74	0.68
Total	15	6.14	1.34					

Table 9.9 Cost Comparison Matrix and Calculation of Normalized Weights for C Contribution of Each Plan

Cost	A	B	C	Cost	A	B	C	Priority Vector
A	1	4	7	A	0.72	0.75	0.64	0.70
B	0.25	1	3	B	0.18	0.19	0.27	0.21
C	0.14	0.33	1	C	0.10	0.06	0.09	0.09
Total	1.39	5.33	11					

* If you have actual data, these can be used in lieu of the subjective comparisons; the math gets a bit more involved than we want for a planning theory book.

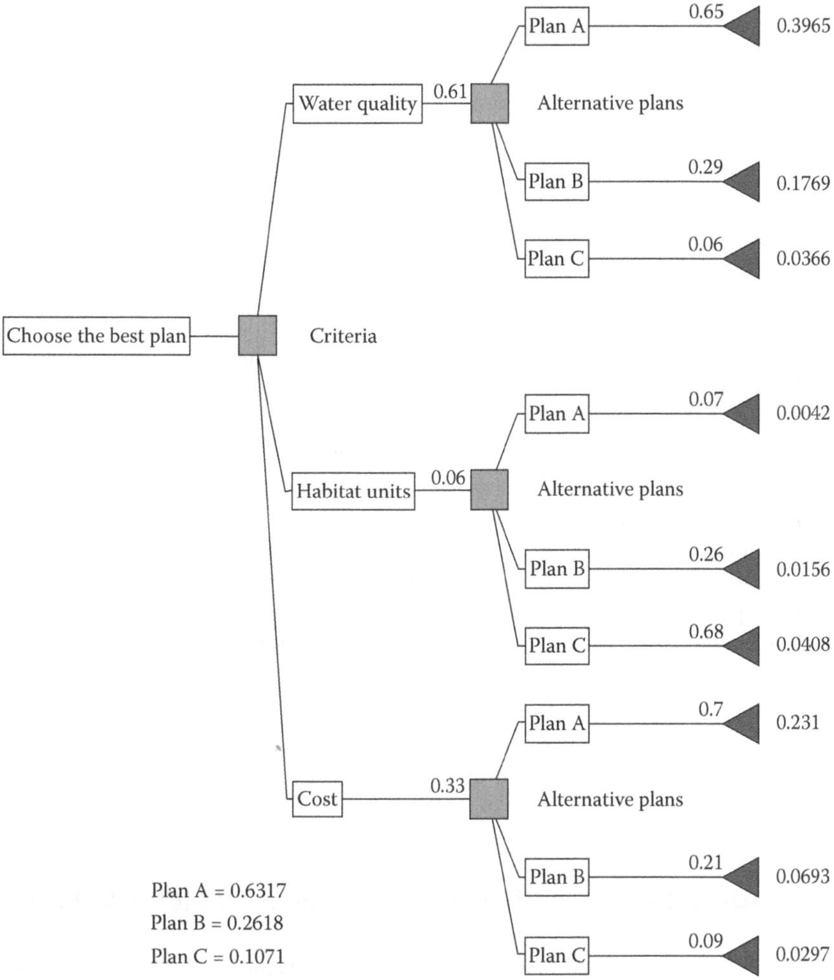

Figure 9.3 Simple hierarchical model with AHP determined weights.

The calculated plan weights follow in the node behind each weighted criterion. The triangle end points show how the criterion weight is distributed among the plans. These values are obtained by multiplying the values along each unique pathway. A final score for each plan is shown at the far right of Figure 9.3, this sums the values for each plan.

Let me offer a nontechnical, soft interpretation of this process. Note the criteria weights sum to 1. These weights, in a sense, are measures of importance. Cost captures 33% of the importance in this decision process. Each node following a criterion has values for the three plans that sum to 1. This shows each plan's share of the total criterion importance. Thus, Plan A captures 70% of the importance of the cost criterion. Multiplying the criterion weight by the node share produces a weighted contribution of the overall importance. So, $0.33 \times 0.7 = 0.231$. Notice the values following the nine triangle end points sum to 1. Thus, Plan A's cost features account for about 23% of the cost criteria. When we add each plan's contributions to the individual criteria, we obtain the final scores. Plan A is the best plan. It meets 63% of all the decision criteria. Plan B meets 26% of the decision criteria and Plan C meets 11%.

The AHP, or any other MCDA technique used, helps planners quantify the judgments used in decision making. Implicit decision making produces decisions, but it often fails to reveal why one alternative is better or preferred to another. MCDA in general and AHP in particular calculate weighted scores for the decision criteria used. These weights are used to calculate a final score for each alternative. This technique enables planners to produce a commensurable quantitative comparison of plans. It also can highlight trade-offs. Look at the weights on the alternative plans nodes. We see Plan A is strong on water quality and cost but not so strong on habitat units.

This suggests several things. First, it begs the question, is there any way to improve Plan A's habitat performance? Doing so makes it an even stronger plan. Second, it causes one to wonder, how important would the habitat criterion have to become before the final sores would produce a different ranking of plans? Although such questions are beyond the scope of this text, MCDA software companies have recognized the commercial value of building such helpful sensitivity analysis tools into their products. If you are likely to adopt one of the MCDA decision-making techniques you would be wise to investigate the uncertainty and sensitivity analysis features of software products you might consider.

Lest the point be lost amidst the details of the examples mentioned earlier, these last two sections have described practical ways for explicitly assigning weights to the selection criteria. Not every planning process is going to do this. Some will rely on the implicit and unstated weights of decision makers.

9.8 ADDENDUM 2: DECISIONS WITH SIMPLE PAIRED RANKING

The determination and handling of subjective weights in decision making is important enough to merit special attention. Let's consider another ranking problem through the lens of a different technique: paired ranking. This is an alternative form of weighted ranking. Following the method of Jones (1998) there are nine steps to this technique. They are as follows:

1. List all criteria for ranking.
2. Rank the criteria pairwise.
3. Assign percentiles to the criteria you will use.
4. Construct a ranking matrix.
5. Rank all items pairwise for each criterion.
6. Calculate weighted criteria ratings.
7. Sum the criteria ratings.
8. Establish the rankings.
9. Conduct a sanity check.

Imagine another planning effort to reconnect an urban area with the natural ecosystem. Suppose four plans (W, X, Y, Z) have been evaluated and qualified as viable solutions. Let the decision criteria for choosing the best plan be the cost of the plan (C), water quality improvements (WQ), and changes in habitat units (HU) as before. Let's apply Jones' paired-ranking technique to this problem, assuming the decision maker wants to select the best of the four plans.

The first step is to identify as many relevant criteria to rank your plans as you can. The second step narrows that list to the most important criteria through pair ranking. For simplicity, this example uses only three criteria, but any number of criteria can be used. By the way, if you need a lot of criteria to provide the ranking of plans, you may need a more sophisticated assessment method or at least some software to handle the calculations for you.

Pairwise ranking works like this. Given three criteria C, WQ, and HU, we ask which is more important: C or WQ, C or HU, or WQ or HU. These choices are presented for all possible pairings of criteria. So, if you have more criteria, you'll have more pairs to consider. Given the decision at hand, assume decision makers have a clear transitive ranking WQ > HU > C. Thus, WQ would receive two most important "votes," (it beats HU and C), HU would receive 1, and C would receive none.

When dealing with a larger number of criteria, the preferences may not always be so obviously transitive. The process can also grow tedious quickly with a large number of criteria. Nonetheless, pairwise ranking can be a useful way to identify the most important criteria from among the list of candidates. The most important criteria are those with the most votes. What lends an air of credibility and confidence to pairwise ranking, however, is when decision makers carefully record the reasons and evidence used to support their rankings.

In the present case, WQ is more important than HU because it is essential to all aspects of ecosystem restoration. HU is more important than C because the success of the reconnection of community and ecosystem rests on a dependable supply of habitat over the long term. Cost is not a trivial consideration; it is just the least important of the three.

The third step, assigning percentiles to the criteria used, is a subjective judgment step. There is no simple deterministic set of values to use. Do not use the votes received as a criterion to calculate these weights. The weights should reflect the decision makers' preferences. If a tool like this is used for making public decisions additional care is going to need to be taken with this step. For this example, assume the weights are WQ = 0.5, HU = 0.3, and C = 0.2. The weights are usually easier to use and for others to understand if they sum to 1 (or 100). Note that equally weighted criteria are not an option with this methodology.

The decision matrix is a simple table, this time with the criteria in columns and the items ranked in rows. (A completed matrix is shown at the end of this example.) Step 4 in the process is another round of pairwise comparisons that is based on evidence. Taking the first criterion, cost, we ask which plan is better on cost: W or X, W or Y, and so on. Table 9.10 summarizes the evidence available for this decision problem. Here, we've used qualitative data; with quantitative data, the process is more transparently obvious.

Using the data from Table 9.10 for the pairwise rankings, Table 9.11 shows the votes received by each plan for each criterion. Plan Y receives three votes on the cost criterion because it is cheaper than the other plans. Thus, when you ask which is better on cost, Y or W, Y gets a vote. The relative weights appear in the column headings.

Weighted criteria rankings are calculated by weighting the number of votes by the criteria weights to obtain the ratings shown in Table 9.12. The ratings are summed and the plans are ranked according to their rating in Table 9.13, which shows a completed ranking matrix. Plan W is the best choice.

This technique can be used to rank items in a simple and straightforward way. It represents the judgments of the people that established the weights and made the pairwise comparisons and no one else's. These judgments are only as good as the evidence they are based upon. When this technique is used to make private decisions about risks, it is a perfectly valid technique to aid decision making.

Table 9.10 Evidence for Criteria Used in Ranking Matrix

Plan	Cost (C)	Water Quality (WQ)	Habitat Units (HU)
W	Most costly	Second best controls	Most reliable
X	Third most costly	Third best controls	Second most reliable
Y	Least costly	Least controls	Third most reliable
Z	Second most costly	Best controls	Least reliable

Table 9.11 Votes Received in Pairwise Comparison of Plans for Each Criterion

Plan	Cost (C) .2	Water Quality (WQ) .5	Habitat Units (HU) .3
W	0	2	3
X	2	1	2
Y	3	0	1
Z	1	3	0

Table 9.12 Calculation of Weighted Ratings for Each Criterion

Plan	Cost (C) .2	Water Quality (WQ) .5	Habitat Units (HU) .3
W	0 × .2 = 0	2 × .5 = 1	3 × .3 = .9
X	2 × .2 = .4	1 × .5 = .5	2 × .3 =.6
Y	3 × .2 = .6	0 × .5 = 0	1 × .3 = .3
Z	1 × .2 = .2	3 × .5 = 1.5	0 × .3 = 0

Table 9.13 Completed Risk Ranking Matrix

Plan	Cost (C) .2	Water Quality (WQ) .5	Habitat Units (HU) .3	Total Score	Rank
W	0	1	.9	1.9	First
Z	.2	1.5	0	1.7	Second
X	.4	.5	.6	1.5	Third
Y	.6	0	.3	.9	Fourth

REFERENCES

Beckhard, R. and R. T. Harris. 1987. *Organizational transitions: Managing complex change*. Boston, MA: Addison-Wesley.

Environmental Protection Agency, Office of Air Quality Planning and Standards. 1985. *Compilation of air pollution emissions factors*. Washington, DC.

Environmental Protection Agency, Risk Assessment Forum. 1992a. *Framework for ecological risk assessment*. Washington, DC.

Environmental Protection Agency. Prepared for the Technical Workgroup, Ostrava (former Czechoslovakia), and EPA by IEC, Inc., and Sullivan Environmental Consulting. 1992b. *Project Silesia: Comparative risk screening analysis*. Washington, DC.

Environmental Protection Agency. Prepared for the Technical Workgroup, Katowice, Poland, and USEPA by IEC, Inc., and Sullivan Environmental Consulting. 1994. *Project Silesia: Comparative risk screening analysis*. Washington, DC.

Johnson, D. W., and F. P. Johnson. 2000. *Joining together: Group theory and group skills*, 7th ed. Boston: Allyn and Bacon.

Jones, Morgan D. 1998. *The Thinker's Toolkit: 14 Powerful Techniques for Problem Solving*. New York: Three Rivers Press.

Mind Tools. 1996–2012. *Team management training: Build and lead an excellent team with Mind Tools!* Accessed June 16, 2012 from https://www.mindtools.com/cgi-bin/sgx2/shop.cgi?page=TeamTools.htm.

Mind Tools Limited. 1996–2012. *The Vroom-Yetton-Jago Decision Model: Deciding how to decide*. Accessed June 16, 2012 from http://www.mindtools.com/pages/article/newTED_91.htm.

Mind Tools Limited. 1996–2012. *The Ladder of Inference: Avoiding "jumping to conclusions."* Accessed June 16, 2012 from http://www.mindtools.com/pages/article/newTMC_91.htm.

Mind Tools Limited. 2008–2010. *Decision-making mistakes and how to avoid them*. Mind Tools Limited e-booklet. Accessed June 16, 2012 from http://www.mindtools.com.

Modern Servant Leader. 2011. Why you need a decision document. Accessed July 28, 2011 from http://modernservantleader.com/resources/why-you-need-a-decision-document/.

Senge P., et al. 1994. *The fifth discipline fieldbook: Strategies and tools for building a learning organization*. New York: Doubleday.

U.S. Agency for International Development. 1990. *Ranking environmental health risks in Bangkok, Thailand*. Washington, DC.

U.S. Agency for International Development. Office of Health, Bureau for Research and Development. 1993a. *Environmental health assessment: A case study conducted in the city of Quito and the county of Pedro Moncayo, Pichincha Province, Ecuador* (WASH Field Report No. 401). Washington, DC.

U.S. Agency for International Development. Environmental Health Division, Office of Nutrition and Health. 1993b. *Environmental health assessment: An integrated methodology for rating environmental health problems*. Washington, DC.

U.S. Agency for International Development. 1994. *Comparing environmental health risks in Cairo, Egypt*. Washington, DC.

CHAPTER 10

Public Involvement

The primary purpose of this ongoing process is to provide people who are affected by and interested in your planning efforts with opportunities to provide input and feedback, as well as to stay generally informed about your efforts.

10.1 INTRODUCTION

Here's the short story: you need people in order to do good planning. People have a right to be involved in decisions that affect them. A few people are going to be very interested and will want to know and be involved in everything. You are going to have a hard time getting the attention of some people who need to know what you are doing. Others won't care until your study is complete.

We talk about the public as if it was a monolithic group, but there is no such thing as "the public." Think of the public as anyone who is not you; there are many publics. Advertisers figured this truth out long ago. Watch how the billboards for a single product change as you move from one part of a city to another. So even though we'll talk about "the public," know that it is not that simple.

Most people get involved by becoming and staying aware. Some will provide input. Some will provide feedback. They need ample opportunities for all of these ways of being involved. Make sure people are provided with input opportunities before all decision-making steps. Let people know how you used their input and their feedback.

We are a busy populace, and there is no one best way to communicate. There are a lot of traditional ways that are still effective. The number of new ways to communicate and participate keeps changing, and planning could do more to use these new technologies to involve people more effectively.

Public involvement is one of two ongoing processes that are part of this book's planning model. Involving the public can make the planner's job harder in the short run because it is time-consuming and it requires resources. In the long run, you get better information and more successful solutions because public support is needed for successful planning. The public lives with the problems and benefits from the opportunities. They are also the people who can oppose you in court when the process turns adversarial. Involve the public effectively or you will regret it.

Here are a few more things to know before leaping into this chapter. There is a rich literature on the value of public involvement. Jim Creighton's *The Public Participation Handbook* (2005) is one of the best portals of entry to this literature and the process. He says public participation incorporates public concerns, needs, and values into government and other public decision-making processes. Quite simply, public support makes for better decisions and public involvement makes for better public support.

10.2 WHAT IS IT?

Public involvement is one of two continuous processes, along with evidence gathering, that is essential to the success of the planning process. Public involvement begins with Step 1, continues throughout all the steps, and even extends beyond the final decision. The International Association for Public Participation defines involvement over a "spectrum" of five levels, as seen in Table 10.1.

Figure 10.1 suggests where these different levels of involvement are most likely to occur in the planning process. Consultative feedback is an essential element of

Table 10.1 International Association for Public Participation Spectrum of Public Participation

Inform	Consult	Involve	Collaborate	Empower
To provide the public with balanced, objective information to assist them in understanding the problem, alternatives, opportunities, and/or solutions.	To obtain feedback on analyses, alternatives and/or decisions.	To work directly with the public throughout the process to ensure that public concerns and aspirations are consistently understood and considered.	To partner with the public in each aspect of the decision including the development of alternatives and the identification of the preferred solution.	To place final decision-making in the hands of the public.

Decision context
1. Inform
2. Consult
3. Involve

Formulate solutions
1. Consult
2. Involve
3. Collaborate

Evaluate solutions
1. Inform
2. Consult

Compare solutions
1. Inform
2. Consult

Make decision
1. Inform
2. Consult
3. Involve
4. Collaborate
5. Empower

Figure 10.1 Public participation levels in the five planning steps.

every planning step. At a minimum, this is providing your various publics with the opportunity to answer the questions, "Did we get this right? What do you suggest?" Providing information and seeking feedback are essential at the outset of a planning study and at each point where decisions are made, especially during formulation and evaluation decisions.

Active involvement of the public will be most crucial in Steps 1, 2, and 5 as well as throughout the evidence gathering process. Collaboration will be most important when solutions are formulated and when a final decision is made. In a few planning situations, the public will be empowered to make the decision. In many, if not most, situations, a planning agency will retain decision-making authority. Creighton (2005) identified six reasons for this common occurrence. First among these is that agencies are often limited by mandates and authorities that constrain what they can do. Second, these agencies are often implementing laws. Third, the agency may be required to meet contractual obligations. Fourth, public consensus may not balance the needs of everyone affected by a decision. Fifth, the public may simply disagree over what ought to be done. Finally, those who participate in the process are more likely to be representing their self-interest than the public interest.

A public participation process is thus a cooperative, consultative, and collaborate process that provides the various participating publics with meaningful and timely opportunities to participate in the planning process. Reasons for practicing public participation are shown in the following text box.

Creighton (2005) identifies the benefits of public participation as:

- Improved quality of decisions
- Minimizing cost and delay
- Consensus building
- Increased ease of implementation
- Avoiding worst-case confrontations
- Maintaining credibility and legitimacy
- Anticipating public concerns and attitudes
- Developing civil society

The public involvement process should be tailored to the study and the needs of its stakeholders and decision makers. A public involvement plan should be prepared at the outset of every planning investigation. It should be clearly linked to when and how decisions are to be made and these linkages need to be carefully communicated to participants. Progress toward decision making should be communicated to participants and the community at large using appropriate methods and technologies. This means providing nontechnical explanations on technical matters when appropriate or when requested to do so by stakeholders and the community.

The planning entity needs to commit to this process. This means providing sufficient time, resources, and expertise to meet that commitment. The process should be inclusive rather than exclusive. Everyone's participation is welcome. Those with a

CORE VALUES FOR THE PRACTICE OF PUBLIC PARTICIPATION

- Public participation is based on the belief that those who are affected by a decision have a right to be involved in the decision-making process.
- Public participation includes the promise that the public's contribution will influence the decision.
- Public participation promotes sustainable decisions by recognizing and communicating the needs and interests of all participants, including decision makers.
- Public participation seeks out and facilitates the involvement of those potentially affected by or interested in a decision.
- Public participation seeks input from participants in designing how they participate.
- Public participation provides participants with the information they need to participate in a meaningful way.
- Public participation communicates to participants how their input affected the decision.

Source: International Association for Public Participation. Accessed June 17, 2012 from http://www.iap2.org/associations/4748/files/CoreValues.pdf.

known interest in the issue need to be identified, invited, and encouraged to be involved in the process. The best process will foster respect for diverse values and interests.

The public deserves a say in decisions about actions that affect their lives. A variety of techniques should be employed to provide opportunities for the public to have input in an open, two-way exchange of information, knowledge, opinions, and perspectives. This public involvement should continue from the beginning of the planning investigation to its end. Communication among all participants should be open, honest, early, and continuous.

The planning team should provide reasonable and timely access to relevant data, reports, models, records, documents, and the like in an understandable and user-friendly way. Education and training requirements to enable this access are provided for in the best involvement programs. A good process provides ample opportunities for understanding issues and decisions and for providing input as well as for receiving feedback. To facilitate the feedback, the process should provide a public account of decisions made and the information required to assure that the public understands how their involvement affected those decisions.

10.3 STAKEHOLDERS

Stakeholders are special members of the public either because they can cause you to fail or because their support is necessary for success. Thus, a stakeholder is someone who can hurt or help your efforts in any given endeavor. Note that this does not include everyone you come into contact with during your study. Anyone with a legitimate interest in your issue can be a stakeholder, however.

10.3.1 Identify Your Stakeholders

It is wise to identify them precisely because they can help you or hurt you. The simplest way to identify your stakeholders is to brainstorm (see the following text box). Think of anyone that might be affected by your study or the solutions that will stem from it. The planning entity has its own stakeholders, the planning process may have another group, and any planned solution may have yet another group of stakeholders.

CONSIDER THESE QUESTIONS TO HELP YOU IDENTIFY STAKEHOLDERS

Whose opinions will matter?
Who is likely to support your efforts?
Who is likely to object or put up obstacles?
Who can make you fail?
Whose support is necessary for success?
Who is not playing but should be?

Who can help you gather necessary evidence?
With whom should you be communicating on a regular basis?
Who has expressed an interest in this issue or issues like this in the past?
Who will be affected by the solution?
Who will you consult before you make decisions?
Who can influence the decision, but is not directly involved with it?
What external groups or organizations are affected by the change? Are there specific contact people?
What people are interested in the success or failure of the plan?
Who will support the study?
Who might be opposed to the study?
Are there political, environmental, or social consequences of the solution?

Stakeholders can include such disparate groups or individuals as your coworkers and boss, elected officials, government agencies, interest groups, residents, academia, and the like. When you identify stakeholder groups, remember you must ultimately identify one or more individuals in the group with whom you can initiate communication.

10.3.2 Prioritize Your Stakeholders

The brainstorming process might produce a long list of people and organizations that could be affected by your study. Because some can help you and others can hurt you it is best to know who they are so they can be appropriately involved in your planning process. Consider the interest/power matrix (Thompson 2012) of Figure 10.2.

The planning process must pay the most attention to stakeholders with a high power to affect your effort. This may violate your sense of social justice, but it remains a simple fact. Your boss will have high interest and power. Your spouse may have high interest and no power. You need to fully engage those with both high interest and high power. Those with less interest you should satisfy but not engage so much that you bore them or turn them off to your study.

Low power, interested people can often be very helpful to you in your evidence gathering efforts. Keep these people well informed. Communicate regularly with them and take care to ensure all their major issues are being addressed as they arise. People with low levels of interest and power should not be ignored. Monitor their interest. Address their concerns but do not bore them with too much communication.

10.3.3 Understand Your Key Stakeholders

Once you identify stakeholders with high levels of interest and power, you have a list of your key stakeholders. You need to learn how they feel about and will react to your study. Then you have to plan to engage them most effectively in your investigation. The best planning studies repeat this process for each group of stakeholders.

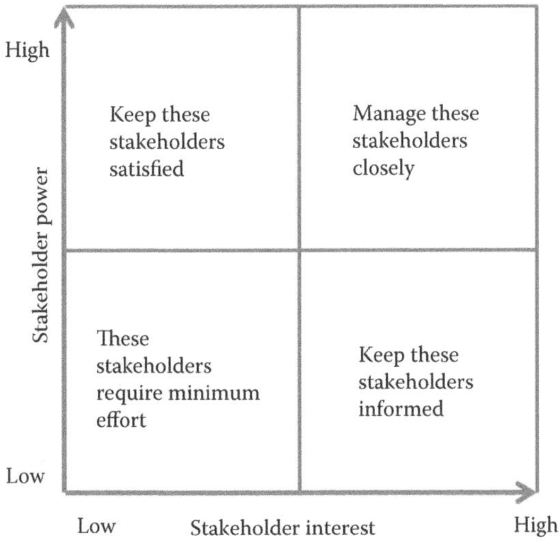

Figure 10.2 Matrix of stakeholder interest and power source. (From Thompson, Rachel. 1996–2012. *Stakeholder analysis winning support for your projects.* Mind Tools. Accessed June 17, 2012 from http://www.mindtools.com/pages/article/newPPM_07.htm. © Mind Tools Ltd, 1996–2012. All rights reserved. With permission.)

Thompson has suggested some key questions to help you understand your stakeholders. Some examples include:

- What financial or emotional interest do they have in the outcome of your work? Is it positive or negative?
- What motivates them most of all?
- What information do they want from you?
- What information is important for them to have?
- How do they want to receive information from you? What is the best way of communicating your message to them?
- What is their current opinion of your work? Is it based on good information? Who influences their opinions generally, and who influences their opinion of you?
- If they are not likely to be positive, what will win them over to support your project?
- If you don't think you will be able to win them over, how will you manage their opposition?
- Who else might be influenced by their opinions?

To get answers to these questions, meet with and talk to your key stakeholders directly. Once you have done this, you can map the stakeholders onto your matrix in a relative manner, as shown conceptually in Figure 10.3. Stakeholders positively inclined toward your efforts are mapped in a Neuropol font, opponents are shown in ALGERIAN. Neutral parties or those yet to take a position have been coded in Bradley Hand. Keep the map current as positions, interest,

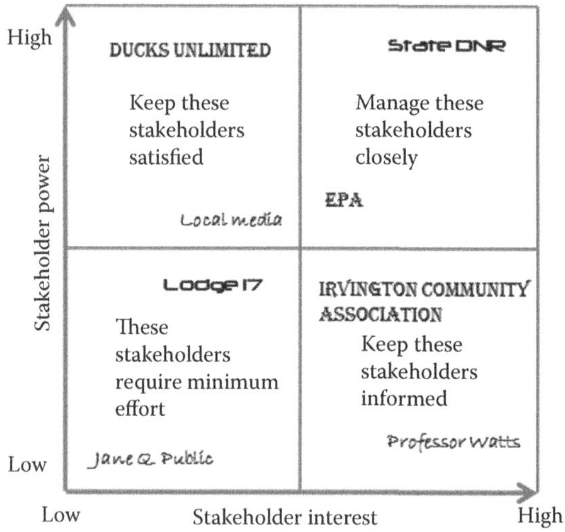

Figure 10.3 Interest/power matrix with stakeholders mapped source. (From Thompson, Rachel. 1996–2012. *Stakeholder analysis winning support for your projects*. Mind Tools. Accessed June 17, 2012 from http://www.mindtools.com/pages/article/newPPM_07.htm. © Mind Tools Ltd, 1996–2012. All rights reserved. With permission.)

and power relationships change throughout your study. As the study progresses, expect more stakeholders to emerge.

Lewinson (2011) suggests strategies for managing each set of stakeholders. The most interested and powerful stakeholders become project collaborators or partners in joint efforts, and they may participate in strategic decision making. Those who are highly interested but have little direct authority to influence the study become your consultants and experts. Powerful stakeholders with less interest are managed as givers of necessary information. All others play a less-active role in the process, receiving less frequent and filtered information.

10.4 INVOLVE THE PUBLIC

There are a wide variety of public involvement techniques available and with technology they continue to grow. Table 10.2 presents examples of four categories of formal techniques for communicating in a public involvement setting. There are traditional methods for getting information to or from stakeholders and the public as well as newer Internet or digital-based tools and techniques for communicating. Several methods appear in both the to and from categories.

Conspicuously absent from Table 10.2 are, hopefully, the most obvious techniques of face-to-face conversation in an informal setting and regular telephone communication. Creighton (2005) is recommended for more details and entrance to the professional literature on these public involvement topics.

Table 10.2 Examples of Traditional and Digitally Based Communication Methods for Public Participation Programs

Getting Information TO People		Getting Information FROM People	
Traditional	**Internet/Digital**	**Traditional**	**Internet/Digital**
Briefings	CDs and DVDs	Advisory group	Blogs
Brochures	Chatrooms	Charette	Bulletin boards
Distribution lists	Computer simulation	Coffee Klatch	Chatrooms
Documents	Discussion boards	Consensus building	Discussion boards
Exhibits and displays	Distance learning	Contests	FAQ and Q&As
Fact sheets	Downloads: data, models, reports	Electronic town meeting	Groups on social networking websites
Feature stories	Electronic newsletters	Field trip	Interactive websites
Forums	Emails, text messages, and SMS blasts	Focus group	Listserv
Interviews	Facebook	Hotlines	RSS feeds
Logo items (e.g., mugs, thumb drives, fans)	Games	Instant voting (clickers)	Role playing
Mailings	Hotlines	Interviews	Shared spaces
Media interviews	Interactive media	Meetings	Site visits
Media kits	Instant messaging	Negotiation and mediation	Surveys, polls, questionnaires
Meetings	Multimedia resources	Public hearings	Texting
News conferences	Online advertising	Shared vision planning	User data analysis
Newsletters	Online videos and simulations		Virtual communication
Newspaper inserts and advertisements	PDF-formatted documents		Wiki spaces
News releases	PowerPoint presentations		Web conferencing
Panels	Project website		
Presentations	Podcasts		
PSAs	Twitter		
Repositories	Videos		
Roundtable discussions	Visualizations		
Symposia	Webcasts		
Visitor centers	Webinars		
Workshops			

10.5 WHEN TO DO PUBLIC INVOLVEMENT?

The commitment to public involvement must be present from the first day of your planning effort. It is not to be an add-on or an afterthought. It is not part of a grand check-off list; held a public meeting, check! It is an essential part of the planning process that touches and is touched by every planning step. Public involvement is as necessary for evidence gathering as it is necessary for decision making.

As noted earlier in the chapter, each planning step provides an opportunity for different levels of public involvement. Different categories of stakeholders will play different roles in a public involvement program. Table 10.3 suggests some recurring and predictable points of input and feedback in the planning steps. Each study is likely to have its own unique public involvement opportunities as well, depending on its decision context, the stakeholders, and their levels of interest.

Dalton and Harter (2012) offer four tips for beginning collaboration. First, the more resource-intensive the public involvement process, the sooner you should communicate your intentions to outside stakeholders. If people and groups are going to require resources to participate in the process, they deserve as much lead time as possible to arrange the availability of those resources.

WORK YOUR PLAN!

- Develop project-specific mailing and contact lists using all appropriate methods (door-to-door personal contact, telephoning, sign-up sheets at local meetings, return postcard mailings).
- Communicate a clearly defined reason (goals and objectives) for public involvement, stressing the value you place on the community's participation.
- Establish communications with the community or affected public using the methods they prefer (electronic mail, printed flyers, mailings, meetings, door-to-door contact, radio, or advertisements in newspapers).
- Advertise your interest through media sources such as the radio, TV, newsletters, and newspapers (use non-English media as well). When possible, include a toll-free contact number.
- Ensure all communications are easy to read (plain English or plain, non-English languages). Include a contact name and number, and provide alternative contacts for non-English speakers.
- Place notices with local government, social, faith-based, and civic organizations and in locations with bulletin boards or kiosks (malls and banks) such as town halls, grocery stores, and senior citizens' centers.
- Advertise and host your own forums, panels, and meetings (at convenient times and locations), and attend community-hosted events to create interest in involvement activities.
- Use surveys, questionnaires, and scientific sampling to identify those who might be concerned or affected by the issues and involvement activities.

Source: EPA Brochure EPA 233-F-03-006. 2011. How to Identify People to Involve. Accessed December 5, 2011 from http://www.epa.gov/publicinvolvement/brochures/peopleinvolv.pdf.

Table 10.3 Points of Input and Feedback for Public Involvement in the Planning Process

Input Opportunities	Feedback Opportunities
Step 1: Decision context	
Help identify problems and opportunities	Confirm problems and opportunities
Help identify planning objectives and constraints	Vet planning objectives and constraints
Help identify decision criteria	Vet decision criteria
Provide input to without condition scenario	Vet without condition scenario
	Vet public involvement process and stakeholder roles
Step 2: Formulate solutions	
Source of solution ideas—measures and plans	React to proposed solutions
Source of community needs, concerns, and values	
Step 3: Evaluate solutions	
Identify candidate evaluation criteria	React to likely effects—beneficial and adverse
	React to solutions eliminated and retained
Step 4: Compare solutions	
May provide insight into important trade-offs	React to differences and trade-offs among solutions
Step 5: Make decision	
May provide weights for decision criteria trade-offs	Keep informed about study progress
May identify preferred solution	React to decision
May make final decision	

Second, the more your process is focused on recommendations, agreements, or stakeholder actions, the sooner you should begin. It is going to take time to identify stakeholders and representatives of stakeholder groups. These representatives are going to need time to gather information, needs, and opinions from their constituencies. If you are going to have to contract out for support for facilitation, logistics, or scientific/technical support, this will also take time.

Third, coordinating the logistics is very time-consuming. Just try to find a time when four people can meet, much less a large group. You need to get an early start on such meetings. Fourth, it always takes longer than you think it is going to take. Dalton and Harter (2012) recommend that you begin talking with stakeholders up to 18 months before you are actually ready to commence a stakeholder involvement process. Several National Research Council recommendations that are especially useful for natural resources planning are seen in the following text box.

RECOMMENDATIONS

The National Research Council Panel on Public Participation in Environmental Assessment and Decision Making reviewed the research and practitioner literature and reported the following conclusions and recommendations regarding public participation:

Conclusion: When done well, public participation improves the quality and legitimacy of a decision and builds the capacity of all involved to engage in the policy process. It can lead to better results in terms of environmental quality and other social objectives. It also can enhance trust and understanding among parties. Achieving these results depends on using practices that address difficulties that specific aspects of the context can present.

Recommendation 1: Public participation should be fully incorporated into environmental assessment and decision-making processes, and it should be recognized by government agencies and other organizers of the processes as a requisite of effective action, not merely a formal procedural requirement.

Recommendation 2: When government agencies engage in public participation, they should do so with clarity of purpose, commitment to use the process to inform their actions, adequate funding and staffing for the duration of the process, appropriate timing in relation to decisions, a focus on how conclusions or policies can be implemented, and commitment to self-assessment and learning from the experience.

Recommendation 3: Agencies undertaking a public participation process should, considering the purposes of the process, design it to address the challenges that arise from particular contexts. Process design should be guided by four principles:

- Inclusiveness of participation
- Collaborative problem formulation and process design
- Transparency of the process
- Good-faith communication

Recommendation 4: Environmental assessments and decisions with substantial scientific content should be supported with collaborative, broadly based, integrated, and interactive analytic-deliberative processes such as those described in *Understanding Risk* (1996) and subsequent National Research Council reports. In designing such processes, the responsible agencies can benefit from following five key principles for effectively melding scientific analysis and public participation:

- Ensuring transparency of decision-relevant information and analysis
- Paying explicit attention to both fact and values
- Promoting explicitness about assumptions and uncertainties

- Including independent review of official analysis and/or engaging in a process of collaborative inquiry with interested and affected parties
- Allowing for iteration to reconsider conditions on the basis of new information

Source: Dietz, T. and Stern, P. C., eds. 2008. *Public Participation in Environmental Assessment and Decision Making*, pp. 1–4. Washington, DC: National Research Council, National Academies Press.

10.6 TAKE AWAYS

Lesson 1: The public is anyone who is not you. There is no such thing as "the" public.
Lesson 2: If you don't involve the public, you will regret it.
Lesson 3: Public involvement is essential throughout the planning process.
Lesson 4: Different stakeholders need to be treated differently.
Lesson 5: Use a rich variety of techniques to involve the public.

REFERENCES

Committee on Risk Characterization, National Research Council. 1996. *Understanding Risk: Informing Decisions in a Democratic Society*. Washington, DC: The National Academies Press.

Creighton, J. L. 2005. *The Public Participation Handbook: Making Better Decisions through Citizen Involvement*. San Francisco: Jossey-Bass.

Dalton, D. and P. Harter. 2012. *Better Decisions through Consultation and Collaboration by EPA's Conflict Prevention and Resolution Center*. Washington, DC: Conflict Prevention and Resolution Center, USEPA. Accessed February 6, 2012 from http://www.ch.doe.gov/eatc-2010/docs/pdf/better_decisions.pdf.

Dietz, T. and Stern, P. C., eds. 2008. *Public Participation in Environmental Assessment and Decision Making,* pp. 1–4. Washington, DC: National Research Council, National Academies Press.

EPA Brochure EPA 233-F-03-006. 2011. How to Identify People to Involve. Accessed December 5, 2011 from http://www.epa.gov/publicinvolvement/brochures/peopleinvolv.pdf.

International Association for Public Participation. Accessed December 5, 2011 from http://www.iap2.org/associations/4748/files/CoreValues.pdf.

Lewinson, M. 2011. Developing a strategy for managing stakeholders: Stakeholder list and stakeholder analysis matrix. Accessed February 6, 2012 from http://www.mymanagementguide.com/developing-a-strategy-for-managing-stakeholders-stakeholder-list-and-stakeholder-analysis-matrix/.

Thompson, R. 2012. Winning support for your projects. Mind Tools Limited. Accessed February 6, 2012 from http://www .mindtools.com/pages/article/newPPM_07.htm.

CHAPTER 11

Tell Your Story

11.1 INTRODUCTION

I've got to tell you, I had never been more proud than when I held a printed copy of the first feasibility study I had ever worked on in my hands. And it was heavy. Words I wrote were scattered throughout the main report and the entire economics appendix was all my work. It was jam packed with facts. I had poured my heart and soul into it. Here is something I have learned since that day: facts are not the best way to get your point across. They do not win people over or persuade them. An emotionally based story can have a greater impact than a traditional argument or a bunch of facts (Ruger 2010). Report writing needs to give way to storytelling. Causative stories are what you want to write: "This caused that and we did this because…"

Why did you do what you did in the study? Why's motivate much more than who's, what's, when's, and where's (although they can be important too). Why's go to motives, feelings, and emotions. Why did you reject the exercise trail, why did you include so many duck boxes, why did you decide what you did?

EIGHT GOOD IDEAS

1. Engaging beginning, interesting middle, satisfying ending.
2. Narrative quality (no data dump, no default formats, no take the last report and add a little to it).
3. Chronology is your friend.
4. No geek speak.
5. Cut to the chase—write it;the way you'd say it; nouns and verbs beat adjectives and adverbs.
6. User-friendly and informative figures and features (let people access data/info in the ways they like)—write on a map or a picture.
7. Tie decisions and judgments to the evidence (say why you did what you did).
8. Tell the truth (no matter where it takes you).

The Central and Southern Florida Comprehensive Review Study Final Integrated Feasibility Report and Programmatic Environmental Impact Statement (United States Army Corps of Engineers 1999) was the basis for the Comprehensive Everglades Restoration Plan (CERP). The report has a total of 4034 pages. The Final Feasibility Report without annexes is 592 pages in length. The story of this comprehensive plan is much, much shorter.

Oh, how we love our data. Figure 11.1 shows an excerpt of a table that took seven pages (pp. 1158–1164) to run in the CERP report (2012).* I know the effort that went into doing this cost-effective analysis of ecosystem management measures. The codes indicate combinations of management measures that numbered 294 in total. Their costs and amounts of water produced are also shown. It was important to do, it was a lot of hard work, and maybe someone somewhere cares about entry 209 on page 1162. But I doubt it. No offense to any of the hardworking analysts out there is intended. I have produced such tables myself. No one really cares about all these numbers and facts.

No one cares as much about your numbers and facts as you do. When telling a story, it is far more important to explain what the numbers mean and why they are important than it is to present the numbers. People care most deeply about the things that touch, move, and inspire them. Facts just can't do that; stories to which people can connect can. When it comes to important decisions, people only care about the facts insofar as they support our feelings about a matter. French novelist Marcel Proust (1957) said, "Facts do not find their way into the world in which our beliefs reside; they did not produce our beliefs, they do not destroy them; they may inflict on them the most constant refutations without weakening them…"

Simmons (2001) says people are up to their eyeballs in information and don't want more information. They want to have faith in your plan and in the story you tell.

* Lest anyone wonder if I am picking on this report, let me assure you I am not. There was any number of candidate plans I could have selected. This was chosen for its high visibility, its availability, and the excellent manner in which the planning team eventually learned how to tell its story. Explore CERP's wonderful, storytelling website at http://www.evergladesplan.org/.

Cost Effective Combinations of Management Measures for Function C

COUNT	CODE													COST ($1,000)	OUTPUT (acre-feet)	AVG. COST ($1,000/acre-foot)
1	A0	B0	D0	E0	F0	G0	H0	I0	J0	K0	Z0	C0	X0	0	0	0.0000
2	A0	B0	D0	E0	F0	G0	H0	I0	J0	K1	Z0	C0	X0	369	582	0.6340
3	A0	B0	D0	E0	F0	G0	H0	I1	J0	K0	Z0	C0	X0	860	8,951	0.0961
4	A0	B0	D0	E0	F0	G0	H0	I0	J1	K0	Z0	C0	X0	972	51,702	0.0188
5	A0	B0	D0	E0	F0	G0	H0	I0	J1	K1	Z0	C0	X0	1,341	52,284	0.0256
6	A0	B0	D0	E0	F0	G0	H0	I1	J1	K0	Z0	C0	X0	1,832	60,653	0.0302
7	A0	B0	D0	E0	F0	G0	H0	I1	J1	K1	Z0	C0	X0	2,201	61,235	0.0359
8	A0	B0	D0	E0	F0	G0	H0	I2	J1	K0	Z0	C0	X0	2,691	67,170	0.0401
9	A0	B0	D0	E0	F0	G0	H0	I2	J1	K1	Z0	C0	X0	3,060	67,752	0.0452
10	A0	B0	D1	E0	F0	G0	H0	I1	J0	K0	Z0	C0	X0	3,095	68,051	0.0455
11	A0	B0	D1	E0	F0	G0	H0	I0	J1	K0	Z0	C0	X0	3,207	110,802	0.0289
12	A0	B0	D1	E0	F0	G0	H0	I0	J1	K1	Z0	C0	X0	3,576	111,384	0.0321
13	A0	B0	D0	E0	F0	G0	H1	I0	J0	K0	Z0	C0	X0	3,995	231,200	0.0173
14	A0	B0	D0	E0	F0	G0	H1	I0	J0	K1	Z0	C0	X0	4,364	231,782	0.0188
15	A0	B0	D0	E0	F0	G0	H1	I1	J0	K0	Z0	C0	X0	4,855	240,151	0.0202

Figure 11.1 Excerpt from seven-page-long report table illustrating planners' tendencies to dump data.

Faith moves mountains, not facts, and facts do not give birth to faith. She suggests that once people make your story their own they will have faith in your plan. This chapter is about how to tell your story.

What are the characteristics of a great report? It is easy to read, it uses plain language, it has maps and graphics, it tells your story, and it is brief. A great report answers all the decision makers' and the public's questions and provides recommendations. It says why things were done as they were and why you recommend what you do. It raises red flags for exceptions to the rules as well as for unique, different, special, or controversial elements.

Here is what you can expect in this chapter. The elements of a good story are the first topic. Reports tend to dump data and follow a process rather than offer a chronological description of the way the action unfolded. It would not be unusual to see a planning report structured around the six-steps of the P&G planning process, for instance. So, how does a story differ from a report? The answer is by its structure. Six classic story structures are offered for your consideration. The hope is one of them and a few of the other story elements described might suit your storytelling needs. Seriously, pick a story structure.

Write simply. It's not that hard. Simple words and short sentences are a good start. This is the message of the chapter's third section. A readability index is described to help you gage how complex your narratives are. A number of rules for writing simply are offered.

Use pictures. Photographs, maps, graphs, charts, and all manner of figures make your story more interesting and more effective. These are the subject of the topic that follows writing simply. The last chapter topic is documenting your story. It differentiates between planning documents and documenting planning. Think of the former as formal reports and the latter as storytelling. They need not be separate tasks, but they can be at times.

Many planning contexts require formal reports. Once, a written report was the only way to tell a story. That is no longer the case. Video reports and multimedia websites with 360 degree virtual tours, animations, games, and puzzles can do the job now. Planners can use YouTube videos, Facebook pages, blogs, tweeting, web conferences, discussion boards, and wiki spaces throughout the planning process. They become part of the living history of the plan and a valuable part of the story or the report, as planners' needs dictate.

11.2 ELEMENTS OF A GOOD STORY

Here is how I learned to prepare a planning report. First, get the last report your office did and do everything they did. Next, add some stuff to it to prove your value. In this time-honored tradition, reports got thicker and thicker while stories got lost and forgotten. Bring back the stories.

Good stories don't just happen. They are more than a series of events strung together. The sequence of events should make sense while developing the story and moving it along. Stories succeed when they evolve according to the audience's expectation of unity. Ruger (2010) speaks eloquently about the role of storytelling and points to Aristotle's *Poetics*, where Aristotle says a good story is unified and focuses on an extended action with a beginning, middle, and end. All good stories have a hero (in planning, perhaps a community, an ecosystem, a species, a physical system, or process) and some type of quest or mission (preservation, restoration, mitigation) that must be completed. Why? Because it works. People are drawn to stories with which they can connect. Ruger (2010) identifies six story structures that people have found universally appealing.

Traditional. This is the three-act play structure. Your story has an engaging beginning ("once upon a time..."), an interesting middle that moves along briskly while building interest (a talking mirror might help), and a satisfying end (they lived happily ever after). To use this in a planning setting, you need a well-defined inciting event (why did you do this study?), a protagonist, and a climax (recommendation). These elements, not the planning process or the last planning report, should drive the story.

A good planning story beginning should accomplish at least three things. First, it should compel the reader to read on. Second, it should provide enough information for the reader to understand your basic findings or recommendation. Finally, it should strike a chord in the reader by revealing to her something that she recognizes as important and in her interest to be informed about.

The middle of your report tells your story. Always bear in mind you do have a story. First, this happened. Then, that happened, and so on. Many planning stories are best told as a narrative of events. When a story is a narrative of events, use time order to organize it. Chronology is your friend. Planning reports that use the steps of the planning process impose a narrative order that is artificial and likely to be confusing to the reader. If the study began when the mayor came to your office and said, "This is what we'd like you to build," then begin your story there. Tell the reader how the study unfolded. Tell the story of how you went from that first day to the recommendation in your study.

Endings need to provide a sense of closure. There should be nothing important left unsaid. All loose threads in the report must be tied up by now. Clear, unambiguous endings are best. Planning studies have a natural ending: the recommendations and steps necessary for implementation. This may be the most practical story structure. If you have a more dramatic flair, there are others.

Hero's quest. This may be the most commonly recognized story structure. The story is driven by the pursuit of a (usually) tangible goal. The hero facing a challenge

PLANOGRAPHY

To this traditional archetype, I add a variation, the invention of my friend, Ken Orth, the "planography." This is the biography or life history of a plan. It's a diary. It begins with what you did the first day of the study, and it ends with the decision. Be sure to include explanations of why things were done and why decisions were made. This is the chronology of the planning process in a rich narrative.

may be a human or other community, an ecosystem, a resource, a natural system, or process. The quest is illuminated by the challenges keeping the hero from the goal. Many planning stories can be told in this structure. The poor urban community that seeks to be reconnected to the natural environment is an example.

The search. The search is driven by the pursuit of an intangible goal, usually the pursuit of meaning. The hero here is often on a mission of discovery of something fundamental about who/what they are or who/what they will become. Sustainability, biodiversity, environmental justice, and social vulnerability are all themes that lend themselves to a search story.

Stranger in a strange land. This is the classic story of change. The hero is put into a new situation (deer in a suburban environment). Everything (e.g., the local rules and customs), is strange and unfamiliar. The hero spends much of the story getting accustomed to their new surroundings (or trying to). In many of the human hero stories of this type, the main character learns we are all the same at heart. A planning story might do well to modify that ending. This story is driven by the experiences of the hero in adapting to conditions in the strange environment. Climate change offers the potential of strange new lands for some investigations.

Boy meets girl. All together now: boy meets girl, boy falls in love, boy loses girl, boy wins girl back. Stories need tension to move forward, so this story is driven by a wonderful little dynamic that is readily adapted to a planning story. "Then what happens…?" is what really drives this type of story. What hurdles must they leap, conflicts must they resolve, or obstacles stand in the way of true love? Reconnecting humans with their natural environment might be a type of story readily told through this genre. Love (or natural values) is hard to find and once found it is hard to keep. This story is driven by conflict and overcoming barriers. Persistent problems with a long history of not-quite-good-enough efforts at solution lend themselves to this story structure.

Coming of age. These are stories about change, transformation, or maturation of the hero. The hero often has to relearn life's lessons as events change them or as they near maturity. The experiences the hero gains as he evolves drive the story. This could be analogous to the changing morphology and function of natural areas. The hero may change significantly or simply figure out they somehow knew the answers all along.

Although the idea may seem uncomfortably silly at first, finding an archetypal story structure that suits your planning story is an excellent starting point for telling a good story well. What kind of story are you telling? Know the answer to that question.

HERO/VILLAIN PAIRS

Fish/acid mine drainage
Indigenous plants/nonindigenous plants
Water quality/phosphorous
The port/sedimentation
Buffalo/brucellosis
Floodplains/development
Development/floods

To engage your audience emotionally, your story needs characters with whom they can identify. When these characters are from the natural environment, it takes a special touch to make them well-rounded without resorting to jargon. But, it is not that difficult to make a forest, an ecosystem, a deer population, or any other warm-and-fuzzy valued resources sympathetic characters. That is not going to be the case in every situation, and finding and developing your heroes is an important part of any planning story.

Your story is also going to need action, something that drives or impels it forward to a climax. Challenges are many in natural resources planning, and challenges create the kind of tension that makes people want to hear your story. Build your story on conflict. Ruger (2010) has identified six sources of conflict, all of which are typically plentiful in natural resource planning studies. These are:

1. Man vs. man includes relational conflicts among people, for example, varying stakeholder groups.
2. Man vs. himself includes the internal conflicts we experience individually and as a community or species, for example, the twin desires of economic development and reducing our footprint.
3. Man vs. society includes social conflict involving a specific group of people, for example, environmental justice, social justice, and social vulnerability issues.
4. Man vs. nature includes external conflicts between the natural and the material worlds, for example, development of floodplains.
5. Man vs. machine includes a conflict with technology, for example, strip mining.
6. Man vs. fate includes internal or external conflict with the moral, ethical, or supernatural, for example, animal welfare issues.

Instead of asking "What else should we put in the report?" ask yourself "What can we do to help people relate to our message?" Focus on connecting with your audience. This is far more important than impressing them with the extent of your knowledge and analysis.

Don't overthink this idea of storytelling. It is not that complicated. Identify your characters before you start writing. Who are the heroes, who are the villains? Jump right into the story at the beginning. Don't clutter things up with "administrative

HOW MANY WRITERS?

Planning reports are often written by various team members, and they read like it too. Not all writers are equally gifted. Not all use the same writing style. Few things are more annoying to a reader trying to understand a complex issue than to try to wade through a poorly organized report written by a dozen people who apparently never spoke to one another or bothered to read what the others had written. That does not mean it *must* be written by one person, but it probably should be.

requirements" that add nothing to your main point, which should be some variation of "this is what we did, this is why we did it, and this is how it turned out." Make it interesting. Don't preach. Let the story grow. Let readers learn the logic of your conclusions for themselves from the power of your story. When you get to the end, stop. Wrap it up quickly (Shephard 1998). The next section offers suggests for telling a good story simply.

11.3 WRITE SIMPLY

Ernest Hemingway is said to have used four rules for writing simply that he learned working for the *Kansas City Star* (quoted in Mitchell 2011). They are:

1. Use short sentences.
2. Use short first paragraphs.
3. Use vigorous English.
4. Be positive, not negative.

Planners could do a lot worse, and frequently do.

Write so your reader understands you. A number of readability indices have been developed to help writers gage how appropriate their writing is for their readers. Gunning (1952) created an easy-to-use and widely applied index that now bears his name, Gunning's Fog Index (or FOG) Readability Formula. Since that time, numerous other indicators of the ease of reading have been developed.

One of these, the Flesch-Kincaid Grade Level Readability Test, is available with Microsoft Word. The Flesch-Kincaid Grade Level Readability Test is described by this rather opaque formula:

$$\text{FKRA} = (0.39 \times \text{ASL}) + (11.8 \times \text{ASW}) - 15.59$$

where FKRA = Flesch-Kincaid reading age, ASL = average sentence length (total number of words divided by the total number of sentences), and ASW = average

WHO IS YOUR READER?

The purpose of a report is to communicate ideas to another person. The first and most important question the writer must ask is who the primary reader of this report is. Once you have identified the reader, empathize with her throughout the writing process. Put yourself in the reader's position. Don't write for the study team or for your own personal glory. Write for that person who is going to be reading. Tell her your story in a way that she can understand. A report written for a technical reviewer or an expert with intimate knowledge of the problem is going to be very different from a report written for Jane Q. Public.

number of syllables per word (total number of syllables divided by the total number of words). The resulting number (FKRA) is the appropriate reading age for the reading material you have tested. A score of 7 means a seventh grader would be expected to read and understand this material. A score of 16.2 indicates material appropriate for a college graduate. In Word 2010, the index is found on the Review tab in the Proofing cluster as part of the Spelling & Grammar function. If it is not already available, it can be added through the Word options found on the File tab. Let's try it on the opening paragraph of the CERP report (2012), which follows:

> The Central & Southern Florida (C&SF) Project extends from south of Orlando to the Florida Keys and is composed of a regional network of canals, levees, storage areas and water control structures. First authorized by Congress in 1948, the project serves multiple purposes. The authorized purposes of the project include flood control, regional water supply for agricultural and urban areas, prevention of salt water intrusion, water supply to Everglades National Park, preservation of fish and wildlife, recreation and navigation. For close to 50 years, the C&SF Project has performed its authorized functions well. However, the project also has had unintended adverse effects on the unique natural environment that constitutes the Everglades and South Florida ecosystem.

Figure 11.2 shows readability statistics generated by Word 2010. The FKRA is 15.1. That means this material is suitable for an advanced college student. That is a much higher grade level than you should target for the general public. An oft-repeated ideal target I was unable to trace to an original source is an index of 7 or 8, that is, a seventh or eighth grade reading level. Anything above 12 is generally regarded as too hard for most people to read.

This index can be used for orally delivered messages as well. It is helpful to note that during low-stress situations, information is processed at an average grade level for the audience. Local newspapers target an eighth-grade reading level. When stress is high, information is processed at four grade levels lower, for example, from eighth grade to fourth grade (Food Insight 2010).

12 SIMPLE WRITING RULES

1. Verbs has to agree with their subjects.
2. Prepositions are not words to end sentences with.
3. And don't start a sentence with a conjunction.
4. It is wrong to ever split an infinitive.
5. Avoid cliches like the plague (they are old hat).
6. Be more or less specific.
8. Parenthetical remarks (however relevant) are (usually) unnecessary.
9. Also, too, never, ever use repetitive redundancies.
10. No sentence fragments.
11. Don't use no double negatives.
12. Proofread carefully to see if you any words out.

Source: Wheeler, S. 2010. Writing rules. Accessed February 13, 2012 from http://steve-wheeler.blogspot.com/2010/04/12-writing-rules.html.

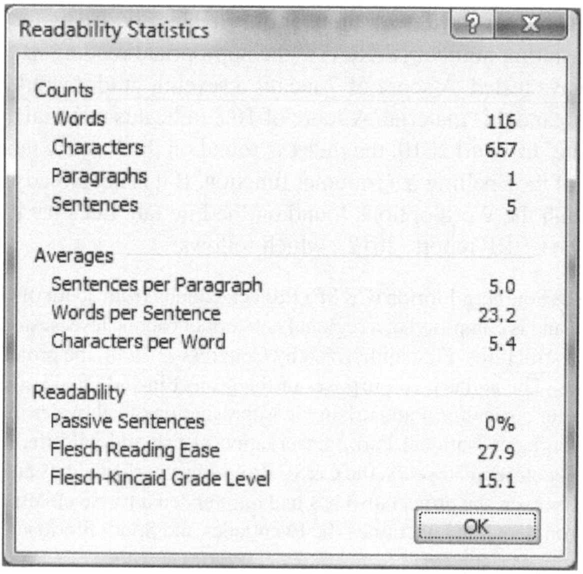

Figure 11.2 Microsoft Word 2010 readability statistics with Flesch-Kincaid reading index score.

11.3.1 A Few More Simple Writing Rules

- *Avoid acronyms and jargon.* You may find acronyms like NEPA, NED, CERCLA, HTRW, USFWS, OMB, and other acronyms second nature. Normal people do not. No acronyms allowed in stories. Likewise, do not use jargon. Habitat units, expected annual damages, potable water, and so on make your message unclear. Avoid words you don't use in everyday speech with your children or nonplanning friends.
- *Fight ambiguity and abstractness.* Words mean different things to different people. If you describe the design level for a wetlands restoration project as –1 foot, we don't know what it means. Sometimes you must get technical and say it is mean sea level, national geodetic vertical datum, mean low water, or mean high water. Be clear.
- *Avoid long words.* Obfuscation does not imply profundity.
- *Kill those euphemisms.* A flood is not a wet water event. A gas station attendant is not a petroleum transfer engineer, and a school bus is not a motorized attendance module.
- *Check the logic of your sentences.* Inane combinations of words often result when we write in haste under a deadline, are distracted, or careless. Make sure all your sentences make common sense.
- *Avoid cop-out phrases.* If an alternative is unacceptable to local interests, say so. Don't say an alternative was suboptimal.
- *Don't mummify your thoughts.* A common mistake in reports and other kinds of writing that are intended to inform and enlighten the reader is to take a good basic idea and then wrap so many meaningless, empty, extraneous, unnecessary words around it that the meaning and very essence of your thought gets lost among the words that exhaust you to read.

- *Don't attempt humor in a report.* Humor is a very personal thing, and you are not likely to hit the target with every reader. See the previous paragraph for an example. Did you find it funny? (I enjoyed it.)
- *Avoid sexist language.* Whether you agree or disagree, it attracts attention these days. On the other hand, avoid nonsensical solutions to this modern sensitivity like (s)he, S/he, or she/he. If it is natural to use gender neutral words, do so. Alternate male and female pronouns now and then if gender neutrality becomes clumsy.
- *Keep explanatory material on target.* If you are telling your story, make sure that what you are writing contributes to the tale you are telling. Concentrate more on telling your story and less on reporting requirements. Get to the point. Eliminate excess words; avoid redundancies.
- *Enumerate.* Enumerate facts, ideas, instructions, questions, and the like. Lists of things have more visibility than text. Lists are a good way to conserve space, and they can be powerful ways to convey ideas.
- *Don't let the subject and verb get too far apart.* This is a simple problem that plagues writers dealing with technical subject matter.
- *Always use spell check.* Then, proofread your work. Never send anything out for editing or review until you have reread what you've written. Ask someone else to read it. Does it make sense to them?

11.4 GRAPHICS

Use graphics in your story. Graphics are visual presentations. They include pictures, maps, charts, tables, drawings, numbers, illustrations, graphs, diagrams, blueprints, artist's conceptions, and virtually any kind of image you can imagine. Every well-chosen graphic tells a story. Graphics can serve different purposes in a planning investigation. Those used to support decision making might be expected to be quite different from graphics used to tell your story to the public.

Show a map. This is where your story takes place. Show the important locations in the story. Do not hesitate to write on your map and to draw an arrow if it helps people understand the story you are telling. Show important places on a map and include ground level photographs as map inserts. Use pictures to illustrate the problem conditions. Doctor photographs to show opportunities to improve conditions. Photoshop is the planner's friend.

If you can illustrate your solution, do so. Conceptual drawings of a finished project juxtaposed to an existing condition can convey a lot of useful information about what your solution is expected to do. Go easy on the blueprints, the public does not understand them.

Show your team at work. Let people see the fieldwork, show them the equipment you used. Photos of you collecting the data will be more interesting to people than your data. Is the elderberry longhorn beetle habitat important to your study?

Show people what the beetle looks like and tell them why people care. Show its ideal habitat, then show the available habitat. Don't just dump facts—tell a story.

When you do have to dump facts, be sure people understand why the facts are important. Give people a reason to care about your data and they will. Then dump them in good graphics. No conversation about good graphics is complete without acknowledging the work of Edward Tufte (1983, 1990, 1997), an expert in the visualization of data and the source of most of the information that follows.

When you use tables, use a horizontal format. The human eye seems trained to look for horizon effects, and it tends to scan side-to-side better than up-and-down. Use no more than nine rows and sort them with the largest numbers at the top. Align your data on the decimal point and show only two significant digits. Use a smaller font than the text and put comparisons of interest in vertically. Use a table anytime you have 20 or fewer numbers.

Tufte has developed a set of principles of graphical excellence that surprise those trained to use pie charts and bar graphs. He says use a well-designed presentation of interesting data. This should include consideration of substance, statistics, and design. The goal is to communicate complex ideas with clarity, precision, and efficiency. This means giving the viewer the greatest number of ideas in the shortest time with the least ink in the smallest space. He notes that excellent graphics are nearly always multivariate. Above all else, an excellent graphic shows the data and tells the truth about it. Minard's graphic of Napoleon's army prepared in the nineteenth century is often cited as an example of an excellent graphic. Images of this graphic are readily found on the Internet. Edward Tufte's website provides many examples of excellent graphics. The examples found at http://www.edwardtufte.com/bboard/q-and-a-fetch-msg?msg_id=0002w4, (accessed June 17, 2012) should stir the imagination of planners who tend to think of PowerPoint and Excel for report graphics.

One of Tufte's recurring recommendations is to maximize the data-ink ratio. This means make the ink on your graphics present information. Do not decorate your data with colors, grids, frames, and shading. Erase all non-data-ink and then erase redundant data-ink.

When you design a graphic, consider size and shape. Do not rely on the default graphs of your software programs. Carefully choose the shape and size of your graphic. Don't settle for default shapes; try different shapes and choose the one that works best. Rectangular graphs are sometimes better than square graphs.

Bigger graphs are not always better than smaller graphs. Sets of thumbnail-sized graphics on a single page can be used to represent aspects of a single phenomenon. The Florida Museum of Natural History provides a great example of multiple small graphs showing shark attacks by location and population size over time. It is reproduced in Figure 11.3 at http://www.flmnh.ufl.edu/fish/sharks/statistics/pop2.htm, (accessed June 17 2012). Multivariate displays like this support comparison and enhance the dimensionality of your story. These multiple minis,

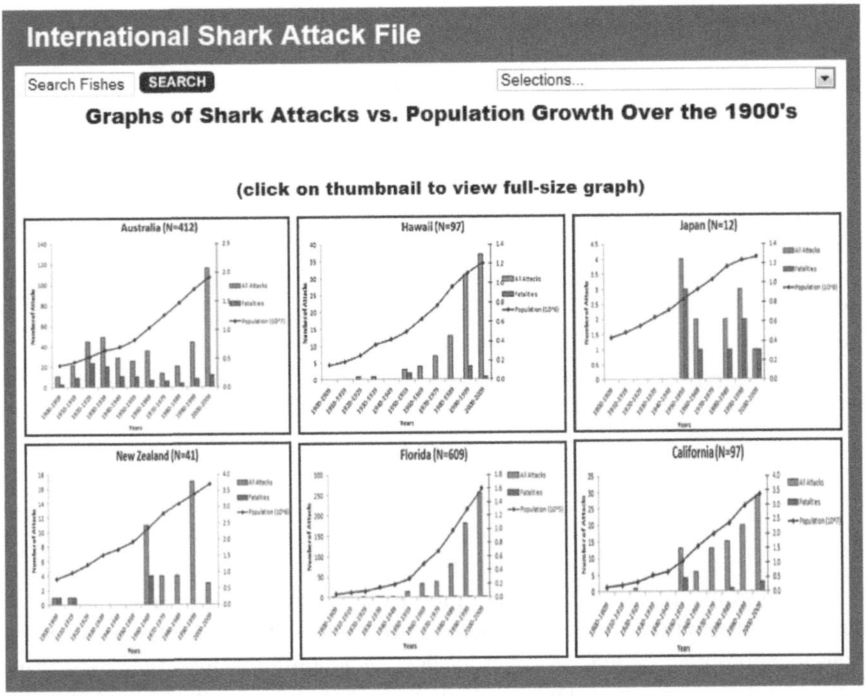

Figure 11.3 An example of multiple minis used to tell the story of shark attacks since the 1900s. (From Florida Museum of Natural History Ichthyology Department. 2012. Accessed February 22 from http://www.flmnh.ufl.edu/fish/sharks/statistics/pop2.htm.)

several small graphics on a page, can tell a rich story compactly. They are well suited to show alternatives or a range of options.

If you're not telling your story in print, the potential for using visual displays literally explodes. You'll find more ideas for visual storytelling in the next section.

11.5 DOCUMENTATION

Think of documentation as a set of different communications that authenticate and support the results of the planning study. The traditional view of documenting a planning study is a planning report. I'd like to propose a somewhat nontraditional view by suggesting that documentation includes both telling your story, as described in Sections 11.2 and 11.3, and reporting your study findings so a decisions can be made. How you document your story and how you document a report could, but need not, be two very different things. Planning documents and documenting planning might require different skill sets.

11.5.1 Planning Documents

Let's begin with the obvious. Someone has asked for and financed a plan, and you owe them some answers. Often a reporting process will be prescribed by the sponsoring agencies. Federal agencies and others following the NEPA process will require either an Environmental Impact Assessment (EIA) and a Finding of No Significant Impact (FONSI) or an Environmental Impact Statement (EIS) and a Record of Decision (ROD). In water resources, planning has historically proceeded in stages. It often begins with a reconnaissance-level study, followed by a feasibility study, and then advanced engineering and design. Each study stage entails an increasing level of detail. Each has its own planning document.

Planning reports are usually prepared as draft documents, which are reviewed and then amended and finalized. Thus, there can be different versions of reports for several planning stages. Formal documentation processes may present you with a long list of must do's, leaving you with few options for innovation. Satisfying the needs of higher authorities is the most important reason for documenting the planning process. When they direct the way you should do it, do it as directed.

Generally a planning report serves two purposes. It is an analytical report and a decision document. As an analytical report, it identifies a problem; analyzes, synthesizes, and interprets pertinent information; and presents conclusions and recommendations for appropriate action. As a decision document, it informs readers of the decision made and the reasons for it. This usually includes a summary of the planning process.

The format for a planning report varies with the magnitude of the study. It can be anything from a letter report of a few pages to a main report with appendices that can run into the thousands of pages (see the IPET text box).

With maps, graphics, editing, writing, printing, distribution, and related costs, documentation can be an extremely expensive part of the planning effort. My experience with report documents, reported above, suggests the most common way of preparing a report is to find the office's most recent example of a report that has successfully navigated the planning review process and replicate it as closely as possible.

IPET REPORT

The post-Katrina Performance Evaluation of the New Orleans and Southeast Louisiana Hurricane Protection System Final Report of the Interagency Performance Evaluation Task Force (IPET) of June 2009 has an executive summary and overview with six appendices and 288 pages (United States Army Corps of Engineers 2009). That's right, the executive summary is 10 of those pages. The overview is almost 300 pages long. The entire report is more than 6000 pages long.

Source: United States Army Corps of Engineers. 2009. Performance evaluation of the New Orleans and southeast Louisiana hurricane protection system final report of the interagency performance evaluation task force volume I—Executive summary and overview. Washington, DC: USACE.

My experience with the review of these reports is that most reviewers only look at the technical appendices for which they have technical review responsibility. Virtually no one reads the entire document. If it is true that few, if any, people are reading these voluminous documents, maybe it is time to rethink what documentation is or ought to be.

Who is the reader of this planning documentation? Formal reports are usually written to meet reporting requirements. The readers of these reports are often reviewers and technical experts who are not always likely to read the entire report. It is likely there are many different readers for your report. If so, doesn't it make sense that there might be many different kinds of documentation? Think of documentation as a set of different documents that authenticate and support the findings of the planning effort. The composition of a document depends on the intended audience. For example, technical reviewers are likely to want to have the most detailed documentation. In a flood risk management study, this might mean stream flow data, backwater profiles, rating curves, damage functions for each reach by type of damage, detailed calculations of expected annual damages, interest during construction calculations, and so on. All these things make a dreadfully dull story. These are good candidates for technical appendices or databases available online.

Devoting more effort to carefully documenting project files that can be made available via the web to a relatively limited number of interested parties might be a viable alternative to devoting extensive resources to the preparation of voluminous technical appendices at the cost of better-documented project files. It is not uncommon for report preparation to absorb so much time and energy that little energy is left to carefully organize and document the raw data in project files. Documentation need not mean put it in the report. Documentation could as easily mean keep it in your files and make them available to interested parties.

11.5.2 Documenting Planning

Unlike planning documents, documenting planning is where you have the greatest freedom to tell your story. This story can certainly be told in writing in a traditional report format, but it need not be. Consider using other media, including the major media and others.

EDITING OR REVISING IS QUALITY CONTROL

Never write a report that you or someone else does not edit and revise at least once. Want to know how your report will sound to the reader? Try reading it out loud. Does it hold your interest? If it doesn't, it won't hold the reader's interest either. You know what you want to say—you need to find the best words to say it. Go through the draft literally word by word and ask yourself whether each paragraph is necessary. Cut the ones that are not. Then work on each sentence, applying the same tests. Do the same for each word. Be merciless. Throw out the words, sentences, and paragraphs that do not contribute to the telling of your story.

11.5.2.1 Role of Major Media

Once, all of America got its news from TV, radio, and newspapers. That world is long gone. Where do Americans get their news today? According to a 2010 Pew Research Center survey (Purcell et al. 2010), Americans are accessing news in multiple formats using multiple platforms on many devices. We are no longer loyal to a network or a newscaster. A vast majority of Americans (92%) get news on a typical day from multiple platforms including national TV, local TV, the Internet, local newspapers, radio, national newspapers, and mobile phones. Almost half (46%) of all Americans get news from four to six media platforms a day. Only 7% get their news from a single media platform. Your story might belong on multiple platforms.

The Internet is at the center of this story of change. Most Americans (59%) get news from a combination of online *and* off-line sources, and the Internet is the third most popular news platform, after local and national television news. Americans have turned into foraging opportunists who access news when the urge strikes or the opportunity presents itself. How do you use the Internet to tell your story?

In the new media environment, people get their news in more portable, personalized, and participatory ways. The Pew survey (Purcell et al. 2010) showed news is portable, with one in three mobile phone owners accessing news on their phones. It is more personalized, as 28% of Internet users have news sources of particular interest to them on a customized home page. It is more participatory as well. About 37% of Internet users have commented on the news, contributed to the creation of news, or helped disseminate it via postings on social media sites like Facebook or Twitter. Are you thinking about mobile phone applications for your story?

Broadcast and cable television and local radio remain effective ways of reaching some of your interested audiences. Some 54% of Americans listen to a radio news program at home or in the car (Purcell et al. 2010). Appearances on radio talk show programs can be an effective way to initiate two-way communication with the interested public. Local TV news remains the most common source of news in America: 78% of Americans get news from a local TV station (Purcell et al. 2010). Public service programming on radio and television shows can provide another opportunity to present information to the public. Press conferences can attract media coverage.

Appearances on local television news programs in brief interview segments can be effective. If you cultivate a relationship with local media, it may be possible to generate occasional reports as your story unfolds. The more people know about your story from other sources, the more interested they will be in how it ends.

Press conferences, press releases, and regular contact with reporters can be effective ways of getting newspaper or local magazine coverage of your story. Half of all Americans (50%) read news in the print version of a local newspaper (Purcell et al. 2010). Anyone trying to involve the local media in telling their story would be well advised to coordinate with their public affairs officers well in advance of initiating such contact. There is an art and science to how this is best done, and doing it poorly can be a disaster.

11.5.2.2 Personal Communication as Documentation

Here's a radical idea in this age of technology. Talk to people. Tell them your story. Make it personal. Face-to-face communication as documentation! There are lots of ways to do that. Briefings, large and small group meetings, guest speaker engagements, news conferences, making presentations, manning exhibits and displays at fairs and such, participating in panel discussions, sponsoring field trips, relevant training sessions, walking tours, classes, sponsoring symposia, and holding focus groups are a few ways planners can get out among interested people. Media interviews, hotlines, and talk shows are one step removed from face-to-face encounters. Mailings, newsletters, and news releases are other ways to reach interested individuals. Less-personal communications including newspaper inserts and advertisements and public service announcements overlap with mass media means of documenting your planning study. Establishing repositories of information at local libraries or with interest groups is another way to target your documentation efforts. As you see, it is easy for personal communication to drift into less-personal channels.

11.5.2.3 Computers and the Internet

Digital technology has radically changed the way planners can document their studies and communicate with the public. Imagine computer animations that show interested parties how a plan would function. Puzzles and games provide a great way to communicate with and educate young people. The Everglades restoration website provides some nice examples in their puzzles and "identify that sound" game at http://www.evergladesplan.org/education/puzzles_games.aspx# (accessed January 8, 2012). But let's imagine a little more. Second Life, at http://secondlife.com/ (accessed June 17, 2012), and World of Warcraft, at https://us.battle.net/account/creation/wow/signup/ (accessed June 17, 2012) are great examples of how games can be used to engage many people in virtual exploring, problem solving, and other strategic activities. How long before some daring pioneer creates a game designed to help solve a wicked planning problem?

Virtual tours enable people to visit distant and hard-to-reach places. They are also effective teaching aids for unfamiliar concepts. The Everglades restoration website provides a virtual tour of the Kissimmee River aquifer storage and recovery (ASR) pilot facility at http://www.evergladesplan.org/pm/projects/asr_tour/asr_interactive.aspx (accessed January 8, 2012). Virtual tours are a fantastic tool for documenting conditions, solutions, and so many aspects of a planning study or plan. Virtual tours of the Harris County, Texas floodplain can be found at http://www.hcfcd.org/qtvr/addicks/one.html (accessed June 17, 2012). One of their especially nice virtual tours is of the Greens Bayou wetlands mitigation bank: http://www.hcfcd.org/F_gbwmb.html# (accessed June 17, 2012). If you're not familiar with the possibilities, these sites are well worth a few minutes of your time. The storytelling possibilities literally explode with digital technology.

Interactive maps are a great way to present facts to people. They enable interested individuals to explore the data of interest to them. The Harris County (Texas) Flood Control District (HCFCD), at http://www.hcfcd.org/ME_HarrisCountyFWS.html (accessed June 17, 2012), offers a flood warning system map that shows real-time rainfall data for the Houston area. The point-and-click technology enables interested parties to discover and explore a wealth of data to find the information that is most interesting and useful to them.

Photo galleries enable people to see realistic images of conditions at a planning site. An example is found at the HCFCD website: http://www.hcfcd.org/photogallery.html# (accessed June 17, 2012). There you can identify the general location of interest and then select more specific views from among a series of available photographs. A flood event archive is also made available at this site.

Animations have become comparatively easy to prepare with the advent of software like Macromedia Flash. Animations like this one at the HCFCD website, http://www.hcfcd.org/F_fpexplained.html# (accessed June 17, 2012), explaining floodplains demonstrates the storytelling power of relatively simple animations. Another very nice animation providing a floodplain primer is found at http://www.hcfcd.org/F_floodplaintypes.html# (accessed June 17, 2012).

Photography in the round, 360-degree view is a fantastic way to take people to places that are an important part of your story. A great example of the promise of this technology in storytelling is found at http://www.360cities.net/ (accessed June 17, 2012). I selected this site for the sheer magical power of the impact this technology can have on documenting your planning story. Be sure to check the Strahov Library indoor tour. Use the zoom feature and notice how the details of an individual book can be displayed. Now think about how to use this technology to tell your story.

Interactive databases are another promising technology. A good example is found at The American Council of Science and Health's website: http://riskometer.org/ (accessed June 17, 2012). This site provides users with the opportunity to access health risk data in a manner that best suits their information needs and interests. The alternative ways to explore the data on life-threatening risks in the United States provide a good example of how planners might begin to think less about what table to present and more about how to make the data available so users can access it in ways that are meaningful to them. One of the very best examples of an interactive database is provided by Gapminder http://www.gapminder.org/ (accessed June 17, 2012). This compelling site truly does enable the user to explore the world over time from a desktop. Have a look at these sites and imagine how to adapt these technologies to your planning studies.

By comparison, many of the other tools are not nearly as dramatic, but they are amazingly promising. Let's begin with interactive websites. The CERP site http://www.evergladesplan.org/ (accessed June 17, 2012) and the HCFCD site http://www.hcfcd.org/index.asp (accessed June 17, 2012), are two fantastic examples of what creative planners can do to tell their stories. Don't overlook the obvious. Enabling people to download reports, documents, databases, maps, audio files, video files, photographs, and other forms of digital information from a website can empower your most interested publics to get involved and informed. All of these storytelling aids are commonly found on the Internet.

Consider using web conference capability to hold a virtual meeting or a Q&A session with the planning team. Bringing in an expert to answer the public's questions about an especially lively issue is a great use of the technology. Hosting a panel of experts with varying perspectives can be another way to document the results of your investigations. Making the transcripts available or even recording short meetings is another option. This leads naturally into the use of podcasts as a means of telling your story. A permanent and growing library of strategically created podcasts could make a valuable resource. Youtube is a great place for video presentations and visual documentation of field conditions, planned solutions and explanations of all sorts.

An electronic mailing list or listserv can be created, managed, and used to communicate about topics of interest in an ongoing way. Consider creating a listserv for your study early in the planning process. Members of the listserv can choose to read or ignore regular emails contributed by list members. They can also choose to reply to the entire listserv group or to individuals in the group. A related idea would be a discussion board established to discuss your planning study. Chat rooms can also be provided or set up to aid communication about a study.

Collaborative work spaces could be useful throughout a study process. Wiki spaces enable interested parties to work collaboratively on problem solving and other creative efforts throughout a study process. A wiki space would be a great place to ask the general public to share their ideas on how to solve the problems you face. As part of the documentation of the study effort, wiki spaces can be used to explain and discuss issues collaboratively in an ongoing and interactive way.

As exciting as these new opportunities are, it is important to say a few words about the digital divide. The digital divide refers to the inequalities among groups in terms of their access to, use of, or knowledge about information and communication technologies. It is important to be aware of the digital divide among your study's many publics. Anywhere from a few to many people may be unable to take advantage of these innovations. There is also a potential issue with compliance with Section 508 amendments (1998) to the Rehabilitation Act of 1973. This act requires federal agencies to make their electronic and information technology accessible to people with disabilities.

11.5.2.4 Social Media

Social media are often defined as a group of Internet-based applications that allow the creation and exchange of user-generated content. Social media are used for social interaction. This is generally considered to include but also go beyond social communication. Important features of these media are their ready accessibility and scalability. Both individuals and organizations can make use of these media. What better place to look for examples of social media than Wikipedia, one of the world's foremost examples of what collaborative people with access to technology can do? Wikipedia breaks social media into communication, collaborative, multimedia, reviews and opinions, and entertainment categories. Examples of each follow.

Communication includes blogs, microblogging, Twitter, social networking (e.g., Facebook), events, information aggregators, online advocacy and fundraising, and engagement advertising and monetization. Many resource planning agencies can already be found on Facebook. Creating new accounts for specific studies is the next logical step for all who want to friend or follow a study. As the story unfolds, it can be told with these kinds of media. Consider Twitter. Let people follow your study. A couple of updates every week or two keeps interest alive. Stakeholders can and also do make use of these media to present their own views on issues. A Twitter account for the planning team's use adds opportunities for immediacy and intimacy heretofore impossible to imagine.

Collaborative and authority-building media include things like wikis, social bookmarking, social media gaming, social news, social navigation, and content management systems. Wikis may be one of the more popular and promising collaborative environments for planners to exploit. I like the idea of using wikis throughout the planning process. Publish the team's problems and opportunities statement and ask, "Did we get it right? What's missing that ought to be here? What's here that ought to be missing?" Do the same with objectives, then use the wiki to ask for help identifying solutions. A wiki space that includes that kind of collaboration with the public provides a powerful piece of the story itself.

Multimedia opportunities are exciting. They include photography and art sharing, video sharing, live casting, music and audio sharing, and presentation sharing. Two existing examples are especially exciting. Flickr is a service that enables people to upload and share their photographs. Sharing photographs of study areas has promise. Asking people to share historical photos from an affected area could be a good source of information and data. Sharing those communal visions of the past or of existing conditions is a great way to tell a story. Photo sharing capability can be built into a study website or media on the Internet can be used directly.

The site that is especially interesting to me is YouTube. Think of all the ways you could use video to help tell the story of a planning study. You could ask citizens to make their own videos showing problems or opportunities, or explaining the history or importance of the area. You could sponsor a contest for the best historical video, the best video depicting a problem, or the best video that tells the story of a local species. The possibilities are unlimited.

Reviews and opinions are usually used to review products or businesses. They have also been used as a sort of community Q&A at sites such as wikiHow, eHow, and the like. This is an idea readily adapted to a planning study. Ask people to review your product as you go. Have them review your hydrology if you like, or your EIS. The dialogue becomes an important part of your study's ethnography.

The final major category of social media is entertainment. It includes things like media and entertainment platforms, movie reviews, virtual worlds, and game sharing. Some of these have already been touched on.

11.5.3 Video Reports

Suppose Steven Spielberg offered to tell the story of your plan in a movie. Would you say yes? How many graphs would he use? How long would he spend talking about authorities and such? Steven is not making a movie about your plan, but you still can.

A carefully scripted 10-or 20-minute video can convey a great deal of information effectively. It can be conveyed to a lot of people in a short amount of time. The video offers the options of going on-site and illustrating problems explicitly. A wide variety of points of view can be displayed as well.

Producing effective video reports can be expensive. They can be broadcast quality, corporate quality, or more amateur attempts. If you think writing a report is difficult, try scripting an effective video. This is a job that requires expertise. That expertise can be purchased or acquired through the school of hard knocks.

If you're willing to enroll in that school, start videotaping the study area, the resources, the heroes, the villains, the affected communities, stakeholder groups, team meetings, interviews, field visits, people working, bits of everything you do. Build a library of clips early and throughout the study. Yeah, you'll have to learn about release forms and lighting and angles and you might still need some professional help to pull it all together the first time. But if you have to write a report, why not try writing a script? Printed reports are so yesterday, and you know you want to be so tomorrow.

11.6 TAKE AWAYS

Lesson 1. Planners must report their findings so decisions can be made.
Lesson 2. Always write with your reader clearly in mind.
Lesson 3. Chronology is your friend. Tell your story.
Lesson 4. Write a report that people will read.
Lesson 5: Use technology to help you tell your story.

REFERENCES

Comprehensive Everglades Restoration Plan (CERP). 2012. Lake Okeechobee Watershed. Accessed January 13, 2012 from http://www.evergladesplan.org/pm/projects/proj_01_lake_o_watershed.aspx.

Food Insight. 2010. Risk communicator training for food defense preparedness, response and recovery: Trainer's overview. Accessed January, 12, 2011 from http://www.foodinsight.org/Resources/Detail.aspx?topic=Risk_Communicator_Training_for_Food_Defense_Preparedness_Response_Recovery.

Gunning, R. 1952. *The Technique of Clear Writing.* New York: McGraw-Hill.

Mitchell, A. 2011. Hemingway's 4 rules for writing well. *Wordsmith.* Accessed July 18, 2012, http://wordsmithblog.wordpress.com/2011/04/07/hemingways-4-rules-for-writing-well/.

Proust, Marcel. 1957. *Swann's Way.* Harmondsworth: Penguin Books.

Purcell, K. et al. 2010. Understanding the participatory news consumer. Pew Research Center: Pew Internet. Accessed February 6, 2012 from http://pewinternet.org/Reports/2010/Online-News.aspx.

Ruger, K. 2010. The art of business storytelling. Accessed February 6, 2012 from http://www.themoleskin.com/2010/03/the-art-of-business-storytelling/.

Shephard, A. 1998. What makes a good story? Tips for young authors. Aaron Shephard's Young Author's Page. Accessed June 17, 2012 from http://www.aaronshep.com/youngauthor/elements.html

Simmons, A. 2001. *The Story Factor.* New York: Basic Books.

Tufte, E. R. 1983. *The Visual Display of Quantitative Information.* Cheshire, CT: Graphics Press.

Tufte, E. R. 1990. *Envisioning Information.* Cheshire, CT: Graphics Press.

Tufte, E. R. 1997. *Visual Explanations: Images and Quantities, Evidence and Narrative.* Cheshire, CT: Graphics Press.

United States Army Corps of Engineers. 1999. Central and southern Florida comprehensive review study final integrated feasibility report and programmatic environmental impact statement. Washington, DC: USACE. Accessed February 6, 2012 from http://www.evergladesplan.org/pub/restudy_eis.aspx.

United States Army Corps of Engineers. 2009. Performance evaluation of the New Orleans and southeast Louisiana hurricane protection system final report of the interagency performance evaluation task force volume I—Executive summary and overview. Washington, DC: USACE.

Wheeler, S. 2010. Writing rules. Accessed February 13, 2012 from http://steve-wheeler.blogspot.com/2010/04/12-writingrules.html.

CHAPTER 12

Uncertainty

12.1 INTRODUCTION

Good planners are honest and open about what they know with certainty and what they are uncertain about. The best planners are intentional in the manner in which they address the significant uncertainty in their planning investigations. This is not a trivial challenge because planning investigations are plagued by data gaps and holes in the understanding of the systems with which planners work. Because planning is fundamentally about choosing better futures, the entire planning process is an exercise conducted in uncertainty. Planners may serve society best when they explore the nature and extent of this uncertainty and make us more aware of its implications for the decisions we make today.

Uncertainty, introduced in Chapter 1, at its simplest means we are not sure. It arises at two fundamentally different levels: the macrolevel and the microlevel. At the macrolevel, planners work in a changing and uncertain decision environment. Social relationships are complex and values are varied, changing, and often difficult to discern. This means the systems, processes, social values, and outcomes of concern to planners may be uncertain. At the microlevel, planner's scenarios, models, and quantities are plagued by uncertainty. These are the uncertainties that receive most of the analytical attention in planning studies.

There are always things we know with certainty in natural resources planning. You can measure distances, stream flows, pH, and temperature; our physical world is loaded with facts that can be arrayed. Planners can count populations and know the atomic structures of chemicals. The things you know in any one planning investigation are likely to make a rather impressive "pile"; but every study also comes with a "pile" of things you do not know. It's the planner's job to identify what is in that pile and which of those things are going to matter most to the decisions you have to make.

This chapter focuses on the pile of things that we don't know. If planners are ever going to figure out how best to address the things they don't know they must first understand the nature of that which they don't know. Figure 12.1 sorts the original pile of things we don't know into two distinct sources of not knowing: natural variability and knowledge uncertainty. Knowledge uncertainty is then sorted into three principle piles of things about which our knowledge is uncertain: scenarios and theory, models, and quantities. The quantities are subsequently separated into different types of quantities.

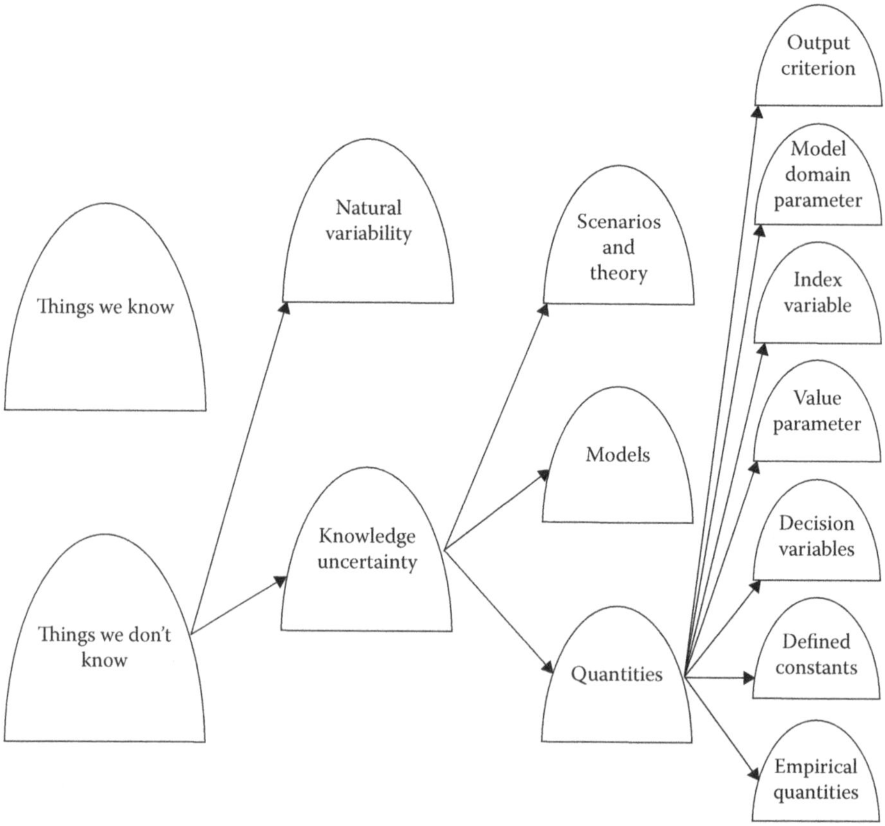

Figure 12.1 Separating what we know from what we do not know and then breaking what we do not know into categories (Yoe 2012, p. 24).

Chapter 1 focused more on the uncertainty in the planner's macroenvironment. This chapter builds on those concepts by parsing microenvironment uncertainty into two major components: knowledge uncertainty and natural variability. Knowledge uncertainty will remain the primary focus as we unpack some details from Figure 12.1 before considering the ways that planners can be intentional about uncertainty in the planning process. The chapter concludes by bridging from the concept of uncertainty to a few words on risk.

12.2 DISTINGUISHING KNOWLEDGE UNCERTAINTY AND VARIABILITY

If an uncertain decision (macro) environment was the only uncertainty a planner had to consider, planning would be far less complex than it is. Uncertainty at the microlevel often presents a much more pressing need. Things are uncertain for basically one of two reasons. First, there are facts or knowledge out there and we just do not have it; we are going to call this knowledge uncertainty. Second, the systems planners deal with are complex and often variable. Even when we are lucky enough to have all the facts, there is often natural variability that keeps us from knowing with certainty how the facts are realized.

Phelge River drains 27,500 square miles, an area the size of South Carolina. Its flow varies from day to day. Its mean daily peak flow in cubic feet per second (cfs) is a constant and a knowable fact. If you do not know that specific fact, that is an example of knowledge uncertainty. Suppose you learn from the U.S. Geological Survey (USGS) that the mean daily peak discharge is 26,700 cfs with a standard deviation of 27,700 cfs. The uncertainty has been removed. Now, a new problem emerges. Even though you know the mean daily peak flow, you have no way of knowing what the peak flow will be on any given day. In fact, you wisely expect the flow to vary from day to day.

Using the very loose definition of uncertainty at the start of this chapter, you say you are not sure what the peak flow will be on any given day, so that must be

NATIONAL RESEARCH COUNCIL (2009)

Uncertainty is a lack or incompleteness of information. Quantitative uncertainty analysis attempts to analyze and describe the degree to which a calculated value may differ from the true value; it sometimes uses probability distributions. Uncertainty depends on the quality, quantity, and relevance of data and on the reliability and relevance of models and assumptions.

Variability refers to true differences in attributes due to heterogeneity or diversity. Variability is usually not reducible by further measurement or study, although it can be better characterized.

uncertainty as well. In a very general sense it is. However, and this is an important "however," this value is uncertain for a very specific, common, and recurring reason: there is natural variability in the universe, specifically in the system that produces stream discharges on the Phelge River.

Natural variability is usually separated out from other causes of uncertainty to preserve that distinction because the techniques and tools used to address uncertainty depend on the cause and type of uncertainty. Hence, we'd say you are no longer uncertain about the mean daily peak flow, but you still do not know the flow on any given day because of natural variability in the universe. For a more formal distinction of these two concepts, let's introduce the terms "epistemic" and "aleatory" uncertainty.

Epistemic uncertainty is a lack of knowledge on the part of the observer. In principle, this uncertainty can be reduced, although it may be difficult or expensive to do so. Epistemic uncertainty arises from incomplete theory and incomplete understanding of a system, modeling limitations, and/or limited data. I prefer the term "knowledge uncertainty" because it is both easier to remember and a more descriptive term.

Generic examples of knowledge uncertainty include lack of data to characterize study area conditions, poor understanding of the linkages between inputs (like water quality and quantity) and outputs (like habitat) in a system, and thinking one value (e.g., the depth of water in different parts of a lake) is greater than another but being unsure of that. More obvious examples include dated, missing, vague, or conflicting information, incorrect methods, faulty models, measurement errors, incorrect assumptions, and so on. Knowledge uncertainty is simply not knowing something that is conceptually knowable. Perhaps the most common example is not knowing a parameter or value we are interested in for model building or decision-making purposes.

Aleatory uncertainty deals with the inherent variability in the physical world. This variability may be attributed to a random process that produces natural variability of a quantity over time and/or space or among members of a population. It arises because of natural, unpredictable variation in the performance of the system under study. In principle, it is irreducible. The variability cannot be altered by obtaining more information, although the additional data may provide a better description of the variability. A larger database provides a more precise estimate of

AN IMPORTANT DISTINCTION

Natural variability cannot be reduced with more or better information. Knowledge uncertainty can be reduced with more and better information through such means as research, data collection, better modeling and measurement, filling gaps in information, and updating out-of-date information, as well as correcting faulty assumptions.

the standard deviation but it doesn't reduce the variability in the population. I like the term "natural variability"; it's easier to use and understand.

Generic examples of natural variability include variation in the strength of winds, the size of rocks, levels of dissolved oxygen in a lake, and the flows in a stream. It can also include variation in the response of an ecosystem to a change in the physical environment, and variation in population counts from year to year. Natural variability is found in virtually any attribute of a population.

Knowledge uncertainty and natural variability are terms used by the National Research Council (2009). It is often convenient to use the term "uncertainty" to encompass both of these concepts and this text adopts that convenience. It takes experience for a planner to be able to comfortably label the reasons why a value may be unknown. Returning to our stream discharge example, at the outset, we are dealing with both knowledge uncertainty and natural variability. Even once knowledge uncertainty is eliminated, natural variability can remain.

12.3 TYPES OF UNCERTAINTY

Our typology of uncertainty begins by differentiating knowledge uncertainty from natural variability. Natural variability tends to apply to quantities only. It is most often addressed through narrative descriptions of the variability, statistics, and probabilistic methods. Knowledge uncertainty is more complex and needs additional sorting. Three major categories of knowledge uncertainty are identified in Figure 12.2: scenarios, models, and inputs.

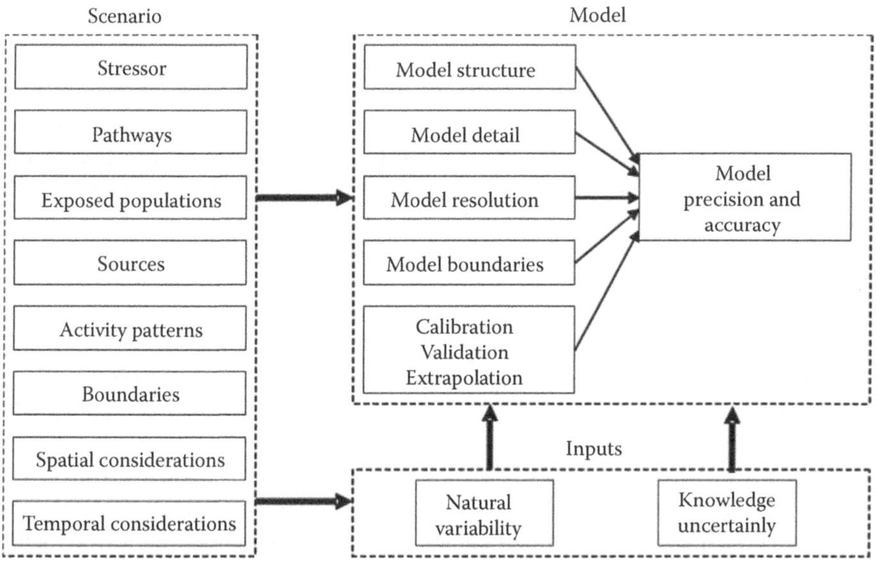

Figure 12.2 Three sources of knowledge uncertainty (Yoe 2012, p. 30).

KNOWLEDGE UNCERTAINTY DEFINITIONS

Scenario uncertainty: Uncertainty in specifying the risk scenario, which is consistent with the scope and purpose of the assessment.

Model uncertainty: Uncertainty due to gaps in scientific knowledge, which hamper an adequate capture of the correct causal relations between risk factors.

Parameter/input uncertainty: Uncertainty involved in the specification of numerical values (be it point values or distributions of values) for the factors that determine the risk.

Source: International Programme on Chemical Safety. 2006. Draft guidance document on characterizing and communicating uncertainty of exposure assessment, draft for public review. Geneva: World Health Organization. Accessed May 15, 2010 from http://www.who.int/ipcs/methods/harmonization/areas/draftundertainty.pdf.

Figure 12.2 shows the types of knowledge uncertainty natural resource planners might encounter. Scenarios are the narratives that describe what planners believe to be most true about the phenomena they study. Models are used to give structure to and to perform calculations for the scenarios based on the inputs provided. Inputs are the quantities used in analysis. You can expect to encounter these three broad types of knowledge uncertainty in planning studies.

Scenario knowledge uncertainty results when the elements of a scenario or their relationships are unknown or incomplete. We might misunderstand the stressors that affect a habitat. Not knowing the relevant activity patterns of a locally threatened species could be another source of scenario uncertainty. Alternatively, we might not understand all the relevant pathways in an ecosystem.

Model knowledge uncertainty reflects the bias or imprecision associated with compromises made or lack of adequate knowledge in specifying the structure, detail, resolution, calibration (parameter estimation), and such of a model. Model structure typically refers to the set of equations or other functional relationships that comprise the specified scenario. Model detail refers to the inclusion or omission of specific phenomena as well as the simplicity or complexity with which they are represented. Model resolution addresses the temporal or spatial scale at which information can be distinguished, for example, minutes versus hours versus years. Model boundaries describe the fidelity with which the desired scenario is captured by the model. Ideally, the precision and accuracy of the model predictions will be assessed as part of the validation exercise. In other words, how well does the model capture reality?

Quantity or input uncertainty is encountered when the appropriate or true values of quantities are not known (knowledge uncertainty) or are variable (natural variability). These quantities are of particular importance to planners and warrant additional discussion. That discussion will enable us to revisit natural variability.

12.4 QUANTITY UNCERTAINTY

TRUE VALUES

The population of a city, number of adult wolves in a pack, percent of channel bottom that is rock, mean strength of materials in a structure, mean daily stream flow, average weight of an adult striped bass, median fuel yields, specificity of a diagnostic test, price of a yard of concrete, and contaminant concentration in a specific exposure are all examples of quantities that have a true value.

The most commonly encountered uncertainty is quantity uncertainty. Quantities can be unknown because of knowledge uncertainty or natural variability. Morgan and Henrion (1990) offer a very useful taxonomy for those seeking to understand the basic types of quantity uncertainty; it is shown in the far right of Figure 12.1. Before considering their taxonomy, we need to make an important distinction. Some quantities have a true or factual value, while others do not. Instead of a true value they have a best or most appropriate value that reflects some subjective judgment. There may be significant consternation about the best or preferred value for these kinds of quantities, but they have no true value that can be objectively discovered. The search for a true value is an objective one, while the search for a best value is subjective.

In general, true values are looked up, measured, or estimated by some means. The means by which quantities with true values are estimated vary and the best choice will depend on the cause of the value's uncertainty as well as the nature of the quantity itself. Best or appropriate values are varied systematically (sometimes called parametric variation or sensitivity analysis) to examine the sensitivity of the model and its outputs to different chosen values.

Planning requires a lot of information. It can involve a great deal of quantitative information that includes many parameters (numerical constants) and variables. The quantities used in planning are frequently a major source of uncertainty. Having a way to think about these quantities and to talk about their uncertainty is critical to the success of any planning effort. Morgan and Henrion's (1990) classification of uncertain quantities includes the following:

- Empirical quantities
- Defined constants
- Decision variables
- Value parameters
- Index variables
- Model domain parameters
- Outcome criteria

Table 12.1 Uncertain Quantity Types with Examples

Types of Quantities	Selected Examples
Empirical quantities	Stream flow, eggs produced daily, vehicles crossing a bridge, temperature, time to complete a task, prevalence of a disease in wildlife
Defined constants	Pi, square feet in an acre, gallons in an acre-foot, speed of light, size of a city, beds in a hospital
Decision variables	Acceptable daily intake, tolerable level of risk, appropriate level of protection, reasonable cost, mitigation goal
Value parameters	Value of a statistical life, discount rate, weights assigned in a multicriteria decision analysis, user-day values
Index variable	A particular year in a multiyear model, the location of an egg on a pallet, a geographic grid in a spatial model
Model domain parameters	Study area, planning horizon, industry segment, climate range, establishing the physical limits for what is possible within the model
Outcome criteria	Mortalities, illness rates, habitat units, infrastructure failures, fragility curves, costs, probabilities, benefit–cost ratios, risk–risk trade-offs

The significance of the objective or subjective nature of the uncertainty, as well as the type of quantity, will become most evident when one chooses a tool, technique, or methodology to treat the uncertainty appropriately. Look at the examples in Table 12.1 and then read the explanations that follow to understand Morgan and Henrion's taxonomy of quantities. The approach used to resolve uncertainty depends very directly on what is uncertain and why it is uncertain.

12.4.1 Empirical Quantities

Empirical quantities are the most common quantities encountered in planning investigations; they have a true value. Empirical quantities are things that can be measured or counted. This includes distances, times, sizes, temperatures, statistics,

DECISION RULE UNCERTAINTY

What is the best endpoint for your purposes? Imprecise or inappropriate operational definitions for desired outcome criteria can be a subtle problem. For example, when we say "risk" precisely what does that mean, how is it measured?

Concerned about a public health risk due to flooding? Should you use lifetime mortality risk, annual risk of mortality, risk to children or other subpopulations, or something entirely different? Concerned about an economic issue? Should you maximize net benefits or minimize costs?

There is no right answer to these questions, only better or worse ones. Someone must decide what the decision criterion or rule will be to resolve this uncertainty.

and any sort of imaginable count. They have exact values that may be unknown, but they are measurable in principle, although it may be difficult to do so in practice. A full range of methods, from narrative descriptions through probabilistic methods, are suitable for addressing uncertainty in these quantities.

12.4.2 Defined Constant

Defined constants have a true value that is fixed by definition. When these values are not known by the analyst, these quantities can end up in the pile of things we do not know. For example, there are 43,560 square feet in 1 acre and 325,851 gallons of water in 1 acre-foot of water. Defined constants provide the perfect opportunity to point out the importance of understanding the nature of your unknowns for the purpose of resolving uncertainty. When you do not know one of these quantities, you do not use sensitivity analysis or probabilistic methods, you look them up.

12.4.3 Decision Variables

This is a quantity that someone must choose or decide. Decision makers exercise direct control over these values; they have no true value. The person deciding this value may or may not be a member of the planning team, depending on the nature of the variable. Policy makers may determine the values of some decision variables to assure uniformity in decision making. An agency may decide it is unacceptable to increase the risk of flooding to anyone, for example. Thus, decision variable values are sometimes set by decision makers external to the planning process.

In other instances, planning team members may make these decisions. Examples could include determining a tolerable level of risk or design characteristics of solutions that differentiate one option from another. Decision variables are subjectively determined. Uncertainty about them is most appropriately addressed through parametric variation and sensitivity analysis.

12.4.4 Value Parameters

These values represent aspects of decision makers' preferences and judgments; they have no true value. They are subjective assessments of social values that can describe the values or preferences of stakeholders, organizations, or decision makers. Like decision variables, some of them may be decided by those external to a specific planning study, while others may be decided by planning team members.

Social values, like the value of a statistical life or society's time preference for consumption, are sometimes established corporately by planning organizations to ensure uniformity in decision making. Establishing decision-specific values, such as assigning relative weights to different decision criteria, may be done by the team. Uncertainty about value parameters is most appropriately addressed through parametric variation and sensitivity analysis.

12.4.5 Index Variables

Index variables identify elements of a model or locations within spatial and temporal domains; a point in time can be referenced as a time step in a model and a grid cell can be referenced using coordinates. If a very specific point in time or place in space is required for some reason, there may be a true value. Random or representative choices of index variables do not have true values and are subjectively determined. Uncertainty in these index variables is most appropriately addressed through parametric variation and sensitivity analysis.

12.4.6 Model Domain Parameters

These values specify and define the scope of the system modeled in planning analyses. These parameters describe the geographic, temporal, and conceptual boundaries (domain) of a model. They define the resolution of its inputs and outputs; they may or may not have true values. Scale characteristics are often chosen by the modeler and most often have no true value in nature. They reflect judgments regarding the model domain and the resolution needed to assess model outputs adequately. Some planning analyses, however, may be restricted to specific facilities, towns, time frames, and so forth. These may have true values. Uncertainty about domain parameters may also be considered a form of model uncertainty. If the domain is the Beastly Wildlife Refuge, it is trivially specific and objective. The hinterland affected by economic activity at the Port of Los Angeles is a much more subjective determination. These kinds of quantities are most appropriately addressed through parametric variation and sensitivity analysis.

12.4.7 Outcome Criteria

Outcome criteria are output variables used to rank or measure the desirability or undesirability of possible model outcomes. Their values are determined by the input quantities and the models that use them. Uncertainty in these values is evaluated by propagating uncertainty from the input variables to the output variables using one of several different methods. Generating the uncertainty about output criteria is the responsibility of the planner addressing it during decision making is the responsibility of the decision maker.

12.5 SOURCES OF UNCERTAINTY IN EMPIRICAL QUANTITIES

Empirical quantities are the most commonly encountered uncertain values that are likely to have true values that must be measured or estimated. Empirical quantities may be uncertain due to knowledge uncertainty or natural variability. When good measurement data are available, there may be little or no knowledge uncertainty about the true value of a parameter or variable. Even when there is no knowledge uncertainty, you may have natural variability to address in planning investigations. It is useful to continue the excellent conceptual framework of Morgan and Henrion (1990) to consider the different sources of uncertainty in empirical quantities. Understanding the reasons you are uncertain about empirical quantities is essential to your ability to choose an effective treatment of that uncertainty in a quantitative risk assessment.

12.5.1 Random Error and Statistical Variation

No measurement can be perfectly exact. Even tiny flaws in observation or in reading measuring instruments can cause variations in measurement from one observation to the next. Then, there is the statistical variation that results from sample bias. If we take measurements on a sample, we only have an estimate of the true value of a population parameter. Classical statistical techniques provide a wide array of techniques and tools for quantifying this kind of uncertainty, including estimators, standard deviations, confidence intervals, hypothesis testing, sampling theory, and probabilistic methods.

12.5.2 Systematic Error and Subjective Judgment

Systematic errors arise when the measurement instrument, the experiment, or the observer are biased. Imprecise calibration of instruments is one cause of this bias. If the scale is not zeroed or the datum point is off, the solution is better calibration of the instrument or data. If the observer tends to over- or underestimate values, a more objective means of measurement is needed or the observer needs to be recalibrated. The challenge to the analyst is to reduce systematic error to a minimum. The best solution is to avoid or correct the bias. When bias can be identified, for example, the scale added 0.1 g to each measurement, it can sometimes be corrected for, that is, recalibrate and remeasure or subtract 0.1 g from each measurement.

The more difficult task is correcting for the biases that are unknown or merely suspected. Estimating the magnitude of these biases is very difficult and often requires a lot of subjective judgment, which, as the following text box notes, can present its own problems. Bias in subjective human estimates of unknown quantities is a topic covered extensively in the literature; see, for example, O'Hagan et al. (2006) or Yoe (2012).

IT FEELS LIKE AN 8 TO ME

Much data are collected outside a laboratory and under less than ideal conditions. Where in the stream does the investigator insert the meter to read dissolved oxygen? How do you estimate how far away a workboat is on the open water? How quickly can you count the deer in a running herd? Subjective judgments like these are notoriously suspect under uncontrolled conditions. Just like with faulty instruments, the solution is better calibrated individuals. Ideally, they are calibrated before the measurements are taken.

12.5.3 Linguistic Imprecision

After all our years on the planet, communication is still humankind's number one challenge. We routinely use the same words to mean different things and different words to mean the same things. This makes communication about complex matters especially challenging. If we say a hazard occurs frequently or a risk is unlikely, what do these words really mean? The problems are more pervasive than that. Tasked with measuring the percentage of mid-day shade on a stream, a group of environmentalists engaged in a lengthy discussion of when mid-day occurs and how dark must a surface be to be considered shade. The best and most obvious solution to this kind of ambiguity is to carefully specify all terms and relationships and to clarify all language as it is used. Using quantitative rather than qualitative terms can also help.

12.5.4 Natural Variability

Many quantities vary over time, space, or from one individual or object in a population to another. This variability is inherent in the system that produces the population of things we measure. Frequency distributions based on samples or probability distributions for populations, if available, can be used to estimate the values of interest. Other probabilistic methods may be used as well.

12.5.5 Randomness and Unpredictability

Inherent randomness is sometimes singled out as a form of uncertainty different from all others, in part because it is irreducible in principle. This cause identifies those uncertainties that are not predictable in practice at the current time. Examples include such things as when the next flood will occur on a stream or where the next wildlife disease outbreak will occur. Such events can be treated as legitimately random processes. The danger here is the personalist view of randomness that could emerge, where randomness is a function of the analyst's knowledge. Phenomena that appear random to one person may be the result of a process well known by a subject matter expert. Strong interdisciplinary planning teams combined with peer involvement and peer review processes provide a reasonable hedge against this sort of problem arising. Uncertainty about such quantities can be addressed by a full range of methods, from narrative descriptions through probabilistic methods.

12.5.6 Disagreement

Organizations and experts do not always see eye-to-eye on matters of uncertainty. Different technical interpretations of the same data can give rise to disagreements, as can widely disparate views of the problem. This is not to mention the real possibility of conscious or unconscious motivational bias.

Disagreements can sometimes be resolved through negotiation and other issue resolution techniques. Allowing disagreements to coexist is another option.

Sensitivity analysis can consider the results of the analysis using each different perspective. An approach for some disagreements is to combine the judgments using subjective weights.

12.5.7 Approximation

Uncertainty due to approximation is similar to model uncertainty. The fact that the model is a simplified version of reality assures uncertainty will remain about the model's outcomes. We are only able to approximate the function of complex systems because scenario, model, and quantity uncertainty exist. Methods for dealing with this source of uncertainty will depend on the specific limitations of the approximation.

12.6 UNCERTAINTY CAUSES RISK

When there is uncertainty in a planning process, there are two kinds of risk planners need to be aware of. There are risks to the people and resources in the study area and there are study risks. To understand these risks, we first have to review and develop a common understanding of some risk language.

Risk is a measure of the probability and consequence of uncertain future events. It is the chance of an undesirable outcome. That outcome could be a loss (fire, flood, disease, death, damage, financial setback, or any sort of hazard) or a potential gain, i.e. opportunity, that is not realized (water quality was not improved, the ecosystem was not restored, biodiversity was not realized, yields did not increase, cost savings were not realized, or any sort of opportunity). What usually creates the "chance" is a lack of information about events that have not yet occurred. We lack information because there are facts we do not know, the future is fundamentally uncertain and the universe is inherently variable. This is the uncertainty of which we've been speaking.

Given the presence of a hazard or an opportunity, there are two important components to a risk: chance or probability and an undesirable outcome or consequence. Risk is often described by this simple equation:

$$\text{Risk} = \text{Probability} \times \text{Consequence} \tag{12.1}$$

HAZARD

In a general sense, a "hazard" is anything that is a potential source of harm to a valued asset (human, animal, natural, economic, social). It includes all biological, chemical, physical, and radiological agents or natural/anthropogenic events capable of causing adverse effects on people, property, economy, culture, social structure, or environment.

Consider this expression as a conceptual model that helps us think about risk rather than an equation that defines it. What this expression conveys is not so much that this is how all risks are calculated (they are not) as much as that both of these elements must be present for there to be a real risk. If an event of any consequence has no probability of occurrence, there is no risk. Likewise, if there is no consequence or undesirable outcome, there is no risk.

A hazard is the thing that has the potential to cause adverse consequence. Storms, hurricanes, floods, forest fires, and earthquakes are examples of natural hazards. When humans and natural resources are exposed to these hazards, there are risks with consequences that include loss of life, property damage, loss of ecosystem function, changes in morphology, and economic loss, to name a few. An opportunity causes the potential for a positive consequence that might not be realized. If a population, an individual, or some asset of interest to us is not exposed to the hazard or opportunity, then there will be no consequence and no risk. The range of possible consequences (loss of habitat, property damage, financial loss or gain, improved environmental conditions, technological success, and the like) is vast because even similar types of consequences can vary in frequency, magnitude, intensity, and duration.

In natural resource contexts, we have pure risks, which are losses with no potential gains and no beneficial result. Speculative or opportunity risks are generally defined as risks with an uncertain degree of gain. These risks are further distinguished by the fact that pure risk events are beyond the decision maker's control; they are the result of uncontrollable circumstances, while speculative risks are the result of conscious choices made in decision making. These two types of risks lead to two distinct risk management strategies: risk avoiding and risk taking. Decision makers select options that will enable them to reduce unacceptable levels of pure risk to acceptable or tolerable levels. Decision makers also choose to take risks when they select one alternative course of action from among a set of alternatives. So, decision makers function as risk avoiders when they decide how best to reduce the adverse consequences of risk and as risk takers when they decide how best to realize potential gains in the future. Uncertainty makes all of this necessary; there is no risk without uncertainty (see the following text box).

A FEW PROPOSITIONS ABOUT RISK

- Risk is everywhere.
- Some risks are more serious than others.
- Zero risk is not an option.
- Risk is unavoidable.

Therefore, we need risk analysis to:

- Describe these risks (risk assessment).
- Talk about them (risk communication).
- Do something about the unacceptable ones (risk management).

Because uncertainty gives rise to risk, the essential purpose of risk analysis is to help us make better decisions under conditions of uncertainty. Risk analysis is a process of decision making under uncertainty that consists of three tasks risk management, risk assessment, and risk communication. Risk analysis examines the whole of a risk by assessing the risk and its related relevant uncertainties for the purpose of efficacious management of the risk, facilitated by effective communication about the risk. It is a systematic way of gathering, recording, and evaluating information that can lead to recommendations for a decision or action in response to an identified hazard or opportunity for gain. Risk analysis is science-based but it is not a science, it is not certain, it is not a solution, and it is not static.

Risk analysis provides a logical framework for making decisions under conditions of uncertainty. We may be uncertain about one or more aspects of the likelihood or the consequence of the risk(s) of concern to us. More troubling, we may be unsure how effective our planned solutions will be. Thus, risk-informed planning (RIP), the interface of risk analysis and planning, is slowly beginning to supplant more traditional planning models.

Risk analysis, like planning, is based on good science. Scientific facts, evidence, and good analytical techniques are hallmarks of both planning and risk analysis. They also consider social values. As important as science is, it is not the sole basis for decision making. Most importantly, they both address uncertainty explicitly. Few, if any, decisions are ever made with complete information and certainty. Lacking complete information and facing sometimes considerable uncertainty does not absolve us of the need to make decisions at times. Risk analysis has evolved explicitly for these kinds of uncertain decision problems, which occur frequently in planning contexts. Risk analysis, like planning, provides information to decision makers; it does not make decisions. It is neither a magic bullet nor a black box.

Experienced planners will recognize the elements of the planning process in the tasks of risk analysis. Risk management is a process of problem identification, requesting information, evaluating risks, and taking action to identify, evaluate, select, implement, monitor, and modify actions to alter levels of unacceptable risk to acceptable or tolerable levels. Informally, risk management is the work one has to do to pose and then answer the following kinds of questions:

- What's the problem?
- What information do we need to solve it, that is, what questions do we want risk assessment to answer?
- What can be done to reduce the impact of the risk described?
- What can be done to reduce the likelihood of the risk described?
- What are the trade-offs of the available options?
- What is the best way to address the described risk?
- (Once implemented) Is it working?

Risk assessment is a systematic process for describing the nature, likelihood, and magnitude of risk associated with some substance, situation, action, or event that includes consideration of relevant uncertainties. Risk assessment can be qualitative, quantitative, or a blend (semiquantitative) of both. It can be informally described by posing and answering the informal questions that build on the Kaplan and Garrick triplet (1981):

- What can go wrong?
- How can it happen?
- What are the consequences?
- How likely is it to happen?

Risk communication is the open, two-way exchange of information and opinion among risk analysts, their stakeholders, and various publics about risks intended to lead to a better understanding of the risks and better risk management decisions. It provides a forum for the interchange of information with all concerned about the nature of the risks, the risk assessment, and how risks should be managed. Risk communication may be informally characterized by its own set of questions (Chess and Hance 1994):

- Why are we communicating?
- Who are our audiences?
- What do our audiences want to know?
- How will we communicate?
- How will we listen?
- How will we respond?
- Who will carry out the plans? When?
- What problems or barriers have we planned for?
- Have we succeeded?

12.7 BEING INTENTIONAL ABOUT UNCERTAINTY IN THE PLANNING PROCESS

Given the ubiquitous nature of knowledge uncertainty and natural variability in natural resource planning, how does one become intentional about addressing this uncertainty in the planning process? One answer is to use risk-informed planning. RIP combines the planning process with risk analysis' focus on decision making under uncertainty. Thus, RIP introduces risk-based decision making to the planning process.

As mentioned in the preceding section, there are two basic kinds of risks that result from the uncertainty encountered in planning. First, there are the hazards and opportunities found in the study area that are more than likely identified among the problems and opportunities that exist. Second, there are planning study risks. Think of these risks as the consequences of planning and its decision-making process that result from not having complete information during the planning process.

The first of these two kinds of risks is addressed by RIP, which is described in Section 12.7.1. Study risks though introduced in this chapter are treated at greatest length in Chapter 15 in the discussion of the risk register.

12.7.1 Risk-Informed Planning

My career as a planner began in the 1970s in the aftermath of Tropical Storm Agnes in the northeast United States. Federal water resource planning was guided by something called the P&S (Principles and Standards) (U.S. Water Resources Council 1973). Paragraph VI.H of the standards is called *Risk and Uncertainty* (1973 p. 88). To the best of my knowledge, this is the first mention of these terms in any official water resource policy document. I was new and so were the P&S. My boss told me to read them and figure out what this risk and uncertainty stuff was all about.

Planning for natural hazards included risk analysis long before anyone invented the term. Over the years, my planning work has led me to risk analysis, where I have done a lot of work. Early I thought was that these folks doing risk analysis were, to an extent, reinventing planning in a new context. It was a paradigm for approaching problem solving that integrated science and social values while differentiating itself by focusing intentionally on the uncertainties that could have an effect on decision making. A sample description of the risk management process is seen in Figure 12.3. Although the five-step process goes by different names, a careful examination of that process will reveal more similarities to the planning process than differences through the first four risk management steps.

The biggest differences between risk analysis and natural resources planning were in the kinds of problems they addressed, the methods used in risk analysis to address uncertainty, and the fact that risk analysis goes beyond the planning stage to include implementation, monitoring, evaluation, and modification of a solution. Otherwise, these two problem-solving approaches are philosophically and conceptually more alike than different. If risk analysis is married to planning, there would be a few notable changes. Let's review them step by planning step.

12.7.1.1 Step 1 RIP Outputs

An RIP process requires the following outputs:

1. A written problems and opportunities statement
2. A written statement of the risk management activity's objectives that anticipates desired outcomes
3. A written list of decision information questions
4. A written list of the decision criteria
5. A narrative description of the risks and/or significant uncertainties of concern to this risk management activity
6. A completed risk profile
7. A decision whether or not to pursue a risk assessment
8. Identification of decision-sharing partners and development of a decision agreement
9. Identification of stakeholders, understanding of their concerns and risk attitudes

Items 1 through 4 are already being produced in the planning process. Item 2 introduces slightly different language, but the objectives here are identical to the

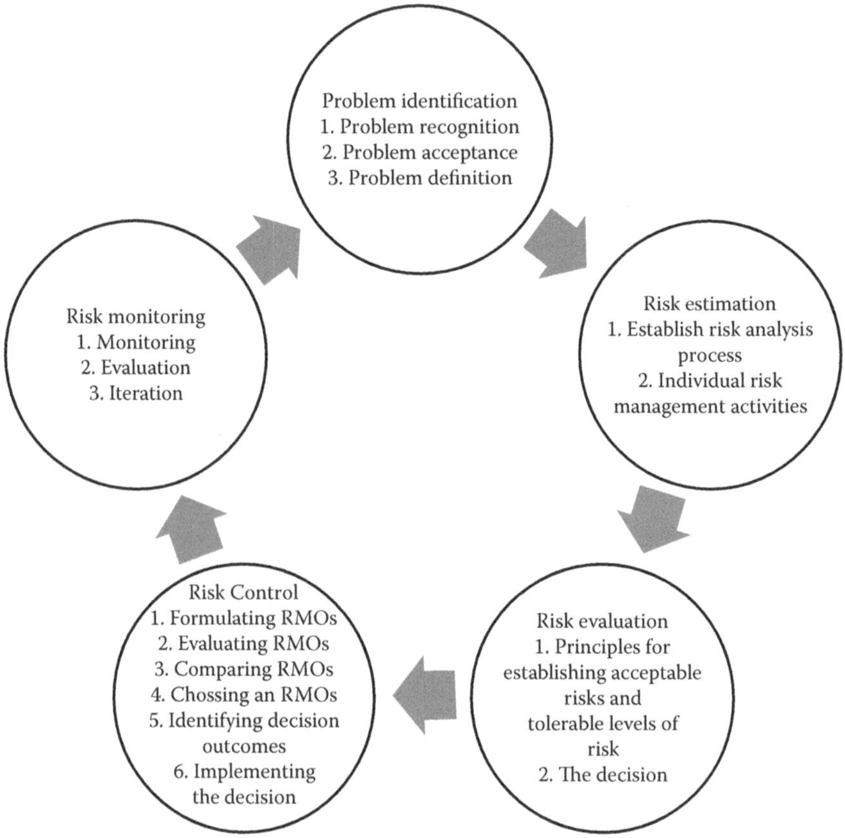

Figure 12.3 A five-step risk management process (Yoe 2012, p. 45).

objectives and constraints discussed in Chapter 4. If specific information is going to be needed to address a risk, item 3 is used to identify that information to planners. Item 5 is added to assure that the relevant risks in the planning study are identified. The problems and opportunities would normally include risks anyway. In RIP, we want to make sure each risk is identified, described in a narrative, and the most significant uncertainties related to that risk are identified.

The risk profile in item 6 is the initial effort to separate what we know about the risks from what we do not know. This is done in a risk profile, which is essentially a situation report that gathers all the readily available information about a risk, before the detailed planning analysis begins. The profile is actually an evidence gathering task but it should occur in Step 1. The idea is to gain a clearer understanding of the most relevant uncertainties. Formal risk assessments will not be part of every planning study. However, there may be times when the identified risks warrant a formal risk assessment, that determination should be made early in the planning process. See Yoe (2012) for a discussion of risk assessment. Items 8 and 9 are part of the public involvement of both planning and risk analysis. A new emphasis on understanding the public's awareness, concerns, and attitudes about the risks identified has been added.

12.7.1.2 Evidence Gathering Process RIP Outputs

The RIP outputs in addition to those already produced in the evidence gathering process include the following:

1. A risk profile and an assessment when appropriate
2. A qualitative or quantitative risk characterization of each significant risk
3. Clear identification of key uncertainties
4. Written answers to the decision information questions

If a formal risk assessment is required to support the planning study, it will be largely completed as part of the evidence gathering process. The process is begun in Step 1's risk profile and continues most intensively through Step 3, where with condition scenario reductions in risk are estimated. Risks identified that do not require a formal risk assessment should be characterized quantitatively if possible during the analysis normally done for Steps 3 and 4. A risk characterization includes one or more numerical risk estimates and an expanded narrative description of the risk that puts the risk estimates in perspective. Later, the risk characterization will include relevant changes in risk attributable to the formulated solutions.

Item 3 focuses specific attention on recognizing and identifying the most important uncertainties identified during the evidence gathering process. If specific data requests are made for risk information, item 4 indicates these questions are to be answered as completely as possible. Additional details may be added to the answers during subsequent planning steps as uncertainty continues to be reduced.

12.7.1.3 Step 2 RIP Outputs

The RIP outputs added at Step 2 include the following:

1. A robust set of alternative solutions, sometimes called risk management options (RMOs)
2. A risk hypothesis for each identified risk
3. Initial adaptive management strategies for plans with uncertain outcomes

Item 1 simply introduces the notion that a solution can also include or function as an RMO. An RMO represents one or more measures that are usually part of a larger plan. A risk hypothesis explains what can go wrong and how it can happen. Each risk should be described by a risk hypothesis that includes the sequence of events that are necessary for harm to occur. This hypothesis provides a blueprint, of sorts, for constructing RMOs that can disrupt the occurrence of a risk. When there is substantial uncertainty about the effectiveness of a formulated solution, that is, when planners are unsure how effective some initial measures may be in managing risk, an adaptive management strategy should be considered one means of addressing this uncertainty. Risk hypotheses are taken up again in Chapter 15.

12.7.1.4 Step 3 RIP Outputs

The evaluation step outputs in RIP can be supplemented by the following:

1. Displays of plan effects that show the significance of the key uncertainties
2. A complete characterization of the risk management effectiveness of each plan (i.e., RMO) to include existing, reduced, residual, transformed, transferred, and new risks
3. Clear identification of key uncertainties and the assessors' confidence in the results
4. A preliminary identification of an upper limit on the tolerable level of risk

A display of effects is already a crucial part of the planning process. In RIP, this display should also convey useful information about uncertainties that matter to decisions being made in the planning process. By this step, the with condition scenarios should be complete for each alternative solution. The risk characterization should be expanded to include the efficacy of each plan in addressing the risks previously identified. In other words, risk management effects need to be included among the evaluation criteria.

Risks need to be broadly construed to include existing risk and other aspects of a risk. To illustrate, consider a flood risk example. Some RMOs reduce the flood risk and the amount by which it is reduced is important to estimate. Residual risk is that part of the existing risk that remains after the RMO is implemented. An RMO might transform a risk. Imagine a community flooded by the slow rise of a river. If that community is protected by a levee or floodwall, the flood problem is transformed from a slow rise to a catastrophic overtopping. In addition, a new risk has been created in the potential failure of the levee itself. Risks can also be transferred. When

ACCEPTABLE VERSUS TOLERABLE RISK

An acceptable risk is a risk whose probability of occurrence is so small or whose consequences are so slight or whose benefits (perceived or real) are so great that individuals or groups in society are willing to take or be subjected to the risk that the event might occur. An acceptable risk requires no risk management; it is, by definition, acceptable.

A tolerable risk is a nonnegligible risk that has not yet been reduced to an acceptable level. The risk is tolerated for one of three reasons: we may be unable to reduce the risk further, the costs of doing so are considered excessive, or the magnitude of the benefits associated with the risky activity are too great to reduce it further. A tolerable risk is not an acceptable risk; it is still unacceptable, but its severity has been reduced to a point where it is tolerated.

a levee is built to protect a town on the left bank of a river, flood flows on the right bank increase exacerbating the risk there.

Item 3 suggests it is not enough to simply identify the key uncertainties and their effects. Analysts should also express their level of confidence in the work they have done so that others may better understand the significance of the uncertainty. If it has become clear that risks cannot be reduced to an acceptable level planners should begin to try to identify the level of risk that affected communities are able to tolerate.

12.7.1.5 Step 4 RIP Outputs

In the comparison step, the same displays of uncertainty, risk estimates, and levels of tolerable risk remain the focus. Displays of plan effects should show the significant differences in risk and uncertainty among the plans. The RIP outputs include the following:

1. Displays that convey the remaining uncertainty and its significance for decision making
2. A complete characterization of the risk management effectiveness of each solution or RMO that includes existing, reduced, residual, transformed, transferred, and new risks
3. A refined notion of the community's tolerable level of risk

12.7.1.6 Step 5 RIP Outputs

The RIP outputs in the final planning step include the following:

1. A tolerable level of risk
2. A selected solution
3. A risk assessment of the selected solution
4. Identification of measurable desired outcomes to evaluate the solution success
5. An adaptive management plan, if required

The selection of a course of action will define what decision makers consider a tolerable level of risk. Sometimes, the tolerable level of risk (TLR) will be established first and then plans will be formulated or selected based on their ability to achieve the desired TLR. Other times, the TLR is what the selected solution provides in a post facto manner.

In RIP, it is highly recommended that a formal risk assessment be conducted for the recommended solution. This is an intentional effort to identify what could go wrong if the selected solution is implemented, how it could happen, what the consequences would be, and the likelihood of those consequences occurring. This assessment may be qualitative or quantitative.

Given that many planning situations are accompanied by substantial uncertainty even at the end of the planning process, it is important to be able to recognize and measure a successful solution. To this end, planners should identify a conceptually measurable outcome of the implemented solution in this last step. That way the solution's success can be monitored. Data can always be collected and evaluated if monitoring is desired.

If the risk assessment for the selected solution suggests significant uncertainty remains, the solution should include an adaptive management component. This would enable responsible parties to design experiments to simultaneously facilitate learning and reduce uncertainty so that phased implementation of the solution's measures can be modified for maximum efficacy.

12.7.1.7 Public Involvement RIP Outputs

The primary addition to the public involvement program under RIP is an effective risk communication process. This will include carefully explaining the risks that do exist and understanding the range of public perceptions concerning the risks identified. The public should be engaged in the determination of a tolerable level of risk. Effective communication about how the RMOs were arrived at and clear training on everyone's role in the risk management process are also required.

12.7.2 The Risk-Informed Planner's Toolbox

The planner's toolbox is a big one. It includes the tools, techniques, and methodologies of a wide variety of disciplines; these are best treated in the literature of those disciplines. Moving efficiently toward RIP requires a few extra tools to be added. The first among these is scenario planning. When the future is so uncertain that a single most likely without condition scenario is inadequate, scenario planning is the most appropriate response to such uncertainties. Scenario planning is described in detail in Chapter 13.

Whether scenario planning is used or not, there is bound to be uncertainty about the facts and data used in the study. The risk-informed tools available to address these sources of uncertainty can be divided into qualitative and quantitative tools.

12.7.2.1 Qualitative Risk Tools

Qualitative risk assessment is distinguished primarily by its lack of reliance on numerical expressions of risk. That means qualitative risk assessment depends on risk descriptions, narratives, and relative values often obtained by ranking or separating risks into descriptive categories such as high, medium, low, and no risk. When the relative values are numeric, but only nominal or ordinal in character, such as when index numbers are used, the risk estimate is said to be semiquantitative, they remain more qualitative than quantitative in character. Quantitative risk assessment relies on numerical expressions of risk.

The qualitative methods used to asses risk continue to grow. A representative list of qualitative tools includes the following:

- Increase or decrease risk
- Risk narratives
- Evidence mapping
- Screening
- Ratings
- Rankings
- Operational risk management (risk matrix)
- Develop a generic process
- Qualitative assessment models
- Multicriteria decision analysis

Details on these and other methods can be found in the literature, for example, see Yoe (2012) for a description of these techniques.

12.7.2.2 Quantitative Risk Tools

In planning, quantitative risk assessment tools are normally going to be preferred when they are available. The array of quantitative tools is as broad as the many disciplines that now practice risk assessment. It is worth noting that the theory, methods, and techniques of good natural and social science methods are the primary tools of risk assessment. A few representative tools are described as follows.

Safety assessment. In the past, getting to "safety" has been the goal of many policies. The problem with a notion like safety is that someone must decide what level of chance or what magnitude of consequence is going to be considered safe. Safety assessments typically require someone to establish a safe level of exposure to a hazard; then the actual exposure is measured and compared to the safety threshold. A maximum daily exposure to a chemical that is considered safe, for example, is called an acceptable daily intake (ADI). As long as the long-term daily exposure for the population at risk is below this value, intake is considered safe. Engineers will define the capacity of a structural member, such as a piece of rebar, and compare the demand on this member to its capacity as a form of safety assessment. Safety assessments are often based on this quotient approach.

Scenario analysis. Scenarios are plausible sequences of events, that is, the stories we tell about what can go wrong and how it can happen. There are a great many tools that can be used to structure scenarios. These range from qualitative narratives through complex quantitative models. Event trees, decisions trees, probability trees, and the like are convenient tools for constructing scenarios.

There are three kinds of scenario analyses: monolithic, deterministic, and probabilistic. Monolithic scenarios were used before uncertainty was widely recognized. Then, it was common practice to develop and present a scenario as if it was a fait accompli, when in fact it was simply one of many possible story lines about a problem or opportunity. Monolithic scenarios are often used by the political system to enact legislation. An idea is proposed and the future it will create is described. Monolithic scenario analysis uses only one scenario for decision making.

Moving from a single scenario to a few selected scenarios involves deterministic scenario analysis (DSA). DSA defines and examines a limited number of specific scenarios. This can be a useful way to organize and simplify an avalanche of data into a small number of possible future states of the system being modeled. The scenarios so identified are usually chosen for specific reasons. They may be exploratory, such as with the worst-case, most likely, or best-case scenarios, or they could be chosen for strategic or tactical reasons.

There are some serious limitations to DSA. First, only a limited number of scenarios can be considered. Second, the likelihoods of these scenarios cannot be estimated with much confidence. Third, this approach is inadequate for describing the full range of potential outcomes.

Probabilistic scenario analysis (PSA) overcomes these limitations by combining probabilistic methods, for example, the Monte Carlo process, with a scenario generation method such as event tree models to produce a PSA. A PSA may produce measures of probability, consequence, or risk, which combines the probability and consequence measures into a single value.

Sensitivity analysis. Expect that uncertainty will never be reduced to zero. This means the risk characterization from your planning studies is at best an estimate. At worst, it may be informed speculation. The outputs of a risk characterization are conditional answers, conditioned on the data and data gaps, assumptions and estimation tools, techniques, and methodologies used to arrive at them.

Sensitivity analysis is the study of how the variation* in a risk-informed output can be apportioned, qualitatively or quantitatively, to different "sources of variation." Analysts need to consider how sensitive a model's output, risk characterization, or other important study outputs subject to uncertainty are to changes or estimation errors that might occur in model inputs, model parameters, assumptions, scenarios, and the functional forms of models. This information must then be effectively conveyed to decision makers so that they can explicitly consider the significance of this sensitivity that arises from knowledge uncertainty and natural variability for their decision making.

Some sensitivity analysis should be an integral component of every planning study. This is the point in a study when the planning team focuses intentionally on better understanding the things they don't know and their importance for decision making. The results of the sensitivity analysis will provide insight into the importance of different sources of uncertainty.

Sensitivity analysis has at times been called "what if" analysis. It can be used to answer questions such as those below, which build on questions first provided by Mokhtari and Frey (2005):

- How might the decision be changed by increases or decreases in selected input values?
- What is the range of values a parameter can assume in a function without changing the decision?
- By how much must a value change to lead to an alternate best decision?

* Variation includes the effects of natural variability and knowledge uncertainty.

- How sensitive is our output to forecast error or other changes in inputs?
- Which inputs contribute most to the variation in the output?
- Which inputs are most responsible for the best (or worst) outcomes of the output?
- What is the rank order of importance among the model inputs?
- Are there two or more inputs to which the output has similar sensitivity, or is it possible to clearly distinguish and discriminate among the inputs with respect to their importance?
- Might changes in our decisions/actions improve our outputs?
- Does the model respond appropriately to changes in assumptions and inputs?

12.7.3 Planning Study Risks

What is an acceptable level of detail in a planning investigation? How complex must the hydraulics and hydrology be? How much design detail is needed to lay out a solution or to estimate its costs? There will always be uncertainty in a planning study, no matter how much detail we may have. At times, planners have no choice but to make some reasonable assumptions and hope they pan out. Other times, planners may choose to assume the risk of an error in an analysis to save time or money. Every time we make an assumption about an uncertainty or proceed with less than the maximum possible analysis we face a risk of making an error. These planning study risks need to be identified and explicitly considered during the scoping process. The planning team will take those risks with which they can live. Study risks with unacceptable consequences must be managed. Consider the following examples.

Imagine that because of an expedited schedule you will be unable to complete the hazardous, toxic, and radioactive waste work for your reconnaissance study. Further, imagine a preliminary search of the land records leads you to suspect some problems, but you don't have enough information to say whether there is a significant impact or not. Consequently, you cannot prepare a finding of no significant impact (FONSI) or an environmental assessment (EA) in this first study stage.

If the plans you formulate, their alignments, dimensions, and all the other attributes are all based on the assumption that there will be no significant hazardous, toxic, and radioactive waste (HTRW) impact, that is, you have assumed an eventual FONSI. If that assumption turns out to be incorrect, this could mean a bust in plan formulation and the need for a full EIS process. A full EIS will slow down the future study phases and the formulated plans may all have to be reformulated and redesigned. The planning team must decide if this risk of delay and additional formulation work is one they can tolerate. If it is, they will assume a FONSI and proceed. If they cannot, they will have to complete the HTRW work in the reconnaissance phase of the study.

A second example would be delaying archaeological fieldwork until the engineering, and design stage of planning (a later iteration of planning) to save time and money in the feasibility study. Proceeding with uncertainty about the existence, number, and location of culturally significant sites presents a risk. This is little risk to the study and its recommendation, but it could delay construction and increase study costs. The number of such planning study risks is varied and virtually limitless.

12.7.3.1 *Risk Register*

The basic solution for these kinds of study risks is informed consent. Planners need to be aware of the potential risks of the assumptions and decisions they make with less than complete certainty. The risk register is a tool for forming that informed consent. A detailed example of a risk register is provided in Chapter 15.

12.8 TAKE AWAYS

Lesson 1: Every study has a pile of things you know and a pile of things you don't know.
Lesson 2: Uncertainty will never be eliminated.
Lesson 3: Uncertainty consists of knowledge uncertainty and natural variability.
Lesson 4: Uncertainty is the source of risk.
Lesson 5: RIP is the best way to be intentional in addressing uncertainty in planning.

REFERENCES

Chess, C. and B. J. Hance. 1994. *Communicating with the public: Ten questions environmental managers should ask.* New Brunswick, NJ: Center for Environmental Communication.

International Programme on Chemical Safety. 2006. *Draft guidance documenton characterizing and communicating uncertainty of exposure assessment, draft for public review.* Geneva: World Health Organization. Accessed May 15, 2010. http://www.who.int/ipcs/methods/harmonization/areas/draftundertainty.pdf

Kaplan, S. and B. J. Garrick. 1981. On the quantitative definition of risk. *Risk Analysis* 1 (1): 11–27.

Mokhtari, A. and H. C. Frey. 2005. Recommended practice regarding selection of sensitivity analysis methods applied to microbial food safety process risk models. *Human and Ecological Risk Assessment* 11 (3): 591–605.

Morgan, M. G. and M. Henrion. 1990. *Uncertainty: A guide to dealing with uncertainty in quantitative risk and policy analysis.* Cambridge: Cambridge University Press.

National Research Council, Committee on Improving Risk Analysis Approaches Used by the U.S. EPA. 2009. *Advancing risk assessment 2009.* Washington, DC: The National Academies Press.

O'Hagan, A., C. E. Buck, A. Daneshkhah, J. R. Eiser, P. H. Garthwaite, D. J. Jenkinson, J. E. Oakley, and T. Rakow. 2006. *Uncertain judgements: Eliciting experts' probabilities.* West Sussex, UK: John Wiley & Sons.

Tainter, J. A. 1996. *Getting down to earth: Practical applications of ecological economics.* Washington, DC: Island Press.

U.S. Water Resources Council. September 10, 1973. *Principles and standards for planning water and related land resources.* 38 *Federal Register* 24778–24862.

Yoe, C. 2012. *Principles of risk analysis: decision making under uncertainty.* Boca Raton, FL: Taylor & Francis.

CHAPTER 13

Scenario Planning

13.1 INTRODUCTION

Good planners separate what is known from what is unknown and then they deal intentionally with the most important unknowns throughout the planning process. The uncertainties are not all equally important. Part of the planners' job is to identify and then address those sources of uncertainty that could significantly affect plan outcomes. Furthermore, planners must convey the significance of these uncertainties to decision makers. These are the uncertainties that can affect decision making.

One of the critical planning tasks in Step 1 is defining the without condition scenario, that is, saying what the future looks like without your plan. Scenario planning, the subject of this chapter, is a viable option for planners when there is too much uncertainty in a study to represent the future with a single most likely scenario. Figure 13.1 divides planning space with two axes: the amount of uncertainty in the study and the consequence of making a wrong decision.

Figure 13.1 Effect of uncertainty and mistake consequences on planning method.

There can be very little to a great deal of uncertainty. Uncertainty increases the likelihood of making a mistake during decision making. The consequences of being wrong can range from trivial to grave. These two notions identify four planning subspaces. When there is little uncertainty and being wrong does not much matter, we can use any kind of routine decision-making processes. As the amount of uncertainty increases but the consequences of a wrong decision are not too serious, traditional deterministic planning processes* are adequate. This process is represented by the common practices of the last 30 years or so.

As the consequences of a wrong decision become more serious, deterministic planning is no longer adequate. Even when there is relatively little uncertainty, if the consequences of a mistake are more serious, then risk-informed planning, which intentionally accounts for uncertainty in all steps of the planning process, including the without condition scenario, is warranted. When the uncertainty is great and the consequences of a wrong decision are serious, scenario planning, that is, multiple without condition scenarios, is needed.

13.2 SCENARIO-PLANNING BACKGROUND

HISTORY OF SCENARIO PLANNING

Herman Kahn introduced the word "scenario" to its planning context at the RAND Corporation in the 1950s. The first applications of scenarios in a planning context are thought to have been in the military strategy studies they did for the U.S. government. By the 1960s, the Wharton School's H. Ozbekahn had used scenarios in an urban-planning project for Paris, France. The theoretical foundations of scenario forecasting, an important component of scenario planning, were principally developed in the 1970s. Royal Dutch Shell is regularly credited with popularizing and modernizing the use of scenario planning for

* This is planning that relies on best guess/representative value point estimates of uncertain knowledge, scenarios, and quantities.

strategic planning in the early 1970s (Wack 1985 a,b). In fact, Wack asserts it was Royal Dutch Shell that came up with the idea of scenario planning. French (Godet 1987) and German (Brauers and Weber 1988) planners have also made early use of these methods. The use of scenario-driven planning spread in the 1970s and by the 1980s, it seemed to have emerged as a distinct field of study with an extensive literature.

All our knowledge is about the past and all our decisions are about the future.

Wilson 1975

What if the future is not like the past? The most common planning analysis, described in earlier chapters, is based on describing a single most likely without condition scenario and comparing effects within it to the plan effects within a single most likely with condition scenario. In actual practice, many future conditions are constructed from individual forecasts and analyses without the benefit of a unifying scenario. Planners often bundle their best efforts and endow them as most likely. That is a distinctly different approach than this text suggests, which is to identify reasonable future scenarios and do your analyses within them. When it is reasonable to choose one of the scenarios as representative of the future, and it often will be, then the key uncertainties can usually be handled through risk-based analytical methods. These include such things as probabilistic-scenario analysis, sensitivity analysis, evidence maps, multi-criteria decision analysis, and the like (Yoe 2012).

What should planners do when it is not possible to identify any one of the alternative scenarios as most representative of the future? One answer is to use scenario planning. If we're brutally frank, we all know any single future scenario will be wrong. Thus, planning is based on what could be, not necessarily what will be, and what could be is wide open to debate. We cannot ignore this fact when the consequences of being wrong may be serious.

Scenario planning was developed in the second half of the twentieth century, mostly in Europe. It grew in popularity largely as a result of the failure of traditional planning as its deterministic view of the future, which failed to capture what really happened, time, and time again. Scenario planning is described in a rich literature, so it is simply summarized in this chapter. More details can be found in the works of Godet (2001), Ringland (2002), Lindgren and Bandhold (2003), or Ralston and Winston (2006). The latter of these is used as the primary reference for the steps described below.

13.3 A SCENARIO-PLANNING PROCESS

Ralston and Wilson (2006) describe the scenario-planning process in 18 steps arranged in four major clusters as follows:

I. Getting started
   1. Develop case for scenarios.
   2. Get executive support and participation.

    3. Define decision focus.
    4. Design process.
    5. Select facilitator.
    6. Form scenario team.
- II. Laying environmental analysis foundation
    7. Gather data and view.
    8. Identify key decision factors (KDFs).
    9. Identify critical forces and drivers.
    10. Conduct focused research on key issues, forces, and drivers.
- III. Creating the scenarios
    11. Assess importance and predictability/uncertainty of forces/drivers.
    12. Identify key axes of uncertainty.
    13. Select scenario logics to cover uncertainties.
    14. Write stories for scenarios.
- IV. Moving from scenarios to decisions
    15. Rehearse future with scenarios.
    16. Make decision recommendations.
    17. Identify signposts to monitor.
    18. Communicate results.

Ralston and Wilson's (2006) "Getting Started" cluster corresponds closely enough with the Step 1 decision context work that it requires no additional discussion here. Step 7 overlaps with the evidence gathering process of the Five by Two planning process, so we will pick up their process with Step 8 and use it as a guide for developing multiple without condition scenarios. Before doing so, it is useful to consider some desirable characteristics of a good without condition scenario. The top five characteristics in no special order are as follows:

1. It tells the story of the study area's problems and opportunities, objectives, and constraints in the future if the planning entity takes no action.
2. It quantifies that story in ways that support planning from formulation through plan selection.
3. It tells the truth.
4. It is based on evidence and good science.
5. It identifies what is uncertain and how that uncertainty might affect decision making.

The case for multiple without condition scenarios is strongest when the following circumstances hold:

- Macro-level uncertainty dominates the planner's concerns.
- The direction of the future is unknown.
- A wide range of future possibilities are already recognized.
- There is widespread disagreement about what the future will be like, each with its own merits.
- Costly surprises have occurred in the past.
- Failure to anticipate the true future could render a solution ineffective.
- Stakeholders want to use scenario planning.

13.3.1 Identify Key Decision Factors

Step 8 in the process above is to identify KDFs. Decision factors are events or outcomes in the future that we would like to know more about in order to improve the quality, relevance, and efficacy of our decisions. KDFs are the most important factors that are outside the control of the decision makers and that will affect their decisions. They are not the decision criteria themselves as much as factors that affect the problems, opportunities, objectives, and constraints. One of the distinguishing characteristics of KDFs is that they are external to the planning study and essentially uncontrollable. This does not mean that internal and controllable factors are unimportant. In fact, these controllable factors are expected to become essential elements of the plans that will be formulated to avoid undesirable futures without action by the study team.

How does the team identify these KDFs? The most effective way is to brainstorm them in a workshop scheduled specifically for this purpose. Assuming everyone is well aware of the decision context, you might begin by asking everyone a seed question. What external issues (i.e., factors beyond your control) concern you the most? If there was an all-seeing oracle, what questions about the future would you like to ask?

You can ask these or other questions for each of the problems, opportunities, and so on. At this point, your goal is to identify as many KDFs (external issues, questions, etc.) as possible. Have participants write their KDFs on an index card, putting any necessary details on the back of the card. Once the idea generation stage of your brainstorming session is completed, examine the answers to your seed questions and begin to cluster them by topic or focus. Give each cluster of ideas a descriptive title. The output of this step is a list of KDFs organized into clusters.

13.3.2 Assess KDFs

Identifying KDFs is an exercise in divergent thinking. It is intended to get as many ideas out on the table as possible. It's not unusual to identify dozens of questions or KDFs in a brainstorming session. Assessing KDFs is a convergent thinking process. The process of clustering the KDFs and naming them is only the first step toward identifying the most significant KDF clusters.

POST-IT NOTE CLUSTERS

Display all the key forces on post-it notes and then let team members move them around into logical groupings. Others can react to what they see and rearrange the notes until a consensus begins to emerge. Remaining points of disagreement are resolved through discussion. The goal is to limit the groupings of forces to four or less. Each grouping is given a name to identify the theme that links all the ideas.

Once you have identified the clusters, the team needs to identify those they believe will have the biggest impact on the success or failure of any decision (i.e., the performance of a plan) they make. The output of this step is a clear identification of the most important KDF clusters. These clusters tell you where your scenarios should focus. Think of them as section titles or headings in your scenario narrative. These clusters will help planners identify key areas of uncertainty that the scenarios need to address. The significant KDF clusters are the external issues the without condition should address. Thus, each scenario for the flood risk management example in the following text box would include a description of flows, property damage, and social vulnerability, among other things.

13.3.3 Identify Forces and Drivers That Determine KDF Outcomes

At this point, the team has identified a set of KDFs that are essential to the decision context. Now, the team needs to know what forces and drivers will determine the outcomes for the KDF clusters they have identified. Table 13.1 provides generic examples of KDFs and their forces and drivers. Think of forces as things beyond the control of stakeholders in the planning investigation and drivers as those things stakeholders can control or at least influence. This task identifies the kind of information that would enable you to answer the questions asked of the oracle.

Once again, the task is to brainstorm the forces and drivers that will shape the future you are trying to describe. These forces and drivers are uncertainties that provide the foundation or underlying structure for the without condition. They include trends, conditions, factors, events, wild cards and shocks, discontinuities, weak signals, and similar things that deserve consideration.

KEY DECISION FACTORS FOR FLOOD RISK MANAGEMENT

Imagine you could brainstorm questions for an oracle to answer about your planning study; some clusters of questions might look like those given below.

Cluster: Flows

What will be the largest flood in the next century?
How many floods will get far enough out of bank to cause damage?
How much of the watershed will be developed in the future?

Cluster: Property Damage

Will the town's floodplain economy thrive or deteriorate?
Will the factory close?
Will the slum areas gentrify?

Cluster: Social Vulnerability

What will the median age look like?
Will per capita income increase or decrease?
Will the minority population grow or decline?

Table 13.1 Generic Examples of Forces and Drivers

Key Decision Factors	Forces and Drivers
Political	Political attitudes and issues Political parties Quality of government Fiscal responsibility Community activism
Demographic	Population growth Age, family size and structure, and ethnicity Education levels Migration patterns Immigration and illegal aliens
Ecosystems	Environmental policy Sustainability and biodiversity Threatened and endangered species Water quantity and quality Hazardous Toxic Radioactive Waste Status and trends Pollution
International considerations	War and terrorism Globalization International trade patterns Strength and role of dollar Protectionism
Natural resources	Energy sources, prices, and policies Resource reserves Resources prices Land use

This task (Ralston and Wilson's Step 9 2006) is akin to the external analysis of strategic planning. There are many methods for conducting a scan of the external environment's forces. The macro-environmental level includes broad national or global contexts in which these forces evolve. The micro-environmental level forces are more study area specific forces. It is better to over-identify these forces and then to winnow them down to key areas of uncertainty than it is to underestimate these forces. A simple way to proceed is to simply ask, "What forces and drivers will influence flows (i.e., substitute KDF_1, KDF_2, etc.)?" Ralston and Wilson (2006) have suggested that as a person identifies a force/driver, they should complete an index card with the following information:

- The name of the force or driver.
- One sentence describing it.
- Possible future outcomes for the force or driver, for example, increasing or decreasing.
- Identify what the force/driver will influence (tie it to a KDF cluster).
- Identify what influences the force/driver.

Give the team time to identify forces/drivers in this way and then begin to ask each person to briefly describe their most significant force/driver, until all the unique forces have been identified. As the forces are identified, the team leader can begin, with the help of the group, to organize them. What you are trying to do in this step is to identify and map the major external forces that will affect the success or failure of your problem solutions and opportunity strategies (i.e., your plans). It is essential that the planning team spend face-to-face time to take advantage of the synergistic synthesis and creativity that is only possible when people spend time working together. Once the team has identified the major forces and drivers that will determine how the KDF clusters will play out, this task is complete.

Table 13.2 shows examples of the kinds of forces and drivers that might emerge in a flood risk management study like the Elliott City example, first introduced in Chapter 4. Many of these forces and drivers will be well known to the planning team. Others will be a mystery. Notice that the forces/drivers have been organized in clusters of related forces.

Table 13.2 Forces and Drivers in a Hypothetical Flood Risk Management Scenario

	Forces and Drivers	
Land Use and Development	**Property Values**	**Hydrology and Hydraulics**
Growth projections (and associated land use conversions)	Depreciated structure values	Future stream conditions
Future land use	Square footage	Precipitation
Flooding potential	First floor elevations	Flooding potential
Urban development/ encroachment	Structure values	Storm frequency events
Future growth rate	Content values	Future changes in H&H without project
Development activities	Land value	Hydrologic estimates of runoff (i.e., modeling)
Economic conditions	Agricultural crop patterns	*n*-values for floodplains
Community follows master plan	Crop yields/productivity	Cross-sectional changes to the channel geometry (changes in hydraulics)
Land use changes	Value of property in floodplain	Percentage impervious surface
Land use mix	Damage surveys	Rainfall
Zoning regulations	Expected annual damages	Runoff
Building codes	Real estate	Hydrographs
Political party in power	Jobs and income	Flow regimes
	National economy	Frequency curves
	Success of new free trade zones	Rating curves
	Education levels in the watershed	Climate change
		Storm frequencies
		Sea level rise
		Channel constrictions
		Erosion rates, land loss
		Vegetation

The outputs of this task are the cards describing the forces/drivers and the group's organizational scheme that has, in effect, separated them into like piles. This provides a basic map of the forces and drivers that will determine the study area's without condition scenario. Drivers and forces are an important enough subject to be taken up in more detail in Addendum 1 of this chapter.

13.3.4 Fill Data Gaps Where Possible

At this point, it is likely that team members will have an uneven understanding of the different forces and drivers. There will likely also be some critical new data gaps. This is the time to prepare simple focus papers* for those forces and drivers that are mysteries to the team. This will require additional analysis by the planning team. Ralston and Wilson (2006) have suggested these papers might address the following:

- The apparent trends in a cluster of forces
- Major uncertainties about how these trends might change in the future
- The impact the force clusters will have on the KDF clusters

This task simultaneously reduces and illuminates the uncertainty in your without condition scenario. The output of this task is a series of focus papers that explain forces that are not clear to all. In so doing, the papers provide not only a common understanding of a force but its attendant uncertainty as well.

13.3.5 Assess the Forces and Drivers

In this step (Ralston and Wilson's Step 11 2006), planners rank and sort all of the forces and drivers. Discerning which of them are most important to determining the future is a fundamentally subjective process. This is important to understand. It is not an objective search for a knowable answer. It requires judgment and the resolution of differences. That is one reason it is important to base the process on as much sound evidence, logic, and transparency as possible.

To this point, the process has focused on identifying environmental forces that might be considered in the without project condition scenario, and judgment of these forces has been suspended. The team may now be looking at dozens of forces and drivers and it is clearly impractical, if not impossible, to consider them all in the without condition scenario. It is equally clear these forces will vary in both the extent of their uncertainty as well as their importance to the future that the planners seek to characterize.

In order to move to the work of developing the without condition, planners must assess the forces and drivers. This is a critical step. A major purpose of the without condition scenario is to make uncertainty about the future explicit. This task will

* The idea of a focus paper is a flexible one. Not every uncertain force/driver will require a focus paper. Some of the uncertainties may be resolved with a phone call. The idea is to resolve or reduce as much of the uncertainty surrounding a force/driver as possible by any means possible, documenting the more important topics.

eventually identify those critical uncertainties that will form the basis for the without condition scenario.

The impact and uncertainty matrix of Figure 13.2 is one tool that has been used to assess the importance of a force or driver to the decision context as well as the degree of uncertainty about the range of future outcomes. To successfully use the matrix, it is critically important that everyone share a common understanding of the matrix's row and column dimensions. Importance or impact is related to the force/driver's influence on future outcomes of the KDFs identified earlier in the scenario-planning process. If stream flows, property damage, and social vulnerability are among the KDFs that will figure in decision making, then you want to identify those external forces and drivers that will have the greatest impact on these three factors.

Uncertainty, as used here, is a little different than it is generally used in this text. For the purpose of this matrix, uncertainty means the extent to which future outcomes of the KDFs are not predictable. If there is widespread agreement on the outcome a force will have on a decision factor, that is low uncertainty. If the team disagrees on what the outcomes might be there is medium or high uncertainty, depending on the extent or range of possible outcomes envisioned by the team. Consider sea level change as an example. Suppose the team's range of opinion is between no change and a 7-m increase—that is high uncertainty. One mindset for rating the uncertainty might be whether you would bet your house on any one particular outcome. If not, there is medium to high uncertainty.

The use of this matrix is quite simple in concept. The team, in another workshop, takes the list of forces and drivers from a list like that in Table 13.2 and sorts them into the cells of the matrix. Thus, sea level change, zoning regulations, the political party in power, the national economy, and every other force/driver identified is pigeonholed in a cell in the matrix. Next, planners take the elements in each cell and rank them. It is important to do good work on this task but it is just as important to bear in mind there is not a right answer for this task and the planning team's judgments can always be vetted by experts or interested stakeholders.

It is recommended that the impact/importance of the force be assessed first. If you assess uncertainty first and a force is judged to be very uncertain because the range of uncertainty is great and includes *no change* as a potential outcome, as was the case in the sea level change example above, some may think no change is the most likely outcome. Then, when the team moves to consider the impact, it may result in a downgraded assessment of the impact. For example, if most people believe the

Level of the impact or importance	Degree of uncertainty		
	High	Medium	Low

Figure 13.2 Impact and uncertainty matrix—a tool for assessing forces and drivers.

likelihood of sea level rise is very small, they may be influenced by this and consider the impact to be low, whereas, if the team considered only the impact of sea level rise, it would surely be high. So, begin by assessing the impact and then the uncertainty.

It can be helpful to establish a threshold for the maximum percentage of high-importance/high-uncertainty forces identified in order to force the team to discriminate carefully in their assessment of the forces. Ralston and Wilson (2006) suggest a maximum of 25%. Only the high-importance and high-impact forces will be used to develop the without condition scenario framework. Some of these will form the basic storyline and its logic. The other forces are not ignored; they can be used to fill in the details of the scenarios or in the analysis done within them later in the technical analyses of the planning process.

One technique for proceeding is to ask each team member to identify no more than five high-impact forces. Going around the room, each planner identifies the forces they chose. Once all have been identified, the team is invited to disagree with any of the nominations. If consensus can be reached on disagreements, the list is adjusted accordingly by demoting forces no longer considered high impact. If not, more formal consensus-building techniques like strength-of-belief voting schemes may be required. An alternative approach when the number of forces is reasonable is to rank each force and discuss them individually with the goal of reaching agreement on an importance and uncertainty rating for each.

The output of this task is a list of the five or so most important high-impact/high-uncertainty forces or drivers. These should be those forces that can lead to the most significant differences in the future, differences that could affect your KDFs and the ultimate choice of a solution.

13.3.6 Identify the Axes of Uncertainty

This (Ralston and Wilson's Steps 12 and 13 2006) is the long-awaited heart of the scenario-building process. It includes identifying the axes of uncertainty and determining the scenario logics. Imagine that we have looked at all the high-impact/high-uncertainty drivers for Elliott City and come up with the regional economy (a combination of land use and development and property values) and the hydrologic regimes as the two axes of uncertainty as shown in Figure 13.3. These two axes reflect the team's judgment about the most important uncertainties that will influence the future without your plan. Now the challenge is to define your scenario logics.

It is common practice to look for the two most significant and different logics to frame your story about the future. Increased runoff and no change in the runoff define one logic. A booming (increased development and rising property values) and a declining (little upstream development and suffering property values) economy define the other logic. If each of these logics is considered an axis in an x–y grid, they will define the four quadrants in Figure 13.3. The two logics form a simple map of the future broadly defining four different scenarios.

The goals of this step are four. First, identify one to four axes that encompass at least the majority of the high-impact/high-uncertainty forces. Second, the scenarios defined by these axes should be distinctly different from one another. If this

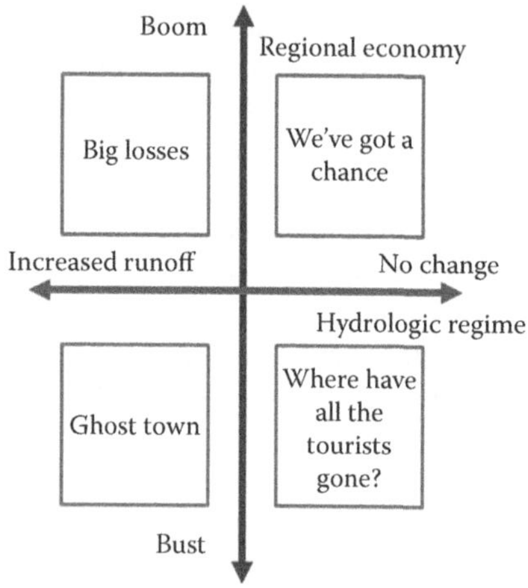

Figure 13.3 Scenario planning logics for Elliot City.

exercise identifies variations of a central theme, that is, when you know you can plan adequately with a single without condition scenario. For example, if you have four floodplain scenarios that are simply four variations on the extent of flooding in the future, all other things equal, or if you have four ecosystem restoration scenarios that differ largely by the size of a wildlife population, then you have variations on a theme.

Third, the axes need to be logical, that is, they should stem transparently from the forces and drivers the team has identified. Fourth, although the probabilities of different scenarios are never estimated explicitly, they must all be plausible. They can surprise and challenge people but they should not strain credulity. Thus, a logic that depends on Star Trek vintage transporting devices to move wildlife may not be a reasonable logic. Likewise, the destruction of the planet by a massive asteroid may be better left to Hollywood to consider. If the pattern is not yet clear, this entire scenario construction method depends on alternating rounds of divergent thinking followed by convergent clustering of related ideas. This is done with KDFs, with forces, and then again with the uncertainty axes.

A logic is defined by Ralston and Wilson (2006) as a hypothesis about the future dynamics of the forces and drivers in the external environment that describes how the world will work in the future without condition. The logic is the central theme or dynamic that explains how the forces in a scenario interrelate. This is plot development for storytelling; it is not analysis and hardcore forecasting. All the work done in the previous steps helps to shape realistic plausible logics. Plausible, as used here, should stretch the team's imagination. The idea is to bound the uncertainty, and planners are sometimes too conservative in their estimates of extremes.

The output of this task is a one-paragraph description of each logic you identify. Thus, there should be one paragraph describing the regional economy's influence on the KDFs and one paragraph describing the hydrologic regime's effects on the KDFs.

13.3.7 Decide How Many Scenarios You Need and Integrate the Details into a Scenario Logic

One of the more important decisions you will have to make is whether you can handle all of the relevant uncertainty adequately within the framework of a single scenario or if you need multiple scenarios to do so. The scenario or scenarios must describe the important trends and impacts of the forces and their associated uncertainty on KDFs. If the associated uncertainty can lead to markedly different outcomes (as opposed to variations on a theme), you are going to need multiple scenarios. Otherwise, a single scenario will do.

Most importantly, the scenario(s) must be useful in decision making. If the team uses one scenario, it must adequately describe the key uncertainties or accommodate them when the scenario is quantified. One scenario will often be sufficient in most planning contexts as long as the analysis within it is risk-based. In other words, the technical analysis done for a single most likely without condition scenario must adequately explore and address the relevant uncertainty that remains. If more than a single scenario is used, those experienced in scenario planning suggest a maximum of four scenarios be used.

There must be enough scenarios to cover the range of relevant possible futures. There need to be few enough scenarios that planners can understand and work with them all. Four axes of uncertainty, for example, produce 16 scenarios, far too many for planners to consider adequately. If two or three scenarios will do the job, use no more scenarios than you need. There could be a single axis of uncertainty, for example, "Will climate change lead to significant sea level rise or not?" gives us a powerful basis for developing two scenarios for the without condition. Using four scenarios has helped planners avoid the good scenario/bad scenario trap and the low/medium/high trap. The number of scenarios is not based on the probability of their occurrence. You can save probabilities for calculations and forecasts made within the scenarios; scenarios are stories not forecasts.

When working with multiple scenarios, it is important to avoid the tendency to anoint one of them as most probable. If one scenario provides you with good coverage of the envelope of uncertainty that characterizes the future use one scenario, as seen in Figure 13.4 (adapted from Ralston and Wilson 2006) on the left, where the future is fairly regular and relatively easy to gage.

There will be times when the map of the future and the envelope of uncertainty that covers it will be too complex for a single scenario. The irregular future on the right of Figure 13.4 represents a greater envelope of uncertainty. The scenarios chosen to represent it are selected to provide as much coverage of that future as possible. The planning team needs enough scenarios to capture all the discussions about significant forces, drivers, and uncertainties that have taken place. In general, do not use any more scenarios than you need to do this well. In the Elliott City example, the four scenarios, although different, are all variations on a theme of more or less flood damage. In that instance, we would use a single without condition scenario and then carefully analyze the range of differences in this critical decision criterion.

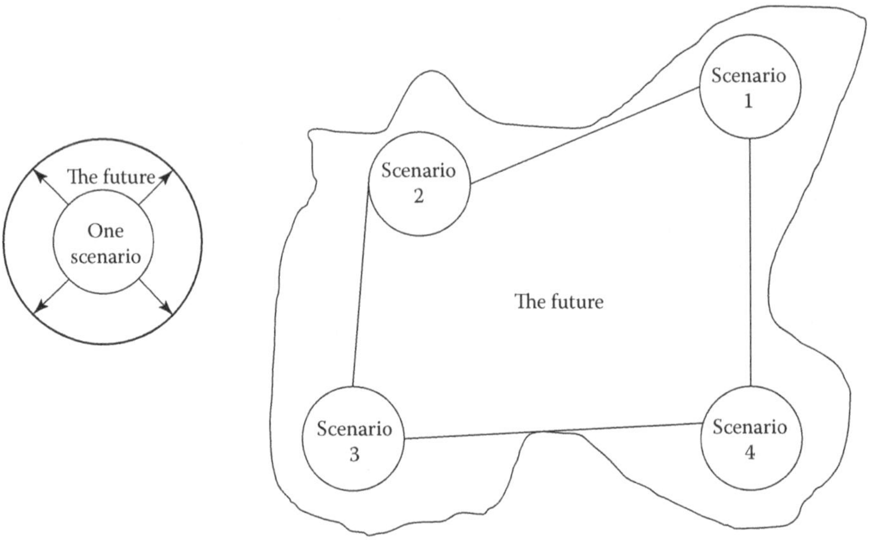

Figure 13.4 Scenarios needed to cover the future's envelope of uncertain.

In contrast, the Lake Okeechobee Watershed project in Florida began with a number of forces and drivers that could affect the nature and success of different solutions. These included runoff, sea level rise, the Lake Okeechobee regulation schedule, storm water treatment area dry outs, ecological response to measures taken, other CERP (2012)* projects, land availability, reservoir phosphorous reduction performance, and phosphorous loads. The axes of uncertainty (see Figure 13.5) identified from these drivers were runoff and phosphorous loads. Four scenarios were identified with the following logics: high-runoff high-phosphorous loadings, high- runoff low-phosphorous loadings, low-runoff high-phosphorous loadings, and low- runoff low-phosphorous loadings. This study would need to consider all four of these scenarios as they encompass different levels of flooding, phosphorous loadings in the water, water supply availability, and significantly different futures for the area.

The outputs from this task are twofold. First, the team should decide on the number of without condition scenarios it will use. Second, all of the previous work on identifying the key uncertainties should be integrated into a clear scenario logic for each of your scenarios.

13.3.8 Write the Scenario

Writing the scenario (Ralston and Wilson's Step 14 2006) is the critical integration step. This is where planners must weave the thread of the drivers and forces into distinctive patterns that describe their impact on your KDF clusters in the

* Comprehensive Everglades Restoration Project (CERP). A description of the Lake Okeechobee Watershed Project can be found at http://www.evergladesplan.org/pm/projects/proj_01_lake_o_watershed.aspx.

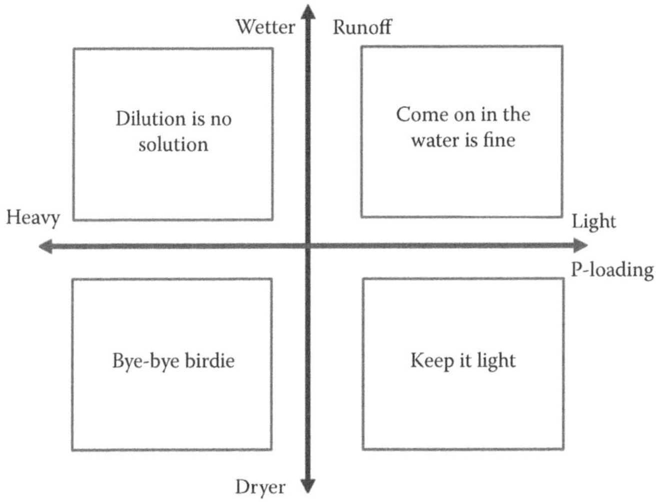

Figure 13.5 Axes of uncertainty forming four scenario logics and four named scenarios for Lake Okeechobee.

future, without condition for the study area. Your scenario should have a plot and a storyline. Cause and effect relationships should be identified and spelled out. The scenario should see and describe the future as a whole rather than as a series of independent and seemingly unrelated trends and events. The scenario describes how the study area problems and opportunities will play out in the future. It is not an analysis.

The logics provide the outline for your story. Elliott City will face a future of more or less flood damages. It is a relatively simple problem. When there are more scenarios, the stories become richer and more distinctive because of the differences among the scenarios. You can, for example, start to tell the story of the Lake Okeechobee watershed if it has a run of dry years with low runoff and high phosphorous loadings to the lake. There will be differences in floods, droughts, and water quality. The story can be expanded quickly in scope and level of detail by relating some cause and effect events you have identified in the work you have done. Focusing on water quality, for example, suppose the state passes a law that requires best management practice on the farm to control phosphorous loads that reach waterways, but farmers are unaware of the law or are unable or unwilling to comply. With just a few cause and effect chains like this the story you tell becomes quite intricate and informative.

The scenario takes on greater meaning when specific developments are described, such as the loss of a super colony of birds because of water quality degradation. Add a dash of conflict between agricultural and environmental interests and you may not have a best seller on your hands but you will have a compelling scenario that will inform decision making. These stories will frame the context for your technical analyses as well.

A scenario is not a forecast. You are not saying this is what will happen. It is a simple statement that suggests given a future that unfolds like this, these are the sorts of things that can happen. Building on the work of Ralston and Wilson (2006), your scenario outputs should include the following:

- Scenario title
- Brief description
- Narrative
- Comparison of scenarios
- Analysis within the scenario

Generally, the team will develop the storyline and major details while one individual will do the actual writing, which the team then vets. The title should be short, descriptive, and, if possible, catchy, so it helps people to understand and remember the storyline (see Figure 13.5). The brief description is a one-paragraph essence of the story. The narrative is the story, the history of the future. It needs to be complete enough to support the necessary analysis and forecasting that will follow. It also has to have enough detail to support the evaluation and decision making of the subsequent steps in the planning process. The scenarios should be no more than a few pages (3–5) in length. Descriptions of the analyses conducted within a scenario can run much longer. Think of the scenario as an informative newspaper article.

The scenarios are, as mentioned, not forecasts, and you can expect their details to change as the team begins to get a better feel for each story. When you use multiple scenarios, it is helpful to include a comparison table to help people understand the significant differences among the scenarios. Do this well before analysis within the scenarios begins.

13.3.9 Conduct Analyses from Within the Scenario

With a scenario in hand, the planning team is ready to make its forecasts and to do its analysis. The scenario frames the team's view of the future in a coherent and unified way that supports technical analysis of an uncertain future. Thus, hydraulic and hydrologic analysis, habitat unit calculations, and water quality analysis, for example, are done under the conditions of scenario logic 1, then they are redone for scenario logic 2, and so on.

The analyses needed to forecast the problems into the uncertain future and to characterize the likely outcomes of opportunities are all conducted within the framework and confines of each without condition scenario. Planners place themselves in each scenario and do the analysis needed to describe what the problems and opportunities will look like. They do the analysis needed to characterize the state of the objectives and constraints and decision criteria that reflect the values that matter to people in each scenario.

13.4 ADDRESSING UNCERTAINTY IN THE WITHOUT CONDITION(S)

The single biggest methodological change I have witnessed since my planning career began has been the growing emphasis on addressing the uncertainty that has always been part of the planning process in a more intentional manner. The evolution of the without condition from a series of forecasts to a scenario inside of which relevant analyses are done is just one piece of evidence of this change.

When the future uncertainty is reasonably manageable, that is, when the most important future differences are variations on a theme rather than distinctly different futures, a single without condition scenario is adequate. The remaining uncertainty, that is, the details of that single future, can be explored using qualitative and quantitative methods in the risk assessment done within the framework of the scenario. Risk assessment techniques, designed to evaluate the effects of uncertainty on decision critical analytical outputs, like those introduced in Chapter 12, will be useful for this purpose.

When the uncertainty driving the future produces significantly different futures, scenario planning should be used. Risk assessment techniques may still be suitable for the analyses done within these multiple scenario frameworks. Identifying the key sources of uncertainty is the critical first step. This step will be of limited value, however, unless planners quantify or otherwise characterize the significance of these uncertainties for decision making. To make this discussion more concrete, let's focus on the expected annual damages (EAD) from the hypothetical Elliott City example used earlier. It is but one of many analytical results that would be important to quantify.

A risk-informed approach should be taken when quantifying values in this uncertain future. Figure 13.6 provides an example of a risk-informed approach to characterizing

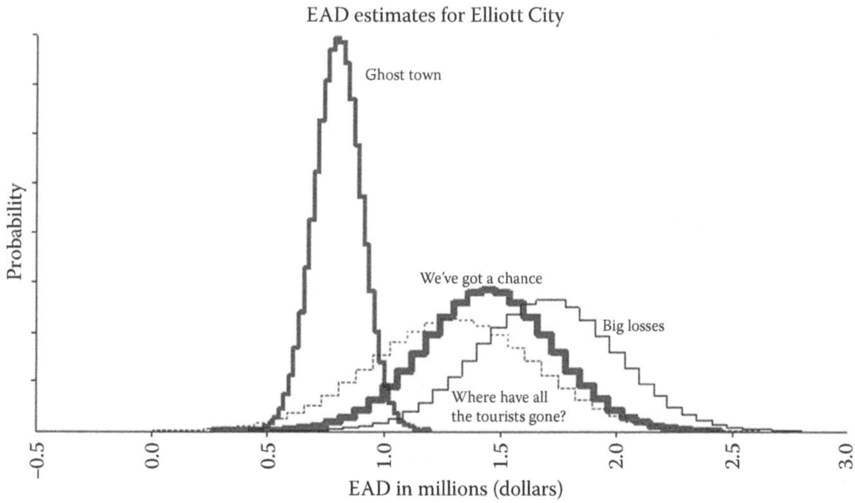

Figure 13.6 Range of expected annual damage estimates reflecting uncertainty in Elliott City without condition scenarios.

Table 13.3 Expected Annual Damages for Elliott City

	Ghost Town	Tourists	Chance	Big Losses
Minimum	$420,490	$2,792	$266,527	$507,033
First quartile	$732,522	$1,063,894	$1,261,015	$1,497,519
Median	$799,957	$1,299,876	$1,449,864	$1,699,942
Third quartile	$867,441	$1,535,998	$1,638,706	$1,902,313
Maximum	$1,190,149	$2,624,839	$2,449,759	$2,796,883

the flood problem via expected annual damages. A different estimate of expected annual damages (EAD) could be produced for each of the four different scenarios identified. These estimates reflect the uncertainty revealed in the scenario development process.

Any one distribution in Figure 13.6 demonstrates the uncertainty (i.e., knowledge uncertainty and natural variability) in an estimate of expected annual damages. The existence of four such distributions represents the knowledge uncertainty about how the future watershed and floodplain development and hydrology will unfold as well as the natural variability in flows and damages. A numerical summary of expected annual damages is provided in Table 13.3.

Such a table clearly conveys the fact that planners really are not sure what the future flood problem, as characterized by expected annual damages, is going to be like without a plan in place. Four scenarios were identified and the analysis of EAD reflects the different logics of these scenarios, whether one or multiple scenarios are ultimately used by the planning team. This is an example of the sort of analysis that can be done to characterize uncertain quantities used to describe problems, opportunities, objectives, constraints, and decision criteria under the without condition scenario.

13.5 OTHER PLANNING TASKS IN SCENARIO PLANNING

Once you have multiple without condition scenarios, what do you do with them? The planning process proceeds as usual, although it is a bit more complex when we are honest about the things we do not know.

13.5.1 Formulation

Formulation can, like quantitation, take place within the scenarios. Planners can *move into* an individual scenario and formulate solutions based on the specifics of the scenario. This approach leads to four sets of solutions: scenario 1 solutions, scenario 2 solutions, and so on. The ideal formulation process would then seek ways to merge solutions from the different scenarios into robust solutions, that is, solutions that will yield desirable outcomes no matter how the future unfolds. An alternative approach is to formulate across scenarios. With this strategy, planners seek to formulate robust solutions from the outset.

When the uncertainty clouding the future is significant and does not lend itself to resolution through robust plans, adaptive management strategies will be useful. Plans are formulated for robustness and flexibility in such a strategy. Consider Table 13.4. Let

Table 13.4 Robust and Flexible Plan Elements for an Uncertain Future

	Measures Comprising Plan				
Plan	**A**	**B**	**C**	**D**	**E**
One	X	X			
Two	X		X		
Three	X			X	
Four	X				X

Plan One be the best plan for scenario one, Plan Two the best for the second scenario, and so on. Robust elements of a plan would form the initial phase of a plan; they would perform well regardless of the manner in which future uncertainty resolves itself. In this case, Measure A might be implemented first. It is part of each best plan and will perform well regardless of the manner in which the future evolves. It is a robust measure.

Flexibility follows in subsequent phases. Research, experiments, and monitoring can be incorporated into a plan in order to reduce the remaining uncertainty. As this uncertainty is resolved and it becomes more clear which of the multiple scenarios the future is most likely to reflect, planners will know which phase 2 option is best. If, for example, the future looks like it will follow the path described in scenario three, then Measure D would be the second phase of the plan.

13.5.2 Assessment

Assessing plans includes the evaluation and comparison steps. Assume with condition scenarios have been completed in a similar manner to the without conditions. The assessment is conceptually simple. Planners can separately evaluate plans formulated separately for each scenario. That means evaluation and comparison is completed for the first scenario, then repeated for the second, and so on. The best plan for each scenario is thus identified.

This is generally inferior to an assessment process that is guided by robustness and flexibility. To do that, first evaluate each plan against every one of the four scenarios to narrow the subset of plans. A qualified plan produces acceptable results in every one of the possible scenarios. Then, compare each of the surviving plans in each of the scenarios to examine the decision criteria estimates the choice decision will be based upon.

13.5.3 Choice

There are many conceptual approaches for selecting a plan once the assessment is completed. Four are offered here. The first looks across all four scenarios to choose the most robust plan, regardless of how they were formulated. Which plan will perform best regardless of the future that is realized? That is the best plan. It may not be the best for any one scenario but it is the most satisfactory given the uncertain path of the future. This approach is based on the implicit assumption that all the described futures are equally likely.

A second approach is to designate one of the alternative scenarios as the most likely without condition and then proceed as usual through the selection process for a single scenario. Once a best plan is chosen, you would then evaluate that plan against the other three remaining scenarios. If this leads to unacceptable results in any scenario, the selected plan could be reformulated to eliminate or render the undesirable effects tolerable. Alternatively, adaptive management features could be added to monitor conditions that would lead to undesirable effects and flexible adaptations to the plan could be triggered at that future point. If the selected plan cannot be fixed for any given future scenario, another plan would be selected. The value of this approach is to anticipate the things that could go wrong, so contingent risk management measures can be formulated in advance.

A third option would base the choice on the regret criterion. Here, regret refers to the situation where we would choose a formulated plan that turned out to be a suboptimal choice for the future that evolves. Consequently, we would regret having made that choice. This approach requires planners to identify the maximum regret associated with the choice of each candidate plan, then choose the plan that minimizes this maximum regret. This decision criterion is sometimes called the minimax criterion. Suppose we have four different possible states of the world (scenarios) and four different plans. To simplify the example, imagine the primary decision criterion is net benefits. Let Table 13.5 summarize the payoff for each plan and state of the world.

From the payoff matrix, a regret matrix is constructed as shown in Table 13.6. Here is how it is derived. Look at Table 13.5. If we choose Plan One and State A is

Table 13.5 Payoff Matrix in Millions of Dollars of Net Benefits for Four Plans and Four States of the World

	State of the World (Scenario Realized) $Millions of Net Benefits			
Plan	**A**	**B**	**C**	**D**
One	10	15	1	−12
Two	5	8	7	4
Three	−1	0	6	20
Four	−5	6	12	2

Table 13.6 Regret Matrix in Millions of Dollars of Net Benefits for Four Plans and Four States of the World

	Regret Matrix $Millions of Net Benefits				
Plan	**Scenario A**	**Scenario B**	**Scenario C**	**Scenario D**	**Maximum Regret**
One	0	0	11	32	32
Two	5	7	5	16	16
Three	11	15	6	0	15
Four	15	9	0	18	18

realized, what is the most regret we would experience? One/A yields \$10 million in net benefits. No other plan yields as much, so there would be no regret. Note the \$0 in Table 13.6 at One/A. If State B is realized, Plan One yields \$15 million, again the maximum payoff, so there is no regret. If State C is realized, we get \$1 million and this is less than the \$12 million of Plan Four. So, we lose and regret the \$11 million loss. If State D is obtained, we lose \$12 million instead of gaining \$20 million with Plan Three. This is a regret of \$32 million. Table 13.6 was generated in this way from Table 13.5.

The maximum regret associated with each plan is shown in the last column. Plan Three minimizes our regret. If we choose Plan Three, the worst outcome would be Scenario B and we would regret the loss of \$15 million in additional net benefits.

A fourth approach uses premise sets. These are the sets of beliefs or assumptions (premises) about how the key uncertainties will resolve themselves. These sets of premises are laid out for decision makers to enable them to choose the set of conditions they believe will prevail. The chosen premise set will then point to the best plan. In a sense, this approach asks decision makers to choose the most likely scenario, but it does so by narrowing the cumulative uncertainty to a few key elements, which will likely be a subset of all the uncertainty.

This approach begins like the first approach above. Plans are formulated individually for each scenario. Imagine that we have identified the best plan for each scenario (Plans R, S, T, and V) and they differ substantially from one another. Choices are presented to decision makers, conditioned on a set of premises as seen in Table 13.7. Consider a hypothetical example like that for Lake Okeechobee in Table 13.7. The decision maker assumes responsibility for resolving the uncertainty when premise sets are used.

Table 13.7 Presenting Choices to Decision Makers Premised on Beliefs about Uncertainty and Planning Objectives

If you believe...	...and want to...	...then choose...
State will pass water quality law Farmers will implement it It is a wet period	Maximize net economic benefits Maximize water quality standards	**Plan R**
Phosphorous runoff will stay the same or increase It is a wet period	Maximize net economic benefits Exceed minimum water quality standards	**Plan S**
State will pass water quality law Farmers will implement it It is a dry period	Maximize net economic benefits Maximize water quality standards	**Plan T**
Phosphorous runoff will stay the same or increase It is a dry period Exceed minimum water quality standards	Produce positive net economic benefits	**Plan V**

13.6 TAKE AWAYS

Lesson 1: Sometimes the future cannot be adequately described with a single scenario.
Lesson 2: Scenario planning produces multiple future scenarios when there is significant uncertainty.
Lesson 3: Scenarios are narrative descriptions of what could be, not forecasts of what will be.
Lesson 4: Analysts occupy a scenario and do their analyses within the confines of its plotline.
Lesson 5: Evaluation, comparison, and decision making are more complex in scenario planning.

13.7 ADDENDUM 1: DRIVERS AND FORCES

Drivers are those things that provide impulse or motivation for change. A driver is something that gives motion to other forces. They are accessible to stakeholders, which can make them more immediate than some forces such as long-term trends. Drivers can also have strong effects on stakeholders. Stakeholders usually adapt to the effects of the drivers, or they may adapt the drivers themselves. Climate change is an example of a force, whereas energy policies are an example of a driver. The drivers for one set of decision factors (e.g., stream flows) may differ from the drivers for another set of decision factors (e.g., social vulnerability of a community). Drivers that have a lot of uncertainty and high potential impact on your KDFs are important to consider in the creation of scenarios. When these drivers or forces could lead to futures that diverge in details that are important to decision making in a planning study, then you should resort to scenario planning. Some examples of drivers include the following:

- Budget priorities
- Office of management and budget policies
- Land use policies
- Water resource policies
- Business cycles
- Climate policies and resource practices
- Renewable energy requirements
- Trade with Cuba
- Enforcement of water quality standards
- Carbon credits or efforts to reduce greenhouse gases

Drivers tend to be more changeable. They can change more often or more quickly that the long-term trends. Stakeholder action and public decision making may influence drivers. Trends are not so easily or readily influenced. Trends, once discerned, may be uncertain in their details but they tend not to lead to divergent futures. The future path of a driver may be very uncertain and capable of having divergent effects on patterns of change and adaptive choices in the future.

Some forces are sudden and others are more gradual and persistent. Trends are more gradual forces that cause changes in the environment or society. Stakeholders in a planning investigation, by themselves, are likely to have little to no power or ability to affect the direction or intensity of the impact of a trend. Trends are often discerned by detectable patterns in behaviors and variables of interest to planners. Some examples of trends include the following:

- Population growth or decline
- Aging population
- Gentrification of property
- Increasing incidence of terrorism
- China's growing economic influence
- Climate change
- Globalization and changing trade patterns
- Increasing interest in sustainability
- The rise of digital technology, including more mobile personal computing and communications
- Increasing political polarization of views on issues

A very different kind of force is the high-consequence low-probability event that is sometimes called a wild card or shock. These events are usually more plausible than likely, but they can have impacts severe enough that even if their likelihood is remote, we need to account for this contingency. We have all experienced situations in our personal lives that we did not foresee. The death of a loved one, a catastrophic fire or flood, loss of health, an unexpected job opportunity, and an inheritance are a few examples. These are powerful forces that can alter our expectations, if not the directions of our lives. The idea of wild cards and shocks should familiar to everyone.

In some without condition scenarios, it may be important to include wild cards that could reshape the trajectories of KDFs. Previously unanticipated problems and opportunities, for which we are totally unprepared, may emerge. What happens if future terrorist attacks result in a decision to offload all cargo into the United States at offshore facilities? What happens to Port Everglades if we normalize trade with Cuba? What happens to U.S. ports if the northern sea passage is opened?

It is often useful to link shocks and trends. What could happen in an emerging trend that would inject a shock into the system? How might a shock affect a trend? Globalization is a trend, and global conflict is a shock that could seriously affect the trend, for example. Some examples of wild cards include the following:

- Dirty nukes found in containerized cargo
- Nuclear bombs
- War
- Cyber collapse
- A large oil spill
- A devastating natural disaster
- A significant leap in human longevity
- A visit by alien intelligence
- Proof of God's existence

The real issue is how to handle wild cards in your scenario. This is an important part of the art of scenario development. When are these wild cards plausible enough to warrant scenario planning and when do we handle them as a significant uncertainty within a single scenario? Scenario planning is warranted when the likelihood or the consequences of such an event are unacceptable.

A third category of forces to consider is discontinuities. They are similar to shocks in that they occur rapidly and can lead to shifts in future trajectories that were largely unanticipated. The major difference is that wild cards and shocks can be imagined and anticipated. Discontinuities themselves are unanticipated sudden sharp breaks that strike with disruptive force; these are unknown unknowns. Containerized cargo, the electronic calculator, spreadsheet software, and the Internet are easy examples of past discontinuities. The printing press changed the world and digital technology is changing it again as physical printing is no longer necessary. Technological discontinuity has rendered many products and services obsolete. Examples of discontinuity could include the following:

- Wiki environments for collaborative working
- The human genome project
- Bioengineering
- Nanotechnology
- Social networking
- Quantum computing
- Telecommuting

A fourth kind of force is the weak signal. Weak signals are the first, often subtle, indications of a change. They usually occur in areas of major uncertainty. You can think of them as the first reductions in uncertainty. They are raw information that serves the purpose of an advanced indication of change. Because the information is incomplete, it is easy to miss, easy to misinterpret, and difficult to read accurately. When we are intentionally looking to reduce uncertainty, however, we may be more inclined to pick up on these weak signals. They rarely exist in isolation and they need to be evaluated in the context of their emergence and other available information.

Saritas and Smith (2010) cite James Hanson's presentation to the U.S. Congress in 1988 about possible global warming on the horizon, and the speed of take up of the *Mosaic* (precursor to the *Netscape*) as the first Internet browser in 1993 as weak signals. They were hard to recognize at the time, but can now be clearly seen as being indicative of things to come. Many weak signals are easier to spot in retrospect, when the change is evident. Saritas and Smith go on to describe weak signals as messages and signs usually associated with early developments in technologies, societal innovations, conflicts, origins of conflicts, and the like that are not easily verifiable from a present-day perspective. The 9/11 Commission Report, for example, found numerous weak signals concerning the attack on the Twin Towers, which were missed. It is easier to see when a neighborhood changed from good to bad when looking backward than it is to see at the time. Look for the weak signals of change in the condition of a natural resource, the use of a port or floodplain, the health of an ecosystem, or the decline in infrastructure performance.

REFERENCES

Brauers, J., and Weber, M. 1988. A new method of scenario analysis for strategic planning. *Journal of Forecasting* 7(1): 31–47.

Comprehensive Everglades Restoration Plan (CERP). 2012. "Lake Okeechobee Watershed." Accessed January 13, 2012 from http://www.evergladesplan.org/pm/projects/proj_01_lake_o_watershed.aspx.

Godet, M. 1987. *Scenarios and strategic management*. London: Butterworth.

Godet, M. 2001. *Creating futures scenario planning as a strategic management tool.* London: Economics.

Lindgren, M. and H. Bandhold. 2003. *Scenario planning the link between future and strategy.* New York: Palgrave Macmillan.

Ralston, B., and I. Wilson. 2006. *The scenario planning handbook.* Mason, OH: Thompson South-Western.

Ringland, G. 2002. *Scenarios in public policy.* New York: John Wiley & Sons.

Saritas, O., and J. E. Smith. 2010. The big picture—trends, drivers, wild cards, discontinuities and weak signals. *Futures* 3287(10): 292–312.

Wack, P. 1985a. Scenarios: Uncharted waters ahead. *Harvard Business Review* 63(5): 72–89.

Wack, P. 1985b. Scenarios: Shooting the rapids. *Harvard Business Review* 63(6): 139–150.

Wilson, I. 1975. From scenario thinking to strategic action. Accessed July 18, 2012, http://horizon.unc.edu/projects/seminars/futurizing/action.asp.

Yoe, C. 2012. *Principles of risk analysis decision making under uncertainty.* Boca Raton, FL: CRC Press.

CHAPTER 14

Economics for Planners

14.1 INTRODUCTION

Economics is for measuring and deciding. In plan formulation, when you are tinkering with attributes and wondering if something should be longer, wider, bigger, thicker, or fancier, economics can help you decide those matters. That is marginal analysis.

Someone is always going to care about cost. Many personal choices come down to how much something is going to cost us. When we hear the cost, we are often able to make an instantaneous assessment of the benefits of the choice and either go ahead or say no, that is too much, based on the cost information. Individual choices are idiosyncratic and personal; we are entitled to make them on any basis we want. When we make a public decision, we have to find a basis for decision making that many people can agree with or at least understand. Economics assists in this role.

There are pros and cons—let's call them benefits and costs—for every decision you encounter in natural resources planning. Benefits and costs comprise a great many of the evaluated effects of plans. In general, we'd like to identify all the benefits of a choice and compare them to all the costs of that choice. Benefit–cost analysis (BCA) is a tool used to compare the pros and cons of a choice in the common metric of money.

Some benefits and costs are readily measured in dollars. The cost of a duck box, the value of a kilowatt of energy, and other goods and services bought and sold in a market are easy to compile. Other values we have to work a little harder for; we know a day of fishing, hiking, or camping has value but it is not bought and sold in a market. Entry fees may not reflect true value as reliably as market prices do. Then there are plan effects like ecosystem services, for example, that can include things that clearly have a value that cannot yet be approximated with reliability. These are some of the challenges faced during BCA. This chapter considers three groupings of economic topics: the economic principles of marginal analysis and incentives; cost concepts including economic costs, project costs, uncertainty in costs, multipurpose project costs, cost allocation, and cost sharing; and a brief overview of economic analysis techniques.

14.2 ECONOMIC PRINCIPLES

Economic analysis operates on the basis of a number of principles. Two of them are important for planners to understand. They are that rational people make decisions by comparing marginal costs (MCs) and marginal benefits (MBs) and that people respond to economic incentives. Neither is absolutely true, but each is examined briefly in this section because they both add a rationality check to planning.

14.2.1 Marginal Analysis

Economics assumes people are rational, meaning they systematically and purposefully do their best to achieve their objectives, such as maximizing personal utility or business profits. Rational behavior means at some level people weigh the costs and benefits of their actions, if only subjectively. Thinking rationally should be a minimum requirement for planners. If they don't decide rationally, perhaps they

> Daniel Kahneman's (2012) book challenges the economist's view of rational decision making. He examines the decision making of rational beings he calls Econs and contrasts that with the decision making of Humans. A substantial part of his book examines the difference between humans' emotional response to the world around them and a more rational (Econs) response. We each have an experiential system that reacts and guides us readily and a thinking system that kicks in only when forced (by us) to do so. Thus, planners should not expect everyone to respond favorably to the impeccable rationality of their economics or their planning investigations.

should. Planners should also expect that people affected by a plan will respond to it in a way that is rational for them. If planners don't carefully consider the range of rational responses to their plans, the desired planning objectives may not be realized.

Economic theory assumes rational people consider how a decision or change will affect their circumstances and respond in a way that is rational for their personal circumstances. An example will illustrate. Imagine we have a water quality problem and are trying to decide what the quality standard should be. There is a chemical present in the water that can be reduced to varying extents at the costs shown in Table 14.1.* Total cost shows it costs $0.50 to reduce the chemical to 100 parts per million (ppm) per acre-foot of water but $10 to get it down to 50 ppm.

Table 14.1 also shows the cost of each additional increment of water quality improvement in the MC column. Think of MC as the change in cost required to produce one more ppm improvement in water quality. It is defined as the change in total cost divided by the change in output. Going from 100 ppm to 90 ppm costs change by $0.75 while water quality improves by 10 ppm. Thus, $0.75/10 ppm yields the MC of $0.075 for a ppm reduction, say $0.08 for each additional part reduced. MC tends, at some point, to exhibit the increasing trend shown in Table 14.1 due to the diminishing marginal returns that limit the physical universe.

Water quality improvements have benefits for society that are shown in Table 14.2. Total benefits range from $4 for a reduction to 100 ppm to $10.75 for a reduction to 50 ppm. What we gain from each increment of chemical reduction is the difference in the total social benefit divided by the difference in the chemical reduction. This is called a marginal benefit (MB) in Table 14.2. In going from 100 ppm to 90 ppm, benefits rise from $4 to $7, a change of $3. Water quality also changes by 10 ppm, so $3/10 ppm = $0.30 for a ppm, as shown.

MBs are decreasing. This is a common pattern, resulting from the economic principle of diminishing marginal utility. That is, usually, the more you have of something like water quality the less valuable additional improvements in water quality become.

Given these facts, the question for decision makers is what is the most appropriate level of water quality? Because planning decisions can be based on science and social values, any answer is in fact possible. One of the social values someone

Table 14.1 Costs of Different Water Standards for a Chemical

Water Quality Standard (ppm)	Total Social Cost	Marginal Cost
100	$0.50	NA
90	$1.25	$0.08
80	$3.00	$0.18
70	$5.00	$0.20
60	$7.25	$0.23
50	$10.00	$0.28

* If you prefer, consider the dollar values to be in millions of dollars. The magnitude of costs is secondary to the marginal principle being developed.

Table 14.2 Benefits of Different Water Standards for a Chemical

Water Quality Standard (ppm)	Total Social Benefits	Marginal Benefits
100	$4.00	NA
90	$7.00	$0.30
80	$9.00	$0.20
70	$10.00	$0.10
60	$10.50	$0.05
50	$10.75	$0.03

Table 14.3 Total Net Social Benefits for Different Water Quality Standards

Water Quality Standard (ppm)	Total Social Cost	Total Social Benefits	Net Social Benefits
100	$0.50	$4.00	$3.50
90	$1.25	$7.00	$5.75
80	$3.00	$9.00	$6.00
70	$5.00	$10.00	$5.00
60	$7.25	$10.50	$3.25
50	$10.00	$10.75	$0.75

is always going to care about, especially with public decision making, is economic efficiency. Economic efficiency relies on marginal analysis, which is not intuitive for most planners and stakeholders.

Should society pay $10 to improve water quality to 50 ppm? Its value to society is, after all, $10.75 and it only costs $10. Would you trade $10 for $10.75? Most people would say yes. Table 14.3 shows a net profit to society for every level of water quality considered.

The answer changes rather dramatically when we use marginal analysis to choose the optimal level of water quality. Notice, for example, that net social benefits rise to a maximum of $6 at a chemical level of 80 ppm, then decline to $0.75 at 50 ppm. In a world of scarcity, where we cannot do everything, we do not want to waste resources. Choosing 50 ppm wastes resources, as shown in the marginal analysis in Table 14.4.

Notice that Table 14.4 shows a positive level of net social benefits for every level of water quality improvement. Now look at the MC and MB. If you hold the MC in your hand, would you trade that sum of money for the corresponding amount of MB? That is, in its essence, the choice the planner faces on society's behalf. Would you trade $0.08 for $0.30? Of course you would. The net gain is $0.23 (allow for rounding) for every additional ppm reduced; that is a good deal for you and it would be a good deal for society as well. Would you spend $0.18 for a $0.20 gain? Yes. Would it be wise to spend $0.20 to gain $0.10? No. We would not make this deal knowingly. So we would stop at 80 ppm because it is the last level of water quality improvement that pays for itself. Not coincidentally, this is also where net social benefits are maximized.

Table 14.4 Marginal Analysis of Different Water Standards for a Chemical

Water Quality Standard	Total Social Cost	Total Social Benefits	Net Social Benefits	Marginal Cost	Marginal Benefits	Marginal Net Benefits	Do It?
100	$0.50	$4.00	$3.50	NA	NA	NA	Yes
90	$1.25	$7.00	$5.75	$0.08	$0.30	$0.23	Yes
80	$3.00	$9.00	$6.00	$0.18	$0.20	$0.03	Yes
70	$5.00	$10.00	$5.00	$0.20	$0.10	$(0.10)	No
60	$7.25	$10.50	$3.25	$0.23	$0.05	$(0.18)	No
50	$10.00	$10.75	$0.75	$0.28	$0.03	$(0.25)	No

The rule for rational marginal behavior is to undertake any change for which MB $\geq$ MC and avoid any change where MB < MC. Applying this rule will also maximize total net benefits to society. Looking at marginal net benefits, we see they exceed zero for 80 ppm but become negative for any additional reduction. In a world where it is not possible to do everything, it is wise to avoid doing things that require costs in excess of their value to society. Planners should take care to present relevant information in a marginal analytical framework so decision makers and stakeholders can better see the trade-offs among benefits and costs.

This principle of marginal analysis can be applied when planners weigh the benefits of any incremental changes in plan formulation. Should the planted area be 1000 acres or 1200 acres? Should a levee be 15 feet high or 17 feet high? Should a bike path be gravel or porous asphalt? All such questions can be aided by marginal analysis. That they can be does not suggest they should be. Too much marginal analysis can bog a study down in unnecessary details. You don't want to be evaluating the number of nails in a duck box. Your team economist can help you find the balance point.

14.2.2 Incentives

People respond to incentives, that is, the economic rewards or punishments that induce people to act in specific ways. When gas prices go up, we drive less and buy more hybrids and fewer SUVs. Likewise, with planning, people respond to incentives. It is important to understand both the intended and unintended incentives associated with an alternative solution.

Consider the water quality example again. Suppose a baseline study of 10 farms is conducted and it shows all 10 practice some amount of runoff controls, but 5 are not at all effective in their measures. Suppose a plan requires all farms to meet the 80 ppm standard. Planners may think they have solved the problem by requiring compliance as part of their plan. But what will people really do?

This requirement will have an incentive effect on farmers but it may not be the expected one. When requirements change, people are presented with choices that necessitate changes in their behavior. So, imagine that of the five farms not meeting the new standard, two decide to comply, one goes out of business, one chooses to stop trying altogether to lower costs and hopes not to get caught, and another sells his land for houses and moves his farm production to another state with a lower water quality standard. For the last two farms, water quality may have actually gotten worse. A plan can sometimes provide people with a choice of how to respond to the changes it effects. The plan may represent a reward to some and a punishment to others. The five farmers already meeting the standard are rewarded by the increased cost of operation imposed on their competition by the plan. Farmers not already complying are punished by higher costs of operation.

Responses to these incentives should be anticipated by planners. Plans on paper may not realize the desired effects unless the economic incentives produced by the plan are carefully considered. It can be useful, at times, to incentivize plans to assure faster and more complete implementation of the planned measures by those responsible for implementing the measures. Subsidies, rebates, tax breaks, coupons, and such can be effective financial incentives. Helping affected parties understand and then realize the benefits of the measures can also be helpful.

14.3 COST CONCEPTS

Most people equate cost with money; after all, money is the most common metric of costs. There are two critically important concepts for planners to understand about cost. We'll begin with the easier of the two, the level of detail. Cost estimating is an iterative process in most planning studies and planners must always be aware of the level of detail they are working with. The second point is the distinction between economic and financial costs. Both discussions follow.

14.3.1 Level of Detail

Cost information has many uses in a planning study. Cost is used to evaluate and make decisions during plan formulation, especially when looking for the most cost-effective way to obtain an effect. Costs are one of the primary plan effects to evaluate. Cost is an essential input to the economic analysis of a plan. Cost–benefit analysis, cost-effectiveness analysis, and incremental cost analysis are impossible without information about the cost of alternative solutions. Cost estimates are used for budget decisions as well. The costs of a plan may be allocated to different parties to the solution. Once the final cost of a plan is known, that information is critical to the decisions that will determine how the costs are to be financed.

A complicating factor is that cost information is not static and the costs used to make decisions about plans are not likely to look much like the costs that are eventually incurred and financed. This is important information for planners and stakeholders alike. There are different levels of detail in different cost estimates.

This book has invoked the image of a pile of things planners know and a pile of things they do not know when conducting a study. The pile of unknowns is quite large when a solution is first formulated and it takes time and a great deal of analysis to transfer unknowns into the pile of knowns. Consider a simple flood risk management problem. To develop a cost estimate, hydrologic engineers must first estimate flows. Structural engineers must gather information about utilities that may be affected by any effort to manage those flows. Design engineers determine the dimensions and materials required to obtain the desired plan effects with channel modifications, levees, walls, property relocations, and the like. They may do this with or without foundation explorations by geotechnical engineers or real estate analysis for the property and buildings affected. Only then can cost engineers begin to develop detailed cost estimates.

At each planning iteration, every one of these analysts will have a new balance in their own piles of knowns and unknowns to address. The cumulative uncertainty is large early in the planning process and early cost estimates reflect this uncertainty. It is not uncommon to find widely varying levels of detail between iterations. A reconnaissance study, for example, might rely on cost estimates based on unit costs or a 15% design level; in other words, about 15% of the necessary detail is available. A feasibility cost estimate may require at least 30% of the design detail and planning, engineering, and design (PED) gets closer to a 100% level of detail. A cost is never a known fact until after implementation is complete and all claims are settled. That means cost estimates in the early iterations of a planning study will bear little resemblance to the eventual costs of the project. This is a fact that escapes many people, so it is especially important to convey this information to decision makers and interested publics.

Imagine, for example, you are formulating flood risk management solutions for an urban area where you have 19 potential detention reservoir sites (which can be used in endless combinations), three different channel alignments, and multiple scales for these channels. Now suppose policy tasks you to find the solution that maximizes net flood risk management benefits. It is not possible to cost out dozens of detailed cost estimates for every possible combination of measures, much less hundreds (or in complex studies even thousands) of cost estimates. Consequently, planners might use conceptual costs* for screening purposes. Even though the costs are not accurate, if the level of detail/uncertainty is consistent across all the measures considered, these conceptual costs can yield useful information to support decisions. Reasonably accurate ordinal rankings of the costs of solutions that bear no resemblance to the final costs can be useful for early screening decisions.

PLANNING COST VERSUS PLAN COSTS

The costs of conducting a planning investigation, that is, planning costs, can be substantial. Multi-million dollar planning study costs are not unusual. This chapter is not concerned with these planning costs. Instead, it focuses on the cost of a proposed plan of action or solution to the problems and opportunities identified in step one of the process. These costs are sometime called project costs.

* Conceptual costs are average costs based on recent past experience. They reflect little to no actual detail for the specific study area.

Planners and their constituents should know that costs are often iterated at increasing levels of detail that can yield widely varying levels of cost. If these cost data are to become widely available, the public, especially the media, must be carefully educated about the level of uncertainty in the cost estimate. It bears mentioning that this is not a problem in all planning studies. Some studies delay cost estimation until the final solution is identified.

14.3.2 Economic Costs

Economics is the study of how society allocates and distributes its scarce resources. Economists look at costs differently than accountants do. The economic costs and accounting (or financial) costs of an action are not always the same thing. This is a fact that can be difficult for many noneconomists to grasp. To understand economic costs there are two points to consider: people face trade-offs and the cost of any action is measured in terms of foregone opportunities. These are the next two topics.

14.3.2.1 Trade-offs

Resources are limited in relation to society's wants and needs; economists call this situation scarcity. We don't have enough resources to do everything we want or need to do. This simple fact of life holds for individuals, firms, governments at all levels, planning partners, and society at every imaginable level.

If we cannot do everything, then we have to make choices. By their nature, choices require that we do one thing and not do another. Why choose fish instead of beef? Why listen to the Beatles when you could listen to the Stones? We are not always able to articulate why we choose as we do and when we choose for ourselves, there may be no obligation to do so. When we make choices in the public arena in our roles as planners, we owe affected publics an explanation. Whether we can articulate why we do what we do or not, we always are choosing between at least two options (doing nothing or doing something) and choices entail trade-offs. Increased output of ecosystem services increases costs, which must be paid for by consumers or taxpayers. Suppose this increase results in job losses in industry. We have a trade-off to make.

One important recurring social trade-off is that of efficiency versus equity. Efficiency means society is getting the most from its scarce resources, that is, what is sometimes called "getting the biggest bang for the buck." Suppose there are two ways to reduce flood damages to a home. One is to raise the structure at a cost of $30,000 and another is to buy and raze the house at a cost of $300,000. It would be inefficient to raze the house. In fact, you could raise 10 similar houses at the same cost as one purchase. Efficiency is often equated with maximizing net benefits, as in the marginal analysis example given earlier.

Equity refers to how resources are distributed. Imagine that a public decision maker decided the people in the floodplain deserved to be relocated to achieve some level of social justice. Spending $300,000 per house would then be a decision based on equity. As long as there is at least an action/no action choice, all decisions are going to require trade-offs. In democratic decision making, the public is best served by making the nature of the trade-offs explicit. Trade-offs are important to understand opportunity costs.

14.3.2.2 Opportunity Cost

Cost, to an economist, is not always the money price of obtaining something. That idea comes as a surprise to most people. Economists define costs as opportunity cost, which is what you must give up to get something. Opportunity cost is measured in real things sacrificed for a choice. When we choose one thing, we simultaneously forgo the opportunity to have made some other choice. The opportunity given up is what a choice costs us. This is the essence of opportunity cost. Money is a convenient shorthand means of communicating the opportunity cost (see the following text box).

Opportunity always cost is an important concept for planners to understand. It can be confusing because accountants and much of the rest of the world concern themselves primarily with explicit costs. Economists consider both explicit and implicit costs as part of opportunity costs. Here is how it works. Suppose you find the Rolling Stones *Some Girls* vinyl album with the cover that still has Lucille Ball's face on it in an out-of-the-way record shop in Montreal and buy it for $50. Later, in a pub, you take it out to admire and a patron offers you $500 for it. You smile and decline the offer. When you return home a friend asks you what the album cost. How do you answer?

If you said $50, read this carefully, because you are wrong. Your explicit costs were $50. Think of explicit costs as out-of-pocket costs. When the stranger offered you $500 you had a choice: sell your album for $500 or keep it. Had you sold the album, you could have paid back your explicit cost of $50 and made a tidy $450 profit. When you decided to keep the album, you said no to a potential $450 gain. Forgoing this $450 is the implicit cost of your decision. An implicit cost is an opportunity for gain that you decline. Economists make no distinction between actual gains and foregone gains when estimating costs. Your opportunity cost of the album is $500, an explicit cost of $50 plus an implicit cost of $450.

OPPORTUNITY COST EXAMPLE

If you go to the vending machine and come back with a bag of chips and I ask what that cost you, you are likely to reply $1, or whatever the cost was. But if you stood before the machine with your only dollar, weighing the choice of a bag of chips or a bag of pretzels, then choose the chips, that choice cost you the opportunity to have a bag of pretzels. So the economist's answer is the chips cost you some pretzels. Economic cost is a real thing or opportunity foregone.

Others may have chosen M&Ms, a Twix, and so on. It becomes unwieldy to list all the opportunities foregone by a single purchase with so many different people and preferences, so money prices serve a wonderfully efficient shorthand means of communication. When we say it costs $1, that stands for everything real you could have done with that $1.

It is very easy, and quite common, for planners or stakeholders to take an incomplete view of costs and focus only on explicit costs. When government spends money on a natural resources plan (or even just a planning study), one opportunity cost is that the people who pay taxes to support government programs have less money to spend to meet their own wants and needs. In addition, there is an almost limitless menu of public wants and needs we ask government to address in addition to natural resources. Workplace safety, food safety, drug safety, airline safety, terrorism, crime, mental health, and literally every sphere of life is affected by public sector decisions. More natural resources spending means less spending on these other goods and services. More government spending means less private sector spending. This is all part of the cost.

When the government already owns land and a planning study proposes to dedicate it to greenway uses, there may be no money (explicit) cost for the land. There is an opportunity cost, because the land has value in other uses and that value will be sacrificed. If the greenway is located in a new housing development, that implicit cost may be the cost of residential land. If that land was previously used to grow corn, it is the cost of agricultural land. The loss of this alternate use is a cost even if no one pays out-of-pocket for it.

On the other hand, a stream bottom owned by private interests might have to be acquired as part of the study cost. Even though there is a financial outlay of dollars, there is no economic cost at all if the land will remain in use as a stream bottom. Thus, economic costs may be more, less, or the same as accounting or financial costs.

14.3.2.3 Cost Concepts for Solutions

There are a number of other cost considerations of which planners should be generally aware. They are introduced briefly in this section and they include price levels, life cycle cost estimates, constant dollars (real dollars), time value of money (discounting and compounding), and interest during construction.

Learn to pay attention to price levels. Costs may be reported at today's price levels or at price levels from a past or future period. If costs or other dollar values are going to be compared, they must be at the same price level. Would you be better off making $6 per hour in 1966 or $24 an hour in 2012? We don't know, a priori, which is the greater sum of money because the price levels in 1966 were much lower than price levels in 2012. It's impossible to make a direct comparison between $6 in 1966 and $24 in 2012 because of price level differences. When you put together cost estimates, be explicit about the price levels you report. Identifying costs by a month and date, for example, October 2012 price levels, is standard practice. If price levels used in your study differ, they should all be converted to the same month and year price level.

Cost estimates are most useful to decision makers when they are life cycle cost estimates. This includes initial costs and all subsequent costs of a project. The language of cost estimating is somewhat fluid, varying from context to context. With no expectation that the terms used here will be generally acceptable, it is still important to define a few terms. In the past, construction cost or first cost of construction was a term used to identify the cost of implementing a plan. This language was spawned in a time when construction was an essential part of a plan. Natural resource planning need not entail construction. Changing the ways resources are managed are viable solutions to many problems. Thus, implementation cost is a more general term that includes the cost of the initial construction, management, and other changes required to begin to produce the desired outputs and outcomes of a plan.

Operation, maintenance, repair, replacement, and rehabilitation (OMRR&R) costs are the costs of all the activities required to make the solution work as designed throughout its life to realize the benefits identified during the planning phase. Some of them are regular and predictable costs, for example, cutting the grass every month during spring and summer, and some of them may occur at erratic intervals in unpredictable ways. An example of this might be the need to replace a structure damaged by a lightning strike.

OMRR&R costs are often expressed as an average annual equivalent (AAE) cost for purposes of economic efficiency determinations. Because OMRR&R is expected to last the lifetime of the project, often covering many decades, current dollar estimates of OMRR&R costs for those "out years" are not estimated and OMRR&R costs are expressed at the same price levels used to estimate project costs. When one price level is used to estimate costs at different points in time, they are called constant costs as opposed to current or nominal dollar costs.

For example, suppose annual OMRR&R costs of three employees in 2020 are $350,000. That is the nominal cost. To support decision making, these should be converted to October 2012 prices, say $300,000. The nominal cost in 2020 reflects the cumulative effects of inflation. The party responsible for OMRR&R can expect to have to pay $350,000 in 2020, but that is like $300,000 at October 2012 price levels. All costs should be expressed at the same price level and in inflation adjusted constant dollars.

In life cycle costing all of these implementation and OMRR&R costs should be estimated and included in the cost estimate. Life cycle costs are normally defined as the sum of all expenditures associated with a project from its inception to its disposal or salvage. Life cycle costs may be expressed as cumulative net present values or as average annual equivalents.

All dollar values, benefits as well as costs, should be expressed at the same price level. They should also be expressed as equivalent values at a single point in time. Dollars at different points in time have different values because of the time value

Operation and maintenance costs are routine and fairly predictable. Replacement costs are large expenditures that are incurred infrequently but at predictable intervals. For example, a crane needed to install stop logs in a flood protection closure structure may have to be replaced every 20 years or so. This is a regular replacement expenditure that is distinguished from ordinary O&M activities. As projects deteriorate, their performance may fall below design standards, necessitating replacement or a major effort to rehabilitate the project to its original performance conditions. The efforts required to do this and the costs incurred in their performance are considered rehabilitation costs.

of money. Ignoring inflation,* $1,000 today is worth more than $1,000 a year from now because of the time value of money. The simplest way to think about this is to imagine that you could take $1,000 today, put it in your bank, and earn 5% interest on it. So after a year you would have $1,050. Thus, $1,000 today or $1,050 one year from now are equivalent sums, assuming a 5% interest rate. In a planning study dollar sums can occur over a long time span. In order to make valid comparisons, all dollar values must not only be at the same price level but at the same point in time. This common point in time is often called the base.

Let's consider time equivalence in general for a moment. Suppose the base year is 2017. An expenditure of $100 million in 2012 is not the same as an expenditure of $100 million in 2017. The $100 million in 2012 is worth more than $100 million in 2017. To express the 2012 sum in terms that are equivalent to other values in 2017 we must escalate the $100 million in 2012. This process is compounding a cost. The compounding calculation expresses society's time preference for consumption now versus consumption in the future via the interest rate used. For example, if the interest or discount rate is 10%, we would say that society is indifferent between consuming $100 million in resources in 2012 and $161 million in the year 2017. The $61 million† difference between the two values is a time equivalence adjustment.

Benefits and costs that occur after the base year are discounted back to base year values—think in terms of shrinking inflated dollars back to their base year values. Dollar values incurred before the base year are compounded—think in terms of growing or inflating dollar values forward to their base year equivalents. Any good engineering economics or finance text will explain these procedures in detail.

AAE values are a common form of an annuity. Though it is a more common business practice to use net present value as the decision criteria in capital investment decisions, AAEs have frequently been used in many natural resources planning contexts. Natural resource project benefits are measured as income streams that occur over the life of the project. These benefit streams, when regular (and constant), are conveniently described as average annual values. When the stream of benefits is known but irregular, simple discounting techniques can be used to convert the stream to the equivalent of an annual value. When the stream of benefits is random and irregular, probabilistic methods are used to estimate benefit flows, which can be expressed as expected annual flows, which are equivalent to average annual equivalent values.

Costs may occur in more irregular flows of money than do benefits. Most of the costs are incurred during the implementation period. Common practice estimates the cumulative net present value of the implementation costs, this single present value can be converted to an AAE through amortization.

To understand amortization, a mortgage analogy may be helpful. Think of the net present value of the implementation costs of a project as the amount of money to

* If all dollar values are expressed at the same price level, then inflation has been accounted for and it is safe to ignore inflation.

† Compounding is what happens to your money in a savings account. The formula used is: Future Value = Present Value × $(1 + r)n$ where r is the interest rate and n the number of years compounded. The present value is $100 million and the future value is $161 million. When the equation is reversed: Present Value = Future Value/$(1 + r)n$, it can be used for discounting.

be borrowed to finance "construction."* Imagine a single annual payment instead of 12 monthly payments. That annual payment covers the principle and interest of the initial construction. OMRR&R costs are often estimated separately, much as they would be for operating and maintaining a house. The term of your mortgage is equivalent to the period of analysis for your plan. The interest rate you are charged is the interest on your project costs and your fixed annual mortgage payment is the AAE cost.† Once all costs and benefits have been reduced to AAE values with a common price level and base year, it is a simple matter to calculate the benefit-cost ratio and to estimate net benefits. A project with annual benefits of $30 million and annual costs of $10 million has a benefit-cost ratio of 3 and net benefits of $20 million.

Interest during construction is an example of an economic cost that is not actually paid to anyone in most government-financed projects. Suppose we have a project that will take 4 years to build and will cost $100 million. For simplicity, assume the money is borrowed at an interest rate of 8%. Interest during construction adds $16 million‡ to the costs so $116 million is used to calculate the AAE costs. Planning partners have expressed concern, dismay, and confusion about the need to include such costs in the economic evaluation; especially when they learn no one is required to actually pay these costs. If there are no loans during the construction period, there is no obligation for anyone to pay interest during construction and it is not a financial or accounting cost of the project.

Perhaps it is most appropriate to consider interest during construction to be a time equivalence adjustment, particularly when interest during construction is not a money cost of the project. The real resources used by the project have an opportunity cost. That cost is the price of the construction materials, equipment, and labor required to implement the project. If that price is paid before the base year, however, it cannot be considered to be equivalent in time value to the benefits that are expressed in terms of the base year.

14.4 UNCERTAINTY IN COST ESTIMATES

Costs are often estimated for each plan under serious consideration as a solution. They are often initially estimated when data are, at best, preliminary and incomplete. Implementation cost uncertainty is huge at the outset of a planning investigation. It

* Why are there interest charges? There are at least two ways to look at this. Suppose we had to take out a loan to cover the implementation costs. In this case, we would clearly expect to have to pay interest for the loan. When the project is financed without the use of a loan, we as society still incur an opportunity cost. Rather than spend, say, $100 million on a project, we could invest this money in some other way. Suppose we put the money into mass transit that would yield an 8% return on investment. Then, using the money for your project costs society the opportunity to earn this 8% return. Thus, we incur the explicit costs of construction plus the implicit cost of a foregone return, which is conceptually captured by an interest cost.

† The amortization or capital recovery factor is given by the following formula: Amortization = $(r[1 + r]^n)/([1 + r]^n - 1)$, where r is the rate of interest or discount rate and n is the period of analysis.

‡ This value is based on the shortcut computation of interest during construction, IDC = $(c \times n \times r)/2$, where c is construction cost, n is the number of years to construct, and r is the interest rate. More precise methods are used once the schedule of dollar flows is known.

is, however, one of the few values whose uncertainty may be completely resolved by implementation of the plan. All implementation costs are known with certainty at the completion of the project. Complete certainty for other kinds of project outcomes is rare. It can take a long time for project outcomes to become known with certainty and many are never quite known. Cost estimation uncertainty is constantly being reduced as the plan moves toward implementation. However, costs remain a major source of uncertainty in planning investigations.

Consider the simple cost estimate shown in Table 14.5. The cost shows the quantity and unit price for a variety of work items. Total costs are the sum of the individual work items' costs. The example given the cost estimate is about $808,000. This is based on a limited amount of design detail. In the early stages of project planning, much of the cost data are the product of prior experience and educated guessing until, detailed design, and other investigations can be completed, reducing the uncertainty further.

The costs in Table 14.5 appear as point estimates. In fact, each value has been sampled from a distribution of values that represents the uncertainty about the true value of each quantity and unit cost. Figure 14.1 presents the unit price of riprap stone as an uncertain number, in this instance represented by a pert distribution. The distribution in Figure 14.1 conveys the cost estimator's uncertainty about what the true cost of riprap stone is going to be. The distribution is constructed by the analyst's estimate of the minimum possible cost per ton ($41.98) and the maximum cost ($47.23). The only other information provided is the most likely value (mode = $44.34). These three data points and the assumption of a pert distribution enable the analyst to replace a point estimate in the cost calculation with a distribution that expresses the relevant knowledge uncertainty and natural variability for the given cost item.

Each value, except the mobilization quantity, is similarly represented by a distribution. Using the Monte Carlo process, the analyst can calculate a large number of possible costs under a wide range of random conditions. Figure 14.2 shows the distribution that resulted from 10,000 "what if" calculations of construction cost for the project shown in Table 14.5. The cost of the project is not known with certainty but is estimated to be between $0.59 million and $1.05 million, with an expected cost of $0.81 million. Based on these simulation results, there is a 90% chance the actual cost will be between $0.71 and $0.91 million.

Table 14.5 Project Cost Summary for a Small Water Control Structure Apron

Account	Item	Quantity	Unit	Unit Price	Amount
15.0.a.-	Mobilization, demobilization and preparatory work overflow	1	Each	$65,205	$65,205
15.0.1.-	Structure				
	Unclassified excavation and backfill	7237	Each	$7	$47,038
	Tremie concrete	260	Each	$200	$51,968
	Sheet pile with anchors	14479	Each	$33	$478,957
	Riprap stone	2585	Tons	$44	$114,847
	Bedding stone	1212	Tons	$41	$50,001
Total construction costs					$808,017

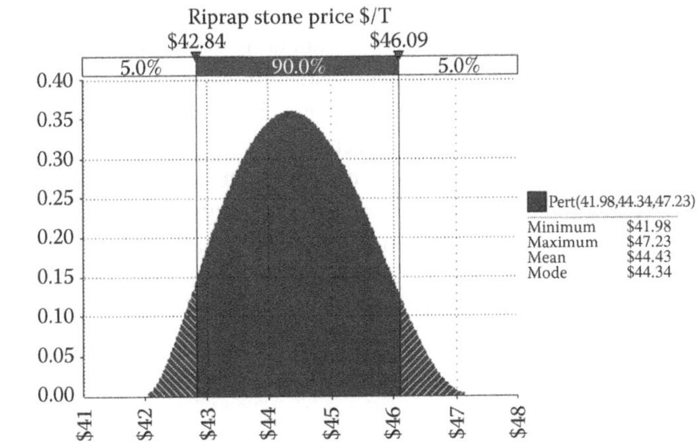

Figure 14.1 Pert distribution for unit price of riprap stone.

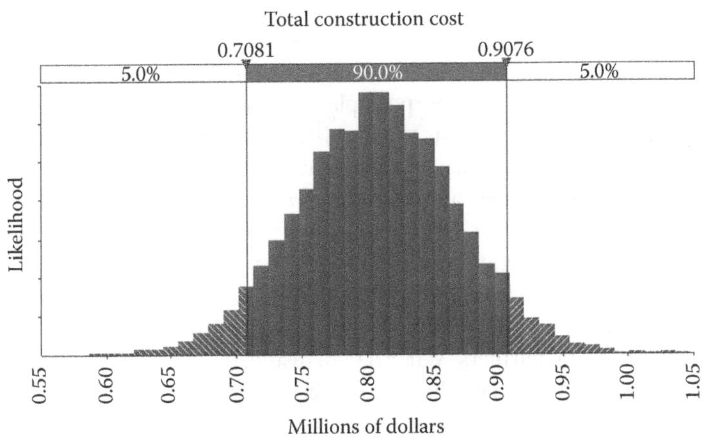

Figure 14.2 Distribution of construction cost estimate representing the knowledge uncertainty and variability in the cost estimate inputs.

MULTIPURPOSE PLANNING

A purpose is a function to be served or an output to be produced by a plan. A multipurpose project includes two or more natural resource related purposes. A purist of the language might make a further distinction between a combined-purpose project and a multipurpose project. A combined project is one that has several individual measures (i.e., physical components or management actions) serving one purpose each, combined in a single plan that serves more than one purpose in total. A multipurpose project has one or more single measures serving several purposes. This means a truly multiple-purpose project has joint costs. Joint costs can present implementation challenges for planners.

As a cost estimate moves closer to actual implementation, uncertainty about quantity and cost estimates is reduced. Subsequent, more sophisticated, iterations of the cost estimate would produce narrower distributions that are also more likely to be centered around the true value of the cost item. In the past, a single cost estimate, including a contingency allowance for uncertainty, was prepared and used throughout the planning study. The uncertainty in cost estimates have always been there, but it is only more recently that it has been addressed in probabilistic ways.

Eventually, one of the values in Figure 14.2 will usually be realized as the final project costs. It is essential that decision makers and stakeholders be kept well informed about the uncertainty in the cost of implementing each plan because deviations from any one cost estimate are to be expected and cost changes get people's attention.

14.5 SORTING OUT THE COSTS OF MULTIPURPOSE PLANS

This is an arcane topic worth preserving. Cost allocation is a rather specialized concept that remains relevant in some special planning contexts. Let's begin by defining a couple of terms used in this discussion. There are two important cost concepts to understand for multipurpose planning. Cost allocation is the process of equitably distributing project costs among different project purposes. This may be important because different purposes may have different cost-sharing formulas. If cost-sharing formulas are the same for all purposes, then this is not an issue. Cost sharing, sometimes called cost apportionment, is the practice of dividing the responsibility for paying the costs of a project among the planning partners. Historically, this has tended to mean federal and nonfederal interests. Collaborative planning and innovative planning partnerships could, however, lead to future cost-sharing relationships with different partners. When a plan results in a multipurpose project, costs must first be allocated and then they are shared based on public policy cost-sharing formulas in effect at the time.

When the money required to implement a solution comes from different sources of funding it is important to allocate the costs of the plan to the proper purpose and then to share the costs of each purpose so everyone pays a fair share of the costs of the solution. The pots may be defined by ownership of the funding sources. Planning

FLOOD CONTROL COST SHARING

In accordance with Section 202 (a) (1) of WRDA 96, the minimum non-Federal cost share for structural flood control projects will be 35%. The maximum non-Federal cost share will continue to be 50%. The non-Federal cost share for nonstructural projects will also be 35%.

Source: Policy Guidance Letter (PGL) No. 51, Flood Control Cost Sharing, July 18, 1997

partnerships may dictate that one partner will participate only in wildlife-related costs and another in drinking water costs and another in yet another kind of cost. Perhaps the best example of cost allocation comes from a long history of state and local governments obtaining federal money to build water resource projects, so let's begin there.

Historically, the federal government has paid part to all of the costs for a federal water resource project. The amount they have paid has depended on the project purpose(s) served by the project. The WRDA 1996, for example, specifies the costs of flood control (now called flood risk management) to be paid by nonfederal interests (see the following text box). Nonfederal interests will pay no less than 35% and no more than 50% of the costs of a structural flood risk management project. If nonstructural measures are used the nonfederal share is a flat 35%. Different project purposes have different project cost-sharing formulas. For example, inland waterway transportation improvements are shared 50/50 between the federal government and the Inland Waterway Trust Fund.

Now imagine a feature of a project, like a dam, that serves both flood risk management and inland navigation. Let that dam cost $100. Make that figure millions or billions if you prefer, we'll work with the simpler small number. Suppose the federal cost share for flood risk management is 60% and the cost share for inland navigation is 50%.* How does the $100 get cost shared? It is a joint cost of two project purposes with different price tags. The answer depends on how much of the $100 is allocated to the flood risk management purpose and how much is allocated to the inland navigation purpose.

COST ALLOCATION LANGUAGE

Separable costs are the costs made necessary because a purpose is included in the multiple-purpose project. It is the minimum amount to be considered for allocation of costs to a given purpose. Specific costs are the costs of project purposes normally serving only one specific project purpose, such as the powerhouse and switchyard in a dam with hydroelectric features.

Joint costs are the total project costs less the separable costs. Joint-use costs are the costs of facilities used for more than one purpose, such as the dam and reservoir.

Alternate costs are the costs of alternative projects with one purpose eliminated to determine separable costs, or the costs of single-purpose projects necessary to obtain the same benefits for the corresponding purpose as the multiple-purpose project.

Source: Department of the Army U.S. Army Corps of Engineers. 1958. EC 1105-2-219 Cost Allocation for Multipurpose Projects Including Ecosystem Restoration. Washington, DC.

* Lest the point is missed let's be clear: The nonfederal partner gets dollars for flood risk management for 40¢ each. A navigation dollar costs 50¢. The nonfederal partner wants as much of that dam cost to be allocated to flood risk management as possible because their share of those costs is 20% lower.

To complete the example, imagine the cost is allocated as $60 for flood risk management and $40 for inland transportation. Now the math is simple:

$$60\% \times \$60 = \$36$$
$$50\% \times \$40 = \underline{\$20}$$
$56 is the federal share

$44 is the nonfederal share

Splitting the $100 into $60 and $40 is cost allocation. Calculating the federal share is cost sharing.

The necessary conditions for cost allocation are that a plan serves two or more purposes and these purposes have different cost-sharing arrangements. When the costs of a plan can be clearly and unambiguously allocated to one purpose or another, cost allocation is a simple manner. One just identifies the separable costs for each purpose. The problem arises when there are components of a plan that serve multiple purposes. These joint costs must be apportioned to the appropriate project purpose before they can be shared.

Cost allocation may seem an arcane art to natural resources planners. The future of collaborative planning and innovative partnerships, combined with growing numbers of interest groups and increasing pressure on government organizations to cut costs, may well conspire at some point to revive the need for this ancient art. Cost sharing then builds on the cost allocation when different project purposes have different cost-sharing arrangements.

14.5.1 A Cost Allocation and Cost-Sharing Example

Here is your fair warning. The ability to do cost allocations is disappearing, and that may or may not be good. This section is included solely and specifically to preserve some knowledge of that ability, because everything old is new again at some point.

How does one go about allocating costs to various project purposes? Historically, this has been done in the United States by using the Separable Cost Remaining Benefits (SCRB) method. This method was adopted by a March 12, 1954 interagency agreement among the Department of Interior, Department of the Army, and Federal Power Commissioner as the preferred method (EM 1160-2-101, 1958). The SCRB method was developed in the days of large multipurpose reservoir projects when a limited number of alternative plans were under consideration. Its steps are basically as follows:

1. Estimate the cost of the multipurpose project.
2. Estimate the benefits for each project purpose.
3. Estimate alternate costs of single-purpose projects to obtain the same benefits.
4. Estimate the separable cost of each project purpose.

5. Separable costs are deducted from the lesser of project benefits or the alternate cost.
6. Calculate the joint costs of the project.
7. Calculate remaining benefits.
8. Prorate joint costs based on the ratio of remaining benefits.
9. Allocate costs.

Let this example include flood risk management and ecosystem restoration purposes. For simplicity, assume all dollar values are in thousands of dollars and have been converted to average annual equivalent values. The data for the examples and the calculations are presented in Table 14.6. The three shaded cells indicate specific projects that have to be formulated and for which cost estimates must be prepared.

The total project has AAE costs of $3.9 million. The benefit estimates are shown in the third row from the top. One of the more burdensome tasks of the SCRB method is to identify and cost out alternative plans (row three) by which the benefits could be obtained. The benefits limited by alternate cost (row four) are the lesser of benefits and alternate costs for each project purpose.

The separable costs of flood risk management are the costs of the total project ($3.9 million) less the alternate costs of a stand-alone ecosystem restoration project ($3.2 million), or $0.7 million. The separable costs of ecosystem restoration are $3.9 million – $1.6 million = $2.3 million, as shown (row five). Remaining benefits are the difference between benefits limited by alternate cost and the separable costs, $0.9 and 0.7 million (row six).

Joint or residual costs of $0.9 million are obtained by subtracting the sum of the separable costs ($3 million) from the total project costs ($3.9 million). This amount is attributable to both project purposes. It needs to be allocated to the purposes so the costs can be subsequently shared. Total remaining benefits are $1.6 million.

Table 14.6 Multipurpose Project Separable Cost Remaining Benefits Cost Allocation Example ($Thousands)

	Flood Risk Management	Ecosystem Restoration	Total
Total project cost			$3,900
Benefits	$2,700	$3,000	$5,700
Alternate cost	$1,600	$3,200	$4,800
Benefits limited by alternate cost	$1,600	$3,000	$4,600
Separable cost	$700	$2,300	$3,000
Remaining benefits	$900	$700	$1,600
Joint cost to be allocated			$900
Allocation weight	56%	44%	100%
Allocation of joint costs	$506	$394	$900
Total allocation	$1,206	$2,694	$3,900

Fifty-six percent of these costs (0.9/1.6) are attributable to flood risk management, while the remaining 44% (0.7/1.6) are allocated to ecosystem restoration. As a result, \$0.5 million of the joint cost is allocated to flood risk management and \$0.4 million is allocated to ecosystem restoration. In the last allocation step, allocated joint costs are added to separable costs to yield flood risk management costs of \$1,206,000 and ecosystem restoration costs of \$2,694,000 for a total cost of \$3,900,000.

If flood risk management costs are split 50/50 and ecosystem restoration costs are split 35/65 for nonfederal and federal interests, cost sharing for the example is as follows:

$$\text{Nonfederal cost}: 0.5\times\$1,206,000+0.35\times\$2,694,000=\$1,545,940$$
$$\text{Federal cost}: 0.5\times\$1,206,000+0.65\times\$2,694,000=\$2,354,060$$
$$\text{Total cost}\ =\$3,900,000$$

An important point for planning purposes is that to use this method precisely requires analysts to formulate and estimate the cost of plans that serve no purpose other than to support the SCRB method. The alternate cost plans are usually not alternative plans under serious consideration for implementation. They merely support calculations needed for the SCRB method. Plan formulation and cost estimation are time-consuming and expensive. This is an undue burden on planners and the planning budget in the current cost-sharing environment for planning, especially when several alternative multipurpose solutions may be under consideration for implementation.

A two-purpose project would require three cost estimates in total. If 12 two-purpose plans are under consideration, the SCRB method could require formulating and costing 24 additional plans. Clearly this is an unreasonable burden for the planners and a waste of resources. If the SCRB is used at all it should be done only once for the selected plan, in order to limit the burden.

14.5.2 Simplified Cost Allocation Method

The SCRB method is not the only way to allocate costs. There is no "true" allocation method. Historically, the alternative justifiable expenditure method and the use of facilities method have also been used (U.S. Army Corps of Engineers 1958). Different parties tend to favor the allocation method that shifts the largest burden of the costs to others. The only real requirement for an acceptable cost allocation is that it resolves the conflicts of interest among the cost-sharing parties. This means that political resolutions to the question of cost allocation and/or cost sharing are acceptable.

James and Lee (1971) posit eight guiding rules for a good allocation process. First, the allocation should be neither less than its additional cost nor more than its benefits. Second, the sum of the allocated costs must equal total costs. Third, the method should avoid costly data and complex calculations. Fourth, it should be straightforward and easy to understand. Fifth, the allocated costs should be related to pricing. Sixth, the charges resulting from the allocation should be stable and relatively constant over time. Seventh, equity is a reasonable criterion for evaluating the allocation. Eighth, the cost allocation should guide the operation of joint facilities.

These rules and the methods discussed were based on federal policy during the age of multipurpose dam construction in the United States. This is an era that has passed. National values have changed. Although it is not appropriate for a textbook to argue what policy ought to be, it does, however, seem reasonable to argue for common sense. To allocate joint costs, we need a set of weights to decide what percentage of the joint costs is allocated to each project purpose. There are many ways one could imagine doing this. One could use the ratio of separable costs,* the ratio of project benefits,† value judgments, and so on. The example in the previous section used the historically preferred SCRB method, where remaining benefits become the weights.

The computational requirements of this method are extensive. When multipurpose planning began, there was an existing bias in favor of single-purpose plans. Perhaps the SCRB method provided needed assurance that the multipurpose plan was better than a series of single-purpose plans and that each added purpose was indeed warranted. SCRB is not well suited to the smaller scale multipurpose projects of today when dozens or even hundreds of alternative plans may be under consideration. Consequently, an alternative to the classical SCRB method must be found and used during the early iterations of the planning process if the decision criteria are to include costs allocated to a project purpose or the partners' shares of the costs. I will, tongue-in-cheek, call this method the simplified cost allocation method (SCAM).

To facilitate the discussion of such a shortcut, let's use a new example. Imagine a simple multipurpose project with flood risk management and ecosystem restoration purposes that costs $18. The first task is to go item by item and separate those dollar costs into three piles. Dollars for flood risk management only (call these specific costs and consider them analogous to separable costs), dollars for ecosystem restoration only, and dollars for both purposes (call these shared costs and consider them analogous to joint costs). It is time to simplify the language too. The problem becomes how we allocate the shared costs of the project to the component purposes. It may be essential to the planning evaluation step in some instances and the procedure described here can accomplish this task without a SCRB method. Suppose after doing this separation, we have the following three piles of costs:

Specific costs for flood risk management (F) = $10
Specific costs for ecosystem restoration (E) = $3
Shared costs for flood risk management and ecosystem restoration (S) = $5

A simple method for allocating costs to project purposes is summarized in Table 14.7. First, notice the original costs in the first three rows of Table 14.7. The

* The ratio in our example was 700/2300, so 23.3% of joint costs would be allocated to flood risk management and 76.7% to ecosystem restoration.

† The ratio in our example was 900/700, so 56.25% of joint costs would be allocated to flood risk management and 43.75% to ecosystem restoration. Allocation weights can be very sensitive to the method chosen.

Table 14.7 Simplified Cost Apportionment Method

Flood risk management cost (F)	\$10
Ecosystem restoration cost (E)	\$3
Shared cost (S)	\$5
Total cost (TC)	\$18
F share ((F+S)/TC)	83.33%
E share ((E+S)/TC)	44.44%
Normalized F share ((F+S)/TC)	65.22%
Normalized E share ((E+S)/TC)	34.78%
Shared cost allocated to F	\$3.26
Shared cost allocated to E	\$1.74
Total F allocation	\$13.26
Total E allocation	\$4.74

share of total costs represented by flood risk management and shared costs is (\$10 + \$5)/\$18 = 83.33%. When the ecosystem restoration share is calculated in this manner (\$3 + \$5)/\$18 = 44.44%, the sum of these two shares is greater than 100% because shared costs are counted twice in these calculations. To adjust for this, the shares are normalized to sum to 1, that is, 83.33%/(83.33% + 44.44%) = 65.22%, and similarly 44.44% becomes 34.78%.

When the normalized shares are multiplied by shared costs, the allocated portions of the joint cost are obtained. Thus, 65.22% of 5 = \$3.26. Adding the flood risk management share of these shared or joint costs to the specific flood risk management costs yields the allocated cost of \$13.26 shown in the second-to-last row. Ecosystem restoration costs are calculated in an analogous manner.

This method can be extended for any number of project purposes. It has some informal appeal to fairness. All flood risk management costs would represent 83% of total project costs. All ecosystem restoration costs would be 44% of total costs. Flood risk management is a larger part of the project and it seems reasonable that shared costs should be allocated in these larger proportions. The greatest value of this method is its simplicity. No additional plans need be formulated or costed out. It can be readily applied to all plans in the final array of plans under consideration for recommendation. Using this approach on all candidate plans need not replace anyone's ultimate reliance on the SCRB method. The recommended plan can always be subjected to a SCRB if a SCAM is not sufficient for resolving the various parties' conflicts of interests. Thus, SCAM can be used in the early iterations of cost allocation and SCRB can be reserved for use in the final calculation if so desired.

14.6 TYPES OF ECONOMIC ANALYSIS

The most common forms of economic analysis used in natural resources planning studies are summarized in the sections that follow. These are the methods most often used to describe the relevant economic effects of solutions for the evaluation and comparison steps.

14.6.1 Cost-Effectiveness Analysis

Cost-effectiveness analysis is used when you're trying to achieve a fixed objective (like a water quality standard, or a specific mitigation goal) at least cost or when you are optimizing an objective for a fixed amount of money. In a planning context, it might be used to find the least costly way to obtain a given level of risk reduction. For example, what is the least costly way of cutting the flood risk in a community in half? Alternatively, it may be used to find the plan that provides the greatest risk reduction for a given expenditure.

Cost-effectiveness relies on a single measurable output like recreation days, habitat units created, increased tons of commerce, acres affected, water quality improvements, or some other measurable output. It also requires estimates of the costs of obtaining the desired levels of the target output. Cost-effectiveness analysis is most useful when the outcome of a plan cannot be readily or reliably monetized but can be quantified.

14.6.2 Incremental Cost Analysis

This is very similar to the marginal analysis at the beginning of this chapter. When planners are not considering a fixed level of output or a fixed expenditure as they do with cost-effectiveness, incremental cost analysis can be an effective technique. Incremental cost is the cost of a little more (i.e., an increment) of an output. When the increment is one additional unit of output or some other arbitrarily small increment of output, incremental cost is sometimes called marginal cost (MC). It can be used to compare plan outputs when the outputs are measured in like terms.

Consider the creation of habitat units (HUs) in an ecosystem restoration study. Imagine the independently implementable measures with the costs and impacts shown in Table 14.8. In this example, the change in output is habitat units created. The incremental costs show measure C is the best buy, that is, it has the lowest incremental cost. Each HU costs an average of $23,100. It is followed by measure B with a total cost of $30 million, where each additional HU costs $27,300. The two measures if combined have a cumulative plan cost of $45 million.

When incremental costs take an order of magnitude jump, as they do for Measure D, this break point often signals when enough has been spent. Incremental cost analysis helps risk managers obtain the "biggest bang for a buck." Planners might use these results to subjectively suggest it is reasonably cost effective to create 2000 HUs

Table 14.8 Incremental Cost of HU Creation Plans

Measure	Additional Cost	HUs Created	Incremental Cost of HUs Created
A	$10 million	250	$40,000
B	$30 million	1100	$27,300
C	$15 million	650	$23,100
D	$20 million	100	$200,000

at a cost of $55 million but it is not worth spending an extra $20 million for only 100 more HUs. That money may best be spent elsewhere.

14.6.3 Benefit–Cost Analysis

When decision makers seek the plan that will do the most good given the choices available, benefit-cost analysis (BCA) can be a useful tool. BCA, also called cost–benefit analysis, analyzes the advantages and drawbacks of a plan. All the advantages and drawbacks are estimated in monetary terms. If the advantages (benefits) measured in dollars exceed the drawbacks (costs) in dollars the option is said to be economically efficient. This means its value to society exceeds its costs to society. The most desirable plan from a BCA perspective is the one with maximum positive net benefits and this is not necessarily the option with the highest benefit–cost ratio.

14.6.4 Risk-Benefit Analysis

We all accept some risks to realize the benefits associated with the risky behaviors we choose to engage in. Risk-benefit analysis is another comparison method available to planners. It compares the risks of a situation to its related benefits. A tolerable level of risk may be judged to be warranted by the fact that substantial benefits accrue in relation to the risk. A risk-benefit analysis is to some extent the reverse image of cost-effectiveness. In cost-effectiveness, we are unable to express the advantages or benefits in monetized terms, whereas the costs of a plan are easily estimated in dollars. In risk-benefit analysis, the risks (costs) remain quantified in risk-related terms, whereas the benefits that accompany that risk are often monetized or otherwise quantified. Building deep sea oil platforms near a coast may have substantial dollar benefits that might be weighed against the risk of an environmental disaster (e.g., loss of wetland or marine organisms) that cannot be monetized.

14.6.5 Economic Impact Analysis

Economic impact analysis (EIA) studies how the direct benefits and costs of a plan affect the local, regional, or national economy. The economic impacts of plans usually include effects on jobs, incomes, prices, taxes, and possibly measures of economic welfare like consumer and producer surplus. EIA is intended to measure these types of economic effects associated either with the status quo or with particular plans that may be implemented. BCA measures direct benefits and costs of a plan. It typically does not convert these direct effects into their indirect effects on the economy, such as changes in employment, wages, business sales, or land use. This is the role of EIA.

The most common forms of EIA trace spending through an economy and measure the cumulative effects of that spending in the impact region. Many economic impact models also predict impacts on personal income, business production, sales, profits, and tax collections.

14.7 TAKE AWAYS

Lesson 1: Economics aids the rational evaluation and comparison of plans, but rationality may have little to do with decision making.
Lesson 2: Get the costs of solutions as right as you can because someone always cares about costs.
Lesson 3: All money values should be expressed at the same price level and at the same point in time.
Lesson 4: Marginal analysis is a useful technique for evaluating variations in the composition of plans and the attributes of measures.
Lesson 5: A variety of economic analysis techniques are available to support the evaluation and comparison of solutions.

REFERENCES

James, D. L, and R. E. Lee. 1971. *Economics of water resources planning*. Washington: McGraw-Hill.
Kahneman, D. 2012. *Thinking fast and slow*. New York: Farrar, Straus, and Giroux.
U.S. Army Corps of Engineers. 1958. *Cost allocation for multiple purpose projects allocation of cost*. Washington, DC: EM 1160-2-101.

CHAPTER 15

Fast Planning

15.1 INTRODUCTION

A rational, evidence-based planning process that seeks the best plan from among the countless options can be a time-consuming and expensive process. Sometimes what you need is a fast plan. Fast planning is

- Faster
- Possibly cheaper
- Purposeful and structured
- Intuitive
- Concentrated
- Intense and tense
- Uncomfortable and threatening for some

- Fun and rewarding
- Better than you'd otherwise do in the same limited period of time

Fast planning is about getting as much out of the "Five By Two" process, or any other planning process, as quickly as you can. If your study took 506 days, would it have suffered a decline in quality if you had done it in 505 days? How about 504 days? Perhaps you see where this is headed. At what point, moving backward, would we say, "Stop, right there is when the solution was as good as it was going to get." Virtually any planning process can be shortened, and the question is how best to do that. You'll find some thoughts on that in this chapter.

15.2 WHY PLAN FASTER?

The planning process is perfectly scalable. You can go through the steps in an hour, a day, a week, a year, or any amount of time you have. There are many reasons you might find yourself needing to plan faster. Answers are sometimes needed quickly. Time and money may be limited. These are only the most obvious reasons.

People problems are often at the root of many "time" issues. Maybe your team does not spend much time actually working on the study in a typical month. In many organizations planning is just one of the staff functions. Planning may compete for the attention of a technical staff that has more fun doing stuff other than planning. An engineer might prefer construction to planning and your planning stuff waits for his attention. The planning team may be working on too many things. Even planners might have several studies they're working on. Then, of course, things just take too long because people can't get together and get the work done. The reasons to plan faster could also be simpler. Planning take you too long or may be too expensive. You may be collecting more data and doing more analysis than needed to make good planning decisions. If you have partners you collaborate with, they may work faster than you do. People may be impatient. You may be directed to do it faster.

Another reason to plan faster is that your planning process may be bloated and in need of reinvigorating. Planning procedures may have replaced the planning process. Budgets, schedules, conflicts, required coordination and consultations, reviews, and complexity set in and supersede planning. Administering and managing a study can consume more time and resources than conducting it when bureaucracies grow too big. Bureaucracies run slowly, if they can run at all.

Fast and effective starts are essential to good results in many if not most endeavors. It is important to learn what the study team thinks as early in the process as possible. More time and money does not always mean better solutions. You may waste a lot of time and money doing things that do not produce better answers. It may possible to get the information you need, or all the information there is, quickly. A good plan made quickly may be better than the best plan that takes years to make.

Sometimes it is easier to do really good work if the duration of the activity is limited. You may get everyone to plan better over a short intensive period. When everyone gives you their full attention to a single study, you can do better planning in the first month than many studies starving for attention do in several years. A month is not long enough for most of the reasons why you can't do it in a month to take hold! Fast planning will quickly reveal the obvious intuitive solutions. Sometimes they are best. Speed may limit decision making to existing data and models, but the existing data and models may be good enough for planning decisions. These are good things. Public involvement, coordination, and compliance are going to take time, but the planning need not wait.

If none of these arguments compel you, then let me offer a pet theory of mine. No matter how long a planning study takes, you don't get much more than one good, really focused month of everyone's time, anyway. Few people have only one assignment. We spend a lot of time waiting for things and more time waiting for things to happen. Too much time is spent less productively working alone on things rather than in groups with others. Way too much meeting time is spent on administration and management instead of on planning. Oh yeah, we're also good at wasting time, getting lost, chasing our own tails, and refining analyses to the 14th decimal place while key uncertainties go unnoticed. That's why you're going to get one good month at best. So instead of spreading that good month over 3 or 4 years, why not try to get it all at once early in the planning process?

This is not an argument for a 1-month planning process! It is a simple recognition that it may often be possible to do the job faster without sacrificing much in the way of decision quality. To be fair, however, some studies are just going to take as long as they are going to take. Let's set them aside for the moment and consider an alternative idea: the 5-day study.

15.3 FAST TRACK: A 30-DAY PROCESS

If you may only get one good month from each team member over the next few years, then ask for 5 days of everyone's undivided attention as soon as your study begins. Use that time intensively to run as far up the learning curve of your study as you can.

REQUIREMENTS FOR FAST PLANNING

- The fast period must occur first. It's the first month of study, not, for example, a month 90 days after a study is initiated.
- The planning team must be dedicated and work on only this study for the fast period.
- If your planning organization has a hierarchy, the team must be vertically integrated through the organization, that is, the local office through the company headquarters.
- The team must include representatives from your most important stakeholders, that is, agencies, organizations, and other interests who can make you fail or whose support is necessary for success.
- The team must have a dedicated workspace for the fast period.

Source: Ken Orth, Beatles fan and planner (personal communication 1970s)

This means you will have to let go of the notion of planning as an optimizing or maximizing process in favor of planning as a satisficing process. It may not produce the very best plan, but if it produces a very good one, the differences may not matter much.

There are three legs on the fast track stool offered here. The 5-day iteration at the start of your study is the first and most important of these. The second leg is based on a scientific approach to examining and improving your 5-day plan. A plan produced in 5 days is going to leave a lot of questions unanswered. This strategy identifies the chain of events necessary for your 5-day plan to be the best plan or at least a satisfactory plan and converts them into a series of testable hypotheses. The remainder of the study effort tests these hypotheses. This effort is followed by the preparation of a risk register (the third leg of our fast planning stool) to guide the remainder of the study effort and to obtain buy-in from the vertical team and influential stakeholders. This is where the team identifies the things it's not going to do as well as the risks of not doing them. Each leg is discussed next.

15.4 FIVE-DAY ITERATION

Have the entire planning team spend the first 5 days of a new study initiative together completing the first iteration of the planning process. This does not mean the team should spend any 5 days together. They should spend the first 5 days or as close to that as is humanly possible. Here's your agenda:

Day 1—team field trip of study area with diaries.
Day 2—determine the decision context and produce those six pieces of paper.
Day 3—identify the without condition scenario and gather as much evidence as you can; brainstorm a list of measures.
Day 4—formulate plans all day long begin to evaluate them.
Day 5—get to an answer.

15.4.1 The Entire Team

Everyone needs to be here for all 5 days. There are no staff meetings to run off to and no other studies to keep going. This requires the undivided attention of each and every team member. If you can afford to do this onsite, that is the ideal situation. If you cannot be onsite, try to get out away from the office. Look for a retreat atmosphere where the team can spend the days working and the evenings relaxing together. If you can't get the team out of town, consider an off-site location so people cannot be buttonholed by supervisors or wander back to their e-mail during the breaks. Arrange for the team to be able to spend as much intensive time together in these 5 days as possible. Make sure some of that intensity is fun. A 5-day process really accelerates those stages of a team mentioned in Chapter 2.

15.4.2 Day 1

You need to see the area you're working on. You need to hear about the problems and opportunities from as many smart people as you can. And you need to do all this together, as a team. The team needs to see as much of the study area together as possible. Have the smartest people knowledgeable about your study area's problems accompany you. Drop the team off and let them walk the area. Drive the area if it is too big to walk. Pile everyone into a couple of buses and drive as much of the study area as possible. If it is too big to drive, arrange a flyover of the area. See it by any means possible, but see it. If it is too wet, sail it. Do whatever you have to do to get the team to spend time together looking at the area. Google Earth is your backup.

Everyone should keep a diary. Everyone should take pictures. The diary is for capturing thoughts, taking notes, and compiling the lists below. The pictures are for capturing problems, opportunities, constraints, objectives, and any visual ideas or things of interest that jump out at you. Have everyone upload their photos to a common share point or website. Set up a wiki space and have everyone enter their lists before the sun sets on the first day. The lists include

- Three landmarks that stick in your mind
- The three most important problems you saw
- The three best opportunities you saw
- The three things that surprised you most
- The three most important stakeholders for this study
- The three most obvious measures needed in a plan
- The three things people are going to want to know about your plan

Have beer and pizza together, then go to bed. Dream of solutions.

15.4.3 Day 2

You need to define the decision you all are going to be making in the next few days. Your entire day's work is to produce the following pieces of paper:

1. A problems and opportunities statement that explains why you are doing this planning study
2. A list of the planning objectives and constraints that state in broad terms what a successful planning solution will do
3. A list of decision criteria that identify what the planning team knows or believes to be important to decision makers
4. A list of questions decision makers would like to have answered as the investigation proceeds
5. A list that singles out for attention those uncertainties that are likely to be most significant in the study
6. The without condition scenario narrative

Post these papers in your workspace and on the share point or web space you've reserved for your study. Everyone should have round-the-clock access to these papers. Keep them up to date. They should change several times over the next 3 days alone.

15.4.4 Day 3

You may not have had time yesterday to prepare a without condition scenario, and this is the sixth piece of paper from Step 1. If not, do it today. Now grab those six pieces of paper and run down as much data as you can. Be sure you begin by pulling apart any previous study efforts you found in the run up to these 5 days. The charge for your team is to go find all the data you're going to need over the rest of this week. Work the phones, cruise the Internet, talk to everyone you can, compile as much literature as possible. Even Wikipedia is in bounds. Answer as many questions the team has raised as possible. Put the resources in a central place, both virtual and real. Spend some time letting people know what you have and where it is.

Before the day ends, the team needs to brainstorm the largest list of potentially viable measures as is humanly possible. Begin the session face to face and let it continue virtually until the morning of Day 4. That means the list has to be posted at your share point before bedtime.

15.4.5 Day 4

When you first meet in the morning, pair every person randomly with a partner. Let each pair choose a plan formulation method and then give them 30 minutes to develop a complete solution to the objectives and constraints you identified on Day 2 using the data and list of measures compiled on Day 3. Partial solutions are not allowed. Have partners present the plans and vote for the best plan. Break the team into small groups and let them spend no more than an hour developing at least three plans that are better than the best plan of the morning. Have everyone report back their plans. Now begin to evaluate them. Divide the plans randomly and in small groups scrutinize the completeness, effectiveness, efficiency, and acceptability of the plans as best you can. At the end of the day, you want to have a list of potential solutions that are complete enough to call plans.

You will not have the data you need, much less the data you want. Do not let that stand in your way. Do the best you can with what you have. Make notes on critical assumptions you make so you can test them later.

15.4.6 Day 5

Here is your task today. By hook or by crook, before the day is done the team must identify the best solution to your problems and opportunities. With luck you begin the day with a list of viable solutions that have passed the first evaluation iteration. Go back to your six pieces of paper and your lists of things people are going to want to know about your plan and produce as much qualitative and quantitative data as you can to evaluate your plans. Comparison is likely to be a very compact step and the evaluation, comparison, and selection steps may all run together in an uncomfortable manner. Don't worry, this is only the first iteration. End the week with an initial selected (ISP) plan. Use the papers you produced, your lists of measures and plans, and the logic you used to arrive at the ISP to write the first draft of your planning report. Keep it to 25 pages or less, just because. The remainder of this 30-day fast track is going to be different.

15.5 25 DAYS OF SATISFICING THROUGH SCIENCE

As you end your 5-day planning iteration, you have identified the best plan you could in the time you had. It may not represent a globally optimized plan, but it should be a good and satisfactory plan. As you begin to move from a very intense 5 days through the remainder of your 30-day fast track planning process, you need a strategy. Here is one.

Assume the plan you have identified is the best solution. Based on the uncertainty you identified and all the work you did, identify the sequence of events that is necessary for your plan to be the best plan. This sequence will vary with the planning environment. It might include such things as it must cost less than $X million, must have economic benefits in excess of costs, and must increase species abundance and diversity. These are not the only things that matter, just the necessary things for your plan to be the best plan.

Those things that are not in the sequence of events are also important at this stage of the strategy. You are not going to pay much if any attention to them. Maybe construction noise is important or avoiding breeding seasons is critical. Even so, they may not be mentioned in your sequence of necessary events. That means it is safe to assume you will find a way around those problems, but you are not devoting resources to them in the preimplementation stages of planning.

With the list of necessary conditions in hand, it is time to formulate what I will call the planning hypotheses, these are statements of the opposite of that sequence. Consider the hypothetical example in Table 15.1. The sequence of necessary events is shown on the left. Planning hypotheses based on these events are shown on the right of Table 15.1.

Here is what is significant about this approach, so don't overlook it. The planning team is not going to spend much time and money on things that are not included among the planning hypotheses. Using a scientific approach, the team will design

QUICK REVIEW OF HYPOTHESIS TESTING

A null hypothesis is a statement of what we believe to be true. We then go in search of evidence to prove the hypothesis is false. Strategically, we go on believing our null hypothesis until we have enough evidence that it is no longer possible to continue believing it.

The justice system says you are innocent. This is what it believes even though it arrested you! The same system now conducts an investigation to gather sufficient evidence to change that belief. In this scientific approach, evidence is used to change beliefs and mere assertion of beliefs is not enough.

Table 15.1 Necessary Events and Their Corresponding Planning Hypotheses

Events Necessary	Planning Hypotheses
Must have positive net NED benefits	Does not have positive net NED benefits
Threatened and endangered species (T&E) compliant	Is not T&E compliant
Reconnects community with ecosystem	Fails to reconnect community with ecosystem
Avoids reduction of species abundance >10%	Causes reduction of species abundance >10%
Reduces flood damages	Does not reduce flood damages
Keeps historic district whole	Fails to keep historic district whole

analyses to test the planning hypotheses. This may represent a major change for many studies that are used to filling their bucket full of facts that may not have much, if any, real effect on the quality of the planning decisions. The planning hypotheses become the team's searchlights, shining on the facts that must be ascertained to know if your plan is best or not.

Given that the first planning hypothesis in your 5-day plan does not have positive net national economic development (NED) benefits, the team will design an analysis to test that hypothesis. This does not mean the team needs to do a full-blown economic analysis, but it must do enough to reject this hypothesis. If that can be done with previous studies or logical argument, so be it. If it requires a complete benefit–cost analysis, so be it. What it will not include is a quantitative optimization of the dimensions and other attributes of your plan.

In a similar manner, each of the other planning hypotheses would be tested. If the analysis completed corroborates your planning hypothesis, then there is reason to believe the plan you have is not the best plan. At that point, it should be reformulated to bring it into compliance with the sequence of necessary conditions.

The purpose of this satisficing strategy is simple. Plan smartly. Only do the things you need to do to make a good decision. Consider data you can use from any previous iterations of your study or similar studies at other locations. What data must you have? What data are nice to have? What data will you never have? Let the planning hypotheses guide you.

15.6 RISK REGISTER

Planning's future orientation assures it is always making decisions under uncertainty. Planning faster means more uncertainty, not less. With uncertainty comes risk. The third leg of this fast track stool is accounting for and managing the risks we create by accelerating the planning process.

There is a generally held belief that more data lead to better decisions. Experience suggests this is not always, or perhaps even often, the case. The data must contribute to the quality of the decision making. Although less uncertainty is preferred to more uncertainty, it is often the case that a little bit of carefully chosen data can reduce the uncertainty just as much as a lot of less-thoughtfully chosen data. More data does not always mean better decisions. In addition, a good decision made in a timely manner may be preferable to a "perfect" or "optimized" decision made years later. Satisficing instead of maximizing is a critical concession required in a faster planning process. Accepting a lesser level of detail can avoid years of protracted data gathering and analysis that provide little in the way of improved decision making. One strategy for doing this is to prioritize data gathering and analyses on areas critical to differentiating among your plan and other potential solutions, as is done with planning hypotheses.

The overarching strategy is to continually ask how additional study detail will affect the next planning decision. That means planners must always know what the next planning decision is and what uncertainty could affect that decision. Then it is important to know what the consequences of a poor decision are and how likely they are to occur. The technique for executing this strategy is to develop and use a risk register.

A risk register is a log in which you record the details of all the risks associated with the planning team's strategy for getting to the next planning decision. Planners rate these risks in terms of their consequences for the study or your plan's outcomes, as well as their likelihood of occurring. The register is ultimately a risk management tool that includes the team's initial plans for managing the greatest risks identified.

The risk register presented in the figures that follow was developed for use by the U.S. Army Corps of Engineers (USACE; 2012) in its effort to implement its SMART Planning Initiative in 2012. A single risk register entry includes the following fields in a simple spreadsheet file:

- Risk number—a unique identifier for each risk entry
- Date—last time entry was updated
- Name—assessors' names
- Task—assumption, strategy, task, decision, problem, question, issue, event, hazard, or opportunity that is to be managed
- Risk and its cause—a description of each risk attributable to the task
- Consequence—how the risk will affect the study schedule and output or project outcomes
- Consequence rating—high, medium, low, none
- Evidence for consequence rating—basis for rating
- Likelihood rating—high, medium, low, none
- Evidence for likelihood rating—basis for rating
- Uncertainty rating—high, medium, low, none
- Risk rating—combination of consequence and likelihood rating
- Decision(s) affected—identifies all decision criteria that could be affected by the risk(s)
- Risk management options—identifies options available for mitigating unacceptable risks
- Recommendation—recommended risk management response

- Study tasks affected—identifies any other study tasks affected by the outcome of the risk identified
- Outcome—effect of course of action
- Notes

Imagine a planning team is preparing to evaluate or screen the plans they have initially formulated. This screening step might precede the availability of quantitative data, and it will be necessary to decide based on less than complete information. The sample entry in the figures that follow illustrates the use of the risk register for the decision to use first-iteration water surface elevations from the hydrology and hydraulics analysis along with topographic mapping to lay out the footprint of levees for a flood risk management plan.

Figure 15.1 shows the first three fields for the register entry. These are simple identification fields. The top row of the register is the field name, the second row provides an explanation of the field, and subequent rows are provided for the individual register entries. Each study discipline could develop its own tab of risks in the spreadsheet. Thus, hydrualics and hydrology, geotechnical, design, economics, environmental, and the like would each have their own page in the risk register.

The first field of Figure 15.2 identifies the item that triggers a risk register entry. It can be a study task or activity, a team decision, an assumption, a strategy, a problem, question, or issue the team is grappling with, an event or a hazard or opportunity that is to be managed. Once this trigger item is identified, the next field is used to identify the risk associated with that trigger. There could be more than one risk associated with a task.

The fields shown in Figure 15.3 present the consequence description, ratings, and their supporting evidence for the risk associated with this task. Note there may be more than one consequence associated with a given risk. The consequences usually focus on study consequences, but when project outcomes can be affected, they may be included as well. The consequence is rated high, medium, low, or none based on

	A	B	C
1	**Item**	**Date**	**Assessors**
2	*Id number*	*Date entry was last updated*	*Name(s) of person(s) assessing the task*
3	1Eng	3-May-12	E. Bunny

Figure 15.1 Risk identification fields in a risk register.

	D	E
1	**Task**	**Risk and its cause**
2	*This is the task, decision, problem, question, issue, event, hazard or opportunity that is to be managed.*	*Briefly identify the risk. Considering the entry in column D, what can go wrong and how can it happen?*
3	Use Water Surface Elevation H&H and topo to determine heights and lengths of levees, floodways, or other features	Those lengths and heights used could be wrong

Figure 15.2 The planning task and its risk fields in a risk register.

	F	G	H
1	**Consequence**	**Consequence rating**	**Evidence for consequence rating**
2	*Describe the consequence of the column D risk. If things do "go wrong" in the way decsribed what is the specific consequence for the study or project outcomes? (List the most significant consequence first if more than one.)*	*If the most significant consequence in coloumn F occurs what is its potential magnitude?*	*Enter specific evidence used to support the consequence rating in coloumn G.*
			There is an inherent margin of error with early iterations of H&H.

Figure 15.3 Consequence fields in a risk register.

the evidence available. When more than one consequence is identified, the rating is provided for the consequence of greatest concern. In this example, the adverse consequences of a mistake in a cost estimate early in the study are judged to be a of medium consequence. Ratings are described later in this section.

Figure 15.4 shows the likelihood rating and its supporting evidence, the uncertainty rating, and the overall risk rating. In this example, the likelihood of an error in cost estimation is rated as high. The uncertainty rating is attributed to the least certain of the consequence and likelihood ratings. In this instance, the assessor has little uncertainty, that is, a lot of confidence, in the ratings provided. The consequence is not going to be great in an early iteration, and mistakes will be made. This is to be expected in early screening efforts. The overall risk rating is based on a medium consequence and high likelihood ratings. The lookup table in Table 15.2 was used to determine the overall risk of the task as high. Risks rated high should be carefully managed. The

	I	J	K	L
1	**Liklihood rating**	**Evidence for liklihood rating**	**Uncertainty rating**	**Risk rating**
2	*What is the liklihood that the most significant consequence in column F will occur?*	*Enter specific evidence used to support the liklihood rating in column I*	*How great is the uncertainty about the either the consequence or liklihood of the risk identified in coloumn E?*	*Qualitative risk rating from lookup table.*
3	High	First iteration. It is likely that H&H WSE will change	Low	High

Figure 15.4 Likelihood, uncertainty, and overall risk rating fields in a risk register.

Table 15.2 Qualitative Determination of Risk-Rating Lookup Table

	Consequence Factor			
Probability	**High**	**Medium**	**Low**	**None**
High	High	High	Medium	None
Medium	High	Medium	Low	None
Low	Medium	Low	Low	None
None	None	None	None	None

team knows errors in cost estimates are very common in early iterations; thus, special care ought to be taken when considering cost information generated from crude water surface elevations and topographic maps. Such information, which is bound to change, ought not be released to anyone outside the team without taking great care to place that information in a proper context. Risks rated low or none can be monitored without management measures enacted. Medium risks can be monitored to see if they move up or down or they may be managed as judgment and resources permit.

Figure 15.5 identifies other study decisions that could be affected by the consequences of this task, along with options for managing the risk and the team's recommended course of action. Here, the team has identified a willingness to live with the consequences of what it considers a highly likely medium consequence. Tolerating the risk is likely to be a common risk management response, especially in an accelerated study. The team may identify a study strategy as a high-risk action and decide it is a risk they are willing to take to save time and money. The transparency of the risk register provides higher authorities and stakeholders with the opportunity to agree or disagree with the team's strategy.

Figure 15.6 shows the final fields of the register. One function of the register is to communicate to the entire team the interests they have in the actions of other team members. The first field shown identifies those other study tasks that could be impacted by the risk identified in the sample entry. The outcome of the study task will be recorded at the time it becomes known. A field for relevant notes is also provided.

	M	N	O
1	**Decision(s) Affected**	**Risk Management Options**	**Recommendation**
2	*Identify all the decision criteria that could be affected by the risk identified. If an important decision not represented among the decision criteria is affected identify it here.*	*Enter options for reducing the risk.*	*Identify any preferred course for managing the risk. Tolerate the risk is the default option.*
3	1. Alternative Screening 2. Affect other recource analysis (cultural, wetlands, etc.)	1. Use better topo data. 2. Cet locations of utility, infrastructure, HTRW, Farmland Trusts, etc.	Accept the risk of the first iteratin and proceed.

Figure 15.5 Decisions affected, risk management options, and recommended course of action fields in a risk register.

	P	Q	R
1	**Study tasks affected**	**Outcome**	**Notes**
2	*For study risks identify any other study tasks that could be affected by the outsome of the risk identified for this entry.*	*Describe the effect of your recommended course of action on study or project outcomes.*	*Make note of any significant information not provided in the other cells.*
3	Cost, Real Estate, Cultural Resources, Environmental, Geotech		

Figure 15.6 Study tasks affected, outcome, and notes fields in a risk register.

Sample descriptions of risk-rating terms are provided below. The definitions should be considered guidelines to help frame the manner in which evidence will be used to determine the rating. A set of sample consequence ratings are as follows:

- High—the consequence of this undesirable result is unacceptable based on evidence.
- Medium—the consequence of this undesirable result is tolerable based on evidence.
- Low—the consequence of this undesirable result is acceptable based on evidence.
- None—there are no undesirable consequences based on evidence.

Risk-rating terms for likelihoods might look like these:

- High—the probability of this undesirable result is unacceptable based on evidence.
- Medium—the probability of this undesirable result is tolerable based on evidence.
- Low—the probability of this result is acceptable based on evidence.
- None—the probability of undesirable results is zero or so low as to be effectively treated as a zero based on evidence.

A sample set of rating terms for uncertainty follow:

- High—there is little to no concrete evidence available.
- Medium—there is some good evidence and some significant data gaps.
- Low—good evidence is available, and data gaps are not significant.
- None—all relevant facts are known.

The register should be prepared as early in the study process as possible. Conceptually, a register could be prepared for the entire study. Many planning studies rely on a plan of work or a project management plan that attempts to detail the study tasks, their sequencing, and schedule. Early experience with the risk register suggests that it may be most useful if it simply documents the tasks to be undertaken to get to the next significant decision in the planning process. Examples of such decisions might include determining the decision context, formulating plans, letting critical contracts, undertaking significant model studies or data collection efforts, evaluating plans, comparing plans quantitatively, and selecting a plan. Every planning study has a number of significant decision points along the way that might be scoped in a risk register.

The register is a living document. It should also be maintained and kept current throughout the planning study. The register will change as uncertainty is reduced and new uncertainties arise. Risks identified may need to be re-ranked in light of new information or the effectiveness of the risk management options.

Once a register is completed for a decision, the risks can be filtered and sorted to identify those of greatest concern. The entire team needs to be aware of the greatest risks to its decision making so that they can be carefully monitored and managed or explored through sensitivity analysis. Everyone needs to be aware of how the strategies used to manage uncertainty can affect their work and responsibilities as well as the quality of their decisions. A good register will at times identify a task that may be too risky to undertake. In these situations, a conscious choice can be made to devote more time and resources to reducing the uncertain in order to reduce or better manage the identified risk. In the example presented previously, the planning team or its higher authorities would have had the option to direct the engineering staff to supplement their approach with a site visit to field truth their information sources, for example.

15.7 IN THE END

In the first 5 days, you will have a candidate best solution. Over the next 25 days, you will test planning hypotheses to refine the solution and to identify the critical analytical requirements for the remainder of your study's duration. Before the 30 days are up, the team will create a risk register that identifies the things you normally do that you won't do for this study. The transparency of the register is to inform and secure the support of the vertical team, including decision makers. After the 30 days, you have the rest of your study period before you. Behind you is at least one complete iteration of the planning process and a plan for reaching the finish line. If that makes you uncomfortable, bear in mind some studies will not get that far in their first year of existence.

15.8 TAKE AWAYS

Lesson 1: Sometimes you may need to plan faster.
Lesson 2: Sometimes you may want to plan faster.
Lesson 3: Fast track planning identifies a solution in 5 days and then spends 25 days refining it.
Lesson 4: Fast track is not any 30 days; it is the first 30 days.
Lesson 5: A risk register is an effective tool to help you decide what data, analysis, and other planning activities are essential to your decision making.

ADDENDUM: SMART PLANNING

On February 8, 2012, Major General Michael J. Walsh, Deputy Commanding General for Civil and Emergency Operations of the U.S. Army Corps of Engineers (2012) issued a memorandum putting into motion a new national water resources planning initiative to deliver quality products in a timely manner. Over the next several months, this initiative began to evolve into something called SMART Planning,* which is a kind of fast planning suited to the needs of the Corps. This addendum describes the state of SMART Planning at the time this book went to print. It is sure to continue to evolve.

Maj. Gen. Walsh identified five imperatives for what has come to be known as SMART Planning. SMART Planning is to

- Manage and balance an appropriate level of detail and acknowledge uncertainty.
- Ensure that vertical integration and engagement of decision makers takes place early and throughout the planning process.
- Identify the federal role in resolving a problem up front.
- Recognize there is no single "best" plan and there are quantitative and qualitative methods of alternative comparison and selection.
- Ensure that all resources needed for study, funding, human resources, data, and information are identified and available for the duration of the study.

SMART Planning is decision-focused planning rather than task-oriented planning. It reorients the planning process away from simply collecting data or completing the tasks on a Gantt chart and refocuses it on doing the work required to reduce uncertainty to the point where the team can make an iterated sequence of planning decisions required to complete a quality study in full compliance with the National Environmental Policy Act (U.S. Environmental Protection Agency 1970).

SMART Planning identifies a series of specific decisions to be made throughout the planning process in a timely manner. It measures progress toward planning objectives and constraints that are realistically attainable. The decisions made are relevant to the planning process, and the objectives pursued are relevant to solving the problems and attaining the opportunities of the study area.

* Maj. Gen. Walsh's initiative was initially known as 3 × 3 × 3 planning, which referred to the directive to complete feasibility planning within 3 years within $3 million and in a report of less than 3 inches thick.

When applied to a new planning study start, SMART Planning is built around the following three specific and intentional iterations of the planning process:

- Five-day iteration
- Existing data iteration
- Targeted data iteration

The first iteration is to be completed within the first 5 days of a 30-day kickoff, similar to that described in this chapter, in which much of that time is devoted to identifying and reducing uncertainty. The second iteration is to be completed in a similarly intensive period once the study team has had time to reduce the existing uncertainty by collecting and analyzing the available data. The third and final iteration of the planning steps is completed when critical gaps in the existing data identified in the previous iteration are targeted and filled.

The vast majority of Corps studies affected by SMART Planning are existing studies.* Some of them were begun decades ago, while others have only been started in recent years. In general, SMART Planning is applied to resource studies that are not making sufficient progress toward completion for any number of reasons. The SMART Planning process may begin at a different point for every existing study depending on the progress they have made since the study was initiated. Each of these studies is to begin its "reset" process with a 3- to 5-day planning charette that is lead by a small team of Washington-level reviewers, an independent facilitator, an expert in the planning process, and a person skilled in the use of a risk register. The purposes of the charette are to introduce the concept of SMART Planning to the study team, to reset the study, and to kick start it along a new decision-focused path.

The process begins by having the study team submit its most current version of seven pieces of paper to the charette team. These papers are the most current versions of the:

- Problems and opportunities statement
- Planning objectives and constraints statement
- List of decision criteria for:
  - Evaluation
  - Comparison
  - Selection
- List of key uncertainties
- Without condition narrative
- List of planning measures already screened/eliminated from consideration
- Names of formulated plans that are under consideration

Examples of the first four of these are shown in the following four text boxes. The charette team reviews this information to ascertain where the Corps study team is in its planning process for the existing study in order to identify the next major planning decision to be made by the planning team. A study that lacks a coherent

* The Corps has 365 feasibility studies identified as active; 68 of these have been ongoing for more than 10 years (2012).

problems and opportunities statement might begin at the beginning of the planning process to develop that statement. A study with a well-developed and up-to-date set of planning papers might begin SMART Planning by focusing on the next significant planning decision needed to move the study forward.

SAMPLE PROBLEMS AND OPPORTUNITIES STATEMENT

1. Urban flooding causes extensive property damage, disruption to the economy, and increases risk to human life, health, and safety.
2. Agricultural floods damage nationally significant crops, including bulbs and seed production.
3. Major interstate commerce transportation routes are disrupted for days at a time by flood waters.
4. Levees built over the last two centuries by a variety of authorities to different levels of protection are maintained to varying standards, are vulnerable to overtopping and breaching, and are in frequent need of inherently risky flood fighting efforts.
5. A fragmented system of reservoirs and levees owned and operated by multiple authorities with varying levels of resource capability have resulted in a flood risk management system that distributes flood risks in a manner viewed as unjust by many.
6. There is a negative perception of planning authority activities, which have adversely impacted the way of life for some stakeholder groups in the basin and thereby resulted in a lack of trust in the agency.
7. Bridges owned by a variety of entities that are located near urban areas cause constriction points that contribute significantly to flood problems.
8. Continued development in the floodplain along with ongoing levee maintenance and repair along the Raven River has caused reduction of wildlife and fish populations and their habitats.

OBJECTIVES AND CONSTRAINTS (PROBLEM(S) ADDRESSED)

1. Improve the reliability and resiliency of the existing levee system over the 50-year planning time frame. (P1, 2, 3, 4)
2. Minimize the transfer of flood risk. (P5, 6)
3. Reduce flood damage in urban areas. (P1)
4. Reduce adverse impacts of flooding in high-risk areas with infrastructure or valuable habitat. (P1, 2, 3, 7)
5. Provide a consistent approach to flood risk management across the basin. (P5, 6)
6. Reduce the risk to human life, health, and safety over the 50-year project life. (P1)
7. Protect and reduce threats to Endangered Species Act listed species, especially salmonid species and their habitat. (P6, 8)
8. Increase floodplain connectivity for habitat. (P6, 8)

9. Limit adverse impacts to natural resources due to measures taken within the Raven River Basin. (P6, 8)
10. Seek solutions that respect the values and interests of stakeholders in the basin. (P4, 5, 6)
11. Avoid or minimize adverse impacts to environment, agriculture, and transportation. (P2, 3)

In this example, let that next significant planning decision the study team needs to make be the screening of formulated plans as part of the evaluation step.* This presumes the study team has formulated a set of plans. To scope this decision, the team would identify the sequence of tasks necessary to make this decision. This would include the following things:

- Identify the specific criteria to be used to evaluate the plans.
- Identify the work needed to develop estimates of these criteria for each plan.
- Develop a risk register for making this decision.
- Identify the individuals responsible for estimating the criteria.
- Schedule a team meeting to conduct the evaluation.
- Document the decision.
- Vet the decision as appropriate with the vertical decision team and the public.
- Establish the most expeditious schedule for accomplishing this work.
- Complete a clear, concise, and complete decision management plan (DMP) for the next decision.

KEY UNCERTAINTIES

- Debris load on railroad bridge and resulting upstream water surface profile
- Project footprint
- Condition of existing levee system unknown
- Climate change
- Subsurface conditions
- Sediment aggradations
- Water surface profile (WSP) with combined measures
- Existing EAD
- Planning authority's levee vegetation policy
- Real estate requirements and issues

* Alternative starting points might include decisions like establishing some aspects of the decision context, formulating plans, evaluating plans, quantitatively comparing plans, and selecting the recommended plan. It could include a more intermediate kind of decision such as how to bring the next important technical analysis, like combined measures water surface profiles or expected annual damage estimates for a flood risk management study or commodity and fleet forecasts for a harbor improvement study, to completion to support decision making. Alternately, it could include how to make a critical planning activity decision such as letting a contract, holding a public involvement event, gathering data, building a model and so on.

The products of the initial planning charette will include

- Fresh, current, and usable versions of as many of the seven pieces of paper as possible
- DMP for the next planning decision
- Risk register to accompany the DMP
- A narrative description of the accomplishments of the charette suitable for including in the final study document

The study team should also accomplish as much of the work required to make the next planning decision as possible during the charette. The resulting DMP should be vetted through the Corps vertical team as soon as possible.

Following the initial charette, the study team proceeds simultaneously along two parallel paths. First, they are to accomplish the tasks identified in the planning charette DMP as expeditiously as possible. Second, they are to begin to scope the decision that will follow the one they are immediately working on. SMART Planning proceeds by always working toward the next decision while scoping the decision after this next one. This process continues until the study culminates in a recommended course of action, which, in the case of the Army Corps of Engineers can include terminating the planning study due to the lack of a federal interest in the planning effort.

Significantly, SMART Planning is risk-informed planning. In order to complete a feasibility study within 3 years and $3 million, SMART Planning must focus on making planning decisions under uncertainty. This means managing uncertainty, hence budgets and schedules, intentionally. The primary means for doing this is the risk register described in Section 15.6.

SAMPLE EVALUATION CRITERIA

- Qualitative EAD reductions
- Qualitative acres of wetland improvement or degradation
- Qualitative stream miles impacted—improvement or degradation
- Qualitative salmonid habitat improvement or degradation
- Reductions in water surface profile
- Reductions in discharge
- Qualitative construction cost
- Qualitative O&M cost
- Total acres of land affected
- Acres of agricultural land lost
- Number of measures in plan that could increase risk
- Number of infrastructure sites affected

REFERENCES

U.S. Army Corps Of Engineers. 2012. February 8, 2012 Memorandum for major subordinate commands. Washington, DC: U.S. Army Corps of Engineers Civil Works Feasibility Study Program Execution and Delivery.

U.S. Environmental Protection Agency. 1970. National environmental policy act. Last updated June 25, 2012, http://www.epa.gov/oecaerth/basics/nepa.html

CHAPTER 16

Practical Tips

16.1 INTRODUCTION

Now that my children are grown, I think I am finally prepared to be a parent. But the job is done and the opportunity has passed, even though I think I could be half decent at it if I began now. Planning is the same way. There are a lot of things you are going to wish you knew on day one of your career. It just doesn't work that way; you are only going to learn these things by planning. There are many things in life we are not prepared for until we are finished doing them. It is the doing that prepares us.

This is your bonus chapter. It has 14 ideas that really didn't fit anywhere else, although some will look kind of familiar. These are things my friend Ken and I decided would have been helpful to know the day they first showed us to our planner's desk. The first one has echoes of fast planning in it. Don't be fooled. This is different. This is for "regular planning." Make sure you do at least one 72-hour iteration of the entire planning process within the first 30 days of your study. Planning studies can waste a lot of time thinking about what they have to do and planning to get started. Just do it.

Here is what the rest of this chapter is about. Spend time in the field. Know your study area. Spend time with the members of your planning team in large and small groups. Take some time to think about what you are doing; don't just do, nevermind what i said above. Be honest about what you know and what you don't know. Politics trumps planning in every game you will ever play. Listen to your elders—they can save you a lot of time. Lists are your friends in planning. Know the relevant planning policies and procedures. Be true to your profession. There is no such thing as bad experience. Be careful of what you say. Write your story as you go. That's it. Enjoy.

16.2 THE 72-HOUR ITERATION (3 IN 30)

Sometime in the first 30 days of your planning study, the entire planning team should meet as a whole for 3 days and complete the first iteration of the planning process. Day 1, you spend on site in the field. Never mind how far away it is or how hard it is to get everyone together. Find a way to do it. See as much of the site as you can. Walk some part of it. See the problems. Visualize the opportunities. Listen to people, ask questions, talk to one another, and take pictures—lots of pictures.

The day after your site visit, assemble the team and complete your first iteration of the planning steps—five steps in 1 day! For Step 1, draft your problems and opportunities and objectives and constraints statements. Preliminarily identify your decision criteria and your stakeholders. Identify what you feel you know well, what you don't know, and what you are going to have to learn in the months ahead. Develop a sketch of the without project condition scenario. Brainstorm ideas and measures that might be useful for achieving your objectives and develop a crude plan or two for Step 2. For Step 3, develop a with condition scenario for each plan and evaluate your first plans qualitatively. Identify the criteria and their metrics that are likely to be most useful for future iterations of this step. Do a quick comparison of your plans for Step 4. Speculate on the gut reactions of stakeholder groups to your plans so far, hold a straw poll on which plan is best, and call it a good day.

On the third day, begin to tell your story. Write your first report in an easy flowing chronological plain language narrative. "The team assembled at the study site at 8 AM and over the next 8 hours this is what we saw and learned..." Say what you did and what it means in writing, pictures, videos, or any other media you have.

TEN MENTAL LOCKS TO AVOID

Roger von Oech (1990) identified 10 attitudes that tend to prevent us from being more creative. They are

1. The right answer.
2. That's not logical.
3. Follow the rules.
4. Be practical.
5. Play is frivolous.
6. That's not my area.
7. Avoid ambiguity.
8. Don't be foolish.
9. To err is wrong.
10. I'm not creative.

See his book for an easy-to-read explanation of any of these mental locks that are not obvious to you.

It is so easy to spend the first month or two looking at work plans, worrying about budgets, thinking about contracts, and all the reasons this study is going to be difficult. Don't. Find 3 days and do a complete iteration of the planning process. That is more planning together than many teams do in a year. It will help you form your team. It will help you understand what you do and don't know. It will force everyone to understand where you are headed. It helps new people understand what planning is all about. A quick study will highlight key issues and uncover key ideas (Zucker 1999). This may be the most fruitful 3 days you will spend doing planning.

16.3 SEEING IS BELIEVING

The field, the field, the field! Make sure you visit the study site early. Visit it as often as possible. There is no substitute for an intimate knowledge of the area you are studying and its people and resources. Walk as much of the study area as is feasible. If it is too big to walk, drive it. If it can't be driven, fly or sail it. Make sure you experience it. Try to become intimate with it. See it in groups. Make time to stand alone and look at things.

Learn about the area. Talk and listen to as many local people as you can about the study area. Ask how places got their names. Learn a little history. Ask everyone, "Who knows this area best?" Find them and talk to them. Take pictures. Jot a few notes to help you remember them. Read the local paper. Buy the local postcards; important stuff often makes it onto those cards.

Listen to people and learn how they feel about this area and these resources. Build relationships while you learn about the study area. Trust will follow. Take advantage of the opportunities planning presents you to know people.

This advice is not just for study managers. It's for every team member. Whatever your planning job is get out in the field and talk to local experts about what you are doing. Spend as much time as you profitably can in the field. You are likely to learn more there than you will in the office. What you learn about yourself in the process is one of the best parts of a planning career.

16.4 LET'S SPEND SOME TIME TOGETHER

Planning is the social solving of society's problems. The planning team needs to spend time together. The entire team needs to convene regularly to work those planning steps. You've got to spend time together on planning activities. Identify problems and opportunities. Create those pieces of paper described in Chapter 4. Review them and refresh them every once in awhile. Share sources of information and good contacts you've made while in the field. Reduce the pile of things you don't know when you gather and meet. Develop those without and with condition scenarios together. Formulate plans together. Evaluate and compare plans when the team is together. Share ideas on how to present your information to others. Develop the team's recommendation. Form, storm, norm, perform, and adjourn.

You do not always need the entire team to spend time together. Economists need a close working relationship with hydrologists when doing flood risk management studies. They need to spend time at each other's cubicles talking about what they need from each other, in what form, and when. Wildlife biologists need to talk to the archeologists to learn where sensitive areas of interest are for ecosystem restoration and wildlife management studies. Plan formulators and design engineers need to spend time on the details of where utilities, bridges, and such things are. You need to spend time with other team members outside of team meetings.

Planning is serious work, but that does not mean it cannot be fun. There is much to be said for having fun at work. We all lead busy lives these days. Do not overlook the value of socializing with your team members. Lunch out every Wednesday, have meals together in the field or an office night at the stadium/bowling alley/pub. Rotating poker games and family picnics can build the team. If your group does not socialize, let that be your contribution. Organize something. Start small and let it grow. We did it, and it worked. When you meet, do planning!

Bonus material. Boring meetings are worse than death by a thousand tiny cuts. Plan your meetings wisely. Do not use them for dispensing information that can be better conveyed in other ways. Few things are more of a buzz kill than to take time out of a busy day to attend a meeting that becomes a series of one-on-one discussions about things of no concern to others. When you meet, do planning!

16.5 TAKE TIME TO THINK

It is very easy to spend your days doing. There are schedules and budgets to keep up, data to gather, calls to return, meetings to attend, models to run, and analyses to do. Make time to think about planning. We get so busy doing that some of us have lost the art of thinking. To many of us, thinking means growing quiet and waiting for something to bubble up out of our subconscious or unconscious selves. We do not always allow ourselves time to go into our minds to find ideas and solutions. This is the deliberate, effortful, and orderly work of slow thinking (Kahneman 2012). Planners need to do more of that. Planning can be hard work.

Spend time thinking about what the problems are and how to best express them. Spend time thinking about how to solve them. Think about the best way to do your own job. Get off autopilot and think about what you are doing and better ways of doing it.

It's a wonderful thing when something just comes to you. Inspiration really is a beautiful thing. More good ideas come from perspiration than inspiration, which is often the subconscious chewing on something you swallowed long ago. You've gained a lot of knowledge in your lifetime. That is the stuff you use to make new ideas, but you've got to spend time working on those new ideas. As Dr. Seuss said, "Oh the THINKS you can think if only you try" (1975).

Take a walk in the woods and solve a problem. Reserve the conference room for just you; get away from your computer and your stuff. Think. When you get an idea, bounce if off a colleague or a friend. Ask them for a reaction or to fix the idea if it is broken. Think by yourself. Then think in pairs and triplets.

Don't confuse planning with analysis. Analysis is part of planning; it is not a substitute for it. Identifying problems and figuring out what to do about them and how to do it—these are the nuts and bolts of planning. Don't make it too complicated. You don't need a lot of data or details to think the thoughts.

16.6 LISTEN TO YOUR ELDERS

I was in upstate New York looking for a creek I could not find in the days before the Internet and iPhones. I called Ken Hartzell at the office. Ken had done the original work here, maybe a decade earlier. He asked where I was and I described the convenience store. He said, "Turn around. Do you see the motel across the street?

The first two chapters of Morgan D. Jones' *The Thinker's Toolkit: 14 Powerful Techniques for Problem Solving* (1998), my favorite book about thinking, are pure genius. You'll find yourself and everyone you know in the delightful descriptions he offers of how and why we think as we do. His book offers 14 chapters, one for each strategy, several of which you might find useful.

Another book that may help you unhinge your thinking is *How to Think Like Leonardo da Vinci* by Michael J. Gelb (1998). He pragmatically discusses seven principles practiced by da Vinci, any one of which could aid your thinking. They are curiosity; commitment to test knowledge; refinement of the senses; willingness to embrace ambiguity; paradox, and uncertainty; striking a balance between science and art and logic and imagination; cultivating grace and ambidexterity; and recognizing the connectedness of all things.

Look behind there" (personal communication 1970s). He did that kind of thing all the time. He knew everything worth knowing.

I hope you are lucky enough to have some wise elders to learn from where you work. Knowledge can be had virtually anywhere. Wisdom, that is, applied knowledge, is harder to find. One of the best teachers of wisdom is experience, and that can be found in a long career. Very often, people who have been around for a good while have seen a lot of things and absorbed a great deal of wisdom in the process.

As I began my planning career, Mr. Joseph Book had just retired from a long career in planning. I never even met the man. There were several boxes of documents he left behind. When no one stepped forward to take them, I did. I now have a very valuable (to me at least) collection of some of the earliest water resources planning documents from the U.S. Army Corps of Engineers. Mr. Book was wise enough to have saved them and I was lucky enough to have read them.

My office had Ken and Milt Cornish and you could ask them anything. They knew the field, they knew the file cabinets, they knew where all the equipment was kept and which pieces didn't work right, they knew the people, they knew the history, they knew how to do things and how to get things done. I had very few assignments that did not begin by me asking someone senior how to do it. I had very few ideas of my own that I did not run by them first.

We all stand on the shoulders of giants and are wise to take advantage of the knowledge those who came before us have gained. Even if you do not work with these people, take some time to get to know them if you can. Drop by their desks, ask for a few minutes of their time. Offer to buy them a lunch for a few words of wisdom. If your organization had a planning hall of fame, find out who would be in it and why. If you get an opportunity to meet those people (annual picnic) or to talk to them (pick up the phone), do so. Listen carefully to what they say.

16.7 LISTS ARE YOUR FRIENDS

If you're going to be in planning for a while, start making and keeping lists. When you leave planning, pass them on to someone who is staying. Lists are your friends. They lend a helping hand when you need one. They are a good source of advice. Make a list of people you have met in the planning process; include other planners and resource people as well as local people. This is your planning network. Add to it whenever you can.

Wouldn't plan formulation for, say, ecosystem restoration, be easier if you could begin with a comprehensive list of measures that have been used before? It sure would be. If that list is going to come into being, chances are you will have to create it. Every time you look at another report, study, or journal article and pick up a new idea, add it to your list. When it gets long enough, you may well have the basis for a nice journal article or at least a web page resource.

You can make lists of words and ways to say things, lists of measures, lists of decision criteria, lists of tools, websites, resources, books and articles, hotels you've stayed at, places to eat, interesting things that happened during the study, you name

Figure 16.1 Nonstructural flood risk management measures.

WORDLES

Create wordles just for your study. Hang them in your team workspace. When you no longer notice them, replace them with new wordles. The wordles in this text were created using Wordle™ found at http://www.wordle.net/ (accessed February 22, 2012).

it. If you plan long enough, you'll use them over and over and they'll become valuable resources more quickly than you can imagine. Once people learn about your lists, you'll also become more popular. Figure 16.1 presents selected nonstructural flood risk management measures in a Wordle list, a cute way to make it available to others.

16.8 KNOW THE RELEVANT POLICIES

Make sure you know how your organization does planning. Read the guidance. Learn where the money comes from, where it goes, and how it gets there. You'll be a better planner. Nothing is more boring than reading policy documents. Few things will serve you better in the long run than reading your boring guidance and policy documents. Find out how your organization came to be in planning; learn its history. Know the policies of your organization and those of your planning partners. Most people do not know these things, and knowledge is power. Know what your

CAN YOU ...

Sign your own letter? Who does sign your letters? Whom can you pick up the phone and call on your own initiative? Whom can't you call? Can you obligate money? Accept a free meal? How binding is your handshake? Can you talk to reporters?

organization can and cannot do. Know what it prevents you from doing. Know what it requires in and of a planning study. Become a policy-knowledgeable planner and more doors will open to you.

Here is the second half of this wisdom. Know these policies well, but do not be bound by them. Never let policy dictate or limit what you think. If you get a good idea but have no authority to implement it, remember that new authorities are created all the time. Even if you can't implement your idea, someone else may, if it is good enough. Know the rules so you can break them intentionally at times. It's also easier to get forgiveness than it is to get permission most of the time. When it is not, it is sometimes easier to get fired than to get permission.

16.9 BE TRUE TO YOUR PROFESSION

My planning career began as an economist for the U.S. Army Corps of Engineers. It was one of the most rewarding and fun periods of my life. The Corps was a great organization to work for. Engineers like to build things. It was, after all, the Corps of Engineers. As a young economist planner, I absorbed the reality that my job was to provide the economic justification for projects that the federal water resources planning system required. No one ever told me that, you just knew it after a few months. Then, after a couple of years, I understood my real job was to tell the truth. Maybe I was a slow learner.

You'll be pushed and pulled in many directions during your career. As a study team member, an employee of an organization, an economist (fill in your discipline or profession here), an ambitious professional, in your many roles, you will have many gods to worship. My advice? Whatever your profession, be a professional at whatever you are first and foremost.

Know the theory. It allows you to think through those unique situations that inevitably arise. It's surprising how many professionals don't know the basic theory of their disciplines. Many of us were busy doing other things in college, like wondering when we are ever going to need to know "this." If you never knew or have forgotten your discipline's theory, don't be ashamed or indifferent, but do mend the situation. Pick up that textbook. Take a theory course.

If you never took a course in something you need to know, take one. No matter how you have to do it, make yourself an expert on those little details of your discipline. Be dead sure of the basics, it is amazing how often they come up and how far they can take you. Once you've mastered the basics, go on to more complex things. You don't have to have a PhD, but there is more to being a professional than having a degree. You've got to be able to learn how to do more things all the time. You have to know your discipline inside and out. You are the one people rely on to know if things are being done correctly or not.

I wish I could tell you that you only need to know your discipline's theory. Unfortunately, that's just the first step. You've got to become something of Renaissance person as well. You need to know something about the disciplines of the people you work with. The neat thing about planning is that it takes place at the intersection of many disciplines. The best economists know some environmental science and some engineering along with some geography, archeology, communications, and decision sciences. Planning is not so much a discipline as it is many mini-disciplines, and it does not require a skill but many mini-skills.

You're never going to know enough statistics. Never mind you chose a major to avoid it. You are a planner now. Statistics are important. You may not have the budget to take a census of a population. That is, you might not have enough money to interview every user of a recreation area or enough to survey each structure in a floodplain. You'll have to take a sample. You need to know how to do that correctly. Worse, you'll need to know what to do with the results. And if you don't have to take samples, you will still need to do some risk and uncertainty analysis. Sooner or later, you'll have to know some statistics and some probability too. Get good at it and make yourself valuable.

There is no substitute for doing your homework. Know your job and know it well. Then, learn the job you want. Nothing will serve you better than being good at what you do, so do your job well. Be a good economist (or whatever you are). Being true to your profession insures your integrity and makes you most valuable to yourself and others. Our natural resources represent a national trust. Your job as a professional is to preserve and protect that trust in the best interests of the people as defined in the laws and rules we live by. That requires people with integrity.

16.10 GAIN EXPERIENCE

There is no such thing as bad experience. Get as much experience as you possibly can. Do every job on a study you can possibly do. Help with an environmental inventory, do a damage survey, read maps, learn how to use surveying equipment or at least a hand level. Find out what software programs planners use and learn how to run as many of them as you can, even if you just look over someone's shoulder. Understand the Marshall-Swift real estate estimation procedure. Learn how to conduct inventories and interviews. Visit businesses in your study area, get a tour of the plant. Ride through the lock with a tug captain or up the channel with the ship's pilot. Find out how things work in that so-called real world. Go into the field with the entomologists to collect insects. Enter data into databases, build the model. Struggle with that future land use forecast. Help develop weights for a multi-criteria decision analysis (MCDA). If a contractor does something for you, make them explain what was done so you could do it. Write a scope of work, fill out a purchase order, serve on a selection committee. Keep track of the budget.

Don't say no to any job in the office. Do every task you can. Know how things get done and what it takes to do them. Some of these things you'll never want to do again, but you will never regret knowing what is involved in accomplishing the work that needs to be done. You'll also develop more respect for the people who do those things.

16.11 PLANNING AND POLITICS

The big "P" of Politics beats the small "p" of planning any day of the week. In the long run, despite your best planning, a political solution may prevail. Don't take it personally. Planning is often involved with political issues, but planners should never be political (Zucker 1999).

Any kind of public decision making is likely to have winners and losers. Winners + losers = politics. Politics can mean the politics of parties, liberal and conservative, native and alien, young and old, rich and poor, hunter and vegan; any way we can be divided, there are political perspectives prepared to embrace it.

Politics can mean pressure from elected officials. Our tiered system of government can give rise to a "teared" system of planning if you take the politics too seriously or not seriously enough. Pay attention to the people who have to smile on your plan. Never lose sight of those who will live with your plan. Congressional representatives, even occasionally presidents, may take an interest in what you are doing. Often agencies like the Environmental Protection Agency (EPA), U.S. Fish and Wildlife Service (USFWS), National Oceanic and Atmospheric Administration (NOAA), and others at the federal level will be interested in your plans at some point. Governors and elected state assembly representatives along with state agency personnel are going to care. Then there are the county administrators, the mayors and council people, homeowners associations, stockholders, and others you need to think about.

Your plans can become political footballs to kick about during election campaigns. They can be treasures to deliver to constituents or evils to protect the public from in the eyes of a politician. Never forget that every elected representative, no matter how genuine and principled they are, is a politician. That is how and why they got the job. Be aware of the politics but remain a planner; do not take sides in political battles.

Sometimes, the strongest political pressures may come from your own organization. There is a good chance your boss or your boss's boss is either in a politically appointed job or hobnobs with the elected politicians. They cannot afford to be politically naïve and the pressure they feel to provide the plan or project the local leaders want may get to them. In the best planning organizations, the bosses will absorb the political pressure and protect the analysts from it. Not everyone works in the best planning organizations. Expect pressure.

You may follow this or some other planning process to a perfect T. You may have impeccably coordinated with the decision makers and identified the decision criteria so you could conscientiously estimate them. Your evaluation and comparison may be perfection itself. Your selection work may be the best a planner has ever done. Then at the last minute, decision makers may pull a criterion no one has even mentioned before from left field and use it as the sole basis for their decision. Worse, you may have no idea how the decision makers decided what they did, after all your team's hard work.

That's the way the world works. If they chose one of your plans, even if it was not the "right" one, take comfort in that. Remember that you function in a political universe, and when a decision is made, it is not your job to be an obstructionist if you disagree with the decision. Competence and personal integrity are the two most valuable resources you could ever have as a planner. On a bad day, even these will not shield you from the criticism and pressure that are your birthright as a planner in a political universe.

16.12 COMMUNICATE AND NETWORK (P.S. BE CAREFUL OF WHAT YOU SAY)

Speaking and writing are, of course, the two most important communication skills you'll ever need. Look for opportunities to speak before groups. If that scares you, then join your local Toastmasters and get over it if at all possible. Speak up at team meetings. Participate in your public meetings. Look for opportunities to speak before stakeholder groups in your studies. Teach a class or lead a seminar. Do whatever you need to do to get comfortable telling a group what you know and what you think. Learn how to make a presentation.

Work on your writing. When writing a report, don't copy the text of a previous report. Planning documents include some of the worst writing that has ever been done in the English language. Look for opportunities to write, then write simply. Work on the newsletters and public notices of your studies. Write issue papers. Prepare working papers for issues that you can circulate among interested parties. Publish your work as often as possible. We know you are not an academic, but publishing helps you and others.

Return all calls promptly. Answer your e-mails. Text back. Think before you tweet or instant message. Keep your mouth shut and your fingers in your pockets when emotions arise. No screens when you are upset. Remember to say please and thank you. Do I sound like a parent? Parents are right sometimes, see listen to your elders, above!

Network. Meet and speak to people. Join the appropriate professional societies for planners and your discipline. Get involved. Join a subcommittee to plan the next annual meeting if you don't know where to begin. Do you own a copy of the directory from your professional society? Use it. You don't need a formal introduction. The people in this directory are your support group. They are dealing with the same types of problems you are and they include some of the world's foremost authorities on those issues. Some of them will even be glad to speak to you! If there is no such directory, start one.

Participate in planner training programs as often as possible. Go to conferences. Volunteer for things that will give you the chance to meet others. When you get where others are don't always stick with the people you know. Eat with and speak to as many people as you can. Make a contact, trade cards. For years, I did none of these things but wish I had!

When you're working on a study, get to know the local people. Go to as many meetings and public functions as you can fit into your schedule. Do some interviews yourself. Get to know the key people, movers, and shakers in your study. Meet the officials, the politicians, the business leaders, and the citizens. They are a valuable source of information.

Never overlook the opportunity to make contacts with academics in your area. They can be really helpful at times. Many are happy to discuss your problem with you if you can focus your questions and not take too much time. Personnel from other agencies, federal, state, and local, can be invaluable contacts. Get to know them.

Learn how to use state-of-the-art communications. A few years ago, the examples would have included fax machines and e-mail. You'll now find them in museums by the horse and buggy exhibit. Now tweeting, texting, Skype, and instant messaging supplant cell phones and e-mail. Social networking media rival the industrial media that once held a monopoly on much social communication. Social media include such things as follows:

- Collaborative working environments like wikis
- Blogs and microblogs like Twitter
- Content communities like YouTube
- Social networking sites like Facebook and LinkedIn
- Virtual game worlds like the World of Warcraft
- Virtual social worlds like Second Life

The technologies used by these types of social media include blogs, video blogs (vlogs), picture sharing, wall postings, e-mail, instant messaging, music sharing, crowdsourcing, podcasts, and voice over IP, to name a few. These technologies provide for interactive dialogues much richer than planners even a decade ago ever imagined.

Imagine a wiki space where anyone can contribute to problem identification. Why isn't there a natural resources planning wiki? Think about a contest that challenges colleges and universities around the world to solve your problems for you. Is your planning team on Facebook? Does your study have a Twitter account that others can follow? Who is going to be the first to use a virtual environment to plan successfully? If you have these communication options, exercise them. If you do not yet have them, then fight to get them. The Internet puts you in touch with the world. That is a powerful workforce to harness. Communication is a must. Finally, learn from the mistakes of others. There are some photos and comments you do not want to share. Think sometimes. I spent a lot of time in the field on one study. A new newspaper reporter spent a lot of time with me. One day, he asked me a hypothetical question. In an unguarded moment I answered it. If such and such did happen, then this project the community wants would be dead in the water. The next day, my comment was the headline. Not just above the fold but the headline. My bosses were all very displeased. Be careful of what you say.

16.13 WRITE AS YOU GO

Open a story account when your study begins and make frequent small deposits to it. You will be surprised at how fast you build up a nice report nest egg. Write the story as it happens and put each piece of narrative in "the bank" and soon enough your report will be largely written. Don't wait until the study is done to document it.

Make sure your story tells people where you are and how you got there. Don't tell more than you know but tell what you know when you know it. Take care to explain uncertainty to the reader. When you use facts and data, make sure the reader knows what every piece of evidence is of. This comes before the evidence itself; they need to know why they need to know it before you give it to them. Always use more reasoning than evidence (words not numbers) when you have a choice in telling your story. Your audience must understand the logic of your arguments. The logic and reasoning for what you did are more important to the audience than the data. If your logic is good, the reader may better understand the data. Make frequent usage of examples, anecdotes, quotations, and comparisons.

Keep it simple. Minimize your use of jargon. If a word is there to impress—cut it. If a word needs defining, just use the definition. If you're tempted to say potable water, but must define potable, just say drinking water. If you must use jargon, introduce the concept before the word. For example, "There is a word we use for the place where an animal lives, we call it habitat. You will see that word come up over and over."

Keep your sentence structure simple and warn your audience about difficult material. "It is not possible to fully understand the nature of your flood problem without talking a little about how often floods occur. The scientific study of how often floods occur is called hydrology and it is difficult for most anyone but a hydrologist to understand." You'll find more in Chapter 11 about telling your story. The important thing is to write as you go.

16.14 TAKEAWAYS

Lesson 1: You learn how to plan by planning.
Lesson 2: There is a lot of stuff you'll wish you'd known back then.
Lesson 3: Listen and think.
Lesson 4: Don't overthink the job.
Lesson 5: Pass on what you learn.

REFERENCES

Gelb, M. J. 1998. *How to think like Leonardo da Vinci: Seven steps to genius every day.* New York: Dell.

Jones, M. D. 1998. The *thinker's toolkit: 14 powerful techniques for problem solving.* New York: Three Rivers Press.

Kahneman, D. 2012. *Thinking, fast and slow*. New York: Farrar, Strauss and Giroux.
Seuss. D. 1975. *Oh, the thinks you can think*! New York: Random House.
Von Oech, R. 1990. *A whack on the side of the head*. New York: Warner Books.
Zucker, P. C. 1999. *What your planning professors forgot to tell you: 117 lessons every planner should know*. Chicago: Planners Press.

Index

I

J

K

L

M

N

O

P

Q

R

For Product Safety Concerns and Information please contact our EU representative GPSR@taylorandfrancis.com Taylor & Francis Verlag GmbH, Kaufingerstraße 24, 80331 München, Germany